$7.50   54415

Composite Structures of Steel and Concrete

Volume 2: Bridges

COMPOSITE STRUCTURES OF STEEL AND CONCRETE

Volume 2: Bridges

Second Edition

R. P. Johnson

Professor of Civil Engineering
University of Warwick

and

R. J. Buckby

Sir William Halcrow and Partners

COLLINS

Collins Professional and Technical Books
William Collins Sons & Co. Ltd.
8 Grafton Street, London W1X 3LA

First published in Great Britain by
Crosby Lockwood Staples 1979
Reprinted by Granada Publishing Limited – Technical Books Division 1980
Second edition published by
Collins Professional and Technical Books 1986

Distributed in the United States of America
by Sheridan House, Inc.

British Library Cataloguing in Publication Data
Johnson, R. P.
Composite structure of steel and concrete.
—2nd ed.
Vol. 2
1. Composite construction   2. Composite
materials
I. Title   II. Buckby, R. J.
624.1'821   TA664

ISBN 0-00-383153-1

Filmset by Eta Services (Typesetters) Ltd, Beccles, Suffolk

Printed and bound in Great Britain by
Mackays of Chatham, Kent

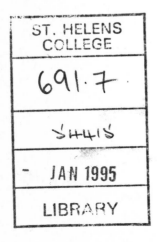

Contents

Foreword

The first volume of *Composite Structures of Steel and Concrete* by Professor Johnson (Beams, Columns, Frames and Applications in Building) was published as a Constrado Monograph in 1975. The second volume (Bridges, with a Commentary on BS 5400 Part 5) was also published as a Constrado Monograph in 1979. This present second edition of Volume 2 has been necessitated, as explained by the authors in the Preface, by the extensive development of relevant codes in recent years, by developments in practice in composite construction, and by progress in research.

The aim of the first edition has been continued, namely to give the designer an understanding of underlying theory, some interpretation and help with the relevant limit-state code, and some practical examples illustrating the design procedure. The changeover to limit-state design has caused particular difficulties for bridge designers in that, for them, the effects of continuity and interaction between structural components have to be allowed for with reasonable accuracy for a range of loading conditions. An enormous amount of time can be saved if the design is carried out in the best order, with a good appreciation of factors likely to be of leading significance at each stage. While it would be impossible for any text to take the place of the build-up of experience acquired by a designer familiar with design procedures developed to satisfy established codes, this volume will certainly help by pointing the way forward.

The design of composite structures to satisfy code requirements presents particular difficulties because of interaction – and in some cases inconsistencies – between specifically composite codes on the one hand and the separate steel and concrete codes on the other. On-going measures aimed at resolving anomalies, the prospect of further revisions of codes, together with the existence of Eurocodes, produces a situation which can cause confusion unless the engineer has some understanding of the theories behind the codes, together with some help in distinguishing where the differences lie. In this revised edition, the authors have aimed at providing such understanding and help.

The design examples have been completely revised to fulfil the intention of illustrating current British practice.

The authors have faced a difficult task in carrying out these revisions but they will be fully rewarded for their labours if readers are thereby assisted in developing, through better knowledge of the principles and how these are related to code requirements, their own techniques and expertise in the safe, economic design of composite bridge structures.

Professor M. R. Horne

Preface to the Second Edition

The fundamentals of composite construction, such as the simpler structural properties of composite cross sections and members, and an account of methods of shear connection, are given in Volume 1, 'Beams, Columns, Frames and Applications in Building', which contains chapters 1 to 5. In Volume 1, the reader is assumed to have no previous knowledge of composite structures. The reader of Volume 2 may need to refer to Volume 1 unless he or she has studied composite structures, or has some experience of designing them.

In this second volume, achievements and trends in Western Europe in the conception and design of bridge superstructures of structural steel and concrete acting compositely are examined, and the available methods of analysis are summarised. The interpretation of limit-state design philosophy is explained, and a detailed presentation and explanation is given of typical design calculations, that should be useful to final-year undergraduates, graduate students, and engineers involved in the design of composite bridges.

SOURCES OF NEW MATERIAL FOR THE SECOND EDITION

No account of this subject can be comprehensive unless it is related to one or more sets of design rules, which can be followed in worked examples. The first edition of this volume was completed in 1978, at about the same time as Part 2 (Loads), Part 4 (Design of concrete bridges), and Part 5 (Design of composite bridges) of British Standard 5400, which will be referred to as 'the Bridge Code'. The exposition referred to those Parts and to the future Part 10 (Fatigue), but was incomplete because the design of a composite superstructure is intimately linked with that of its structural steel components, and Part 3 (Design of steel bridges) was then in an early stage of drafting. Its final form (1982) turned out to be different from what had been anticipated by the drafting committee for Part 5, particularly in respect of the resistance of composite cross sections to bending and to combined bending and shear. In effect, it partly superseded Part 5 and was, in many clauses, incompatible with it. The anomalies have been resolved, pending the preparation of a revised edition of Part 5, by the issue of Departmental Standards by the Department of Transport, relating to the implementation of Parts 3 and 5 for highway structures in the UK.

Revisions have also been made since 1978 to Parts 2 and 4 of the Bridge Code; those to Part 4 being so extensive that it was re-published in 1984. The treatment of fatigue in the first edition of this book was related to a preliminary version of Part 10. This Part was published in 1980, with few revisions to the 1978 version, then or subsequently.

Volume 2 has been extensively rewritten for the second edition, to take account not only of the major developments in the Bridge Code, and of experience of its use, but of seven years of progress in the development of codes of practice for use throughout the European Economic Community, known as 'Eurocodes'. The unofficial *Recommendations for Composite Structures* was published by the European Convention for Constructional Steelwork in 1981, and in 1983–84 an international drafting panel prepared Eurocode 4, *Composite Steel and Concrete Structures*, to which detailed reference is made in section 6.1.

The commentary on specific design methods given in the first edition related mainly to BS 5400: Part 5, with limited reference to Parts 1, 2, 4 and 10. It has been extended in this edition to cover the procedures relevant to composite superstructures given in all Parts of BS 5400 and in the associated Departmental Standards. Comparisons with corresponding clauses in draft Eurocodes 2, 3 and 4 are also given. The treatment of composite columns relates to BS 5400; comparisons with the Eurocodes will be made in the second edition of volume 1.

No clause-by-clause commentary is given on BS 5400: Part 5: 1979, because much of it has been superseded by clauses included in Part 3. There is instead an extended list of contents and a detailed index.

The third source of new material for this edition is recent research. Two colloquia in the USA–Japan series on composite structures have been held, in Tokyo in 1978 and in Seattle in 1984, and a total of 72 papers were presented. Much recent European work was made available to the Drafting Panel for Eurocode 4 and there has, in particular, been progress in our understanding of distortional lateral buckling in hogging moment regions, and of the effects of temperature and shrinkage of concrete.

Finally, there have been developments in design. Examples are composite crossheads of similar depth to the main girders, now common in continuous viaducts, but rare ten years ago; and the widespread use of permanent formwork for concrete decks.

THE WORKED EXAMPLES

The design examples given in Appendices W and X of the first edition have been replaced by a fuller set of design calculations, based on current British practice, for the centre span of a three-span continuous bridge. The design of a footbridge (Appendix Y) has been similarly revised. There has been little development in design procedures for composite columns, which are more conveniently treated in the context of buildings, so the previous Appendix Z has been omitted.

The examples include discussion of particular questions that arise in the calculations. The interpretations given and the calculations themselves are based solely on the authors' understanding of the design documents. They must not be assumed to represent the opinions of the various drafting committees or of the Department of Transport. Despite careful checking, errors are inevitable, so the authors would be glad to be informed of any that may be found.

ACKNOWLEDGEMENTS

The authors are most grateful to the other members of the various code drafting committees with which they have worked, and particularly to Dr J. C. Chapman, Mr H. Mathieu, Professor K.-H. Roik and Mr J. Stark, for the education that such experience provides. Thanks are due also to Mr A. C. G. Hayward of Cass, Hayward and Partners, consulting engineers, for providing material for use in Appendix X; and to the authors' employers, the University of Warwick and Sir William Halcrow and Partners, for facilities provided. Finally, the authors are grateful to their wives, Diana and Joy, for their unfailing support.

R. P. Johnson
R. J. Buckby

Notation

The symbols used in this book follow wherever possible the practice of BS 5400[1], which was based partly on Appendix F of CP 110: 1972. Where reference is made to the European codes,[3, 4, 5] their symbols are used. These are based mainly on ISO standard IS 3898. 'Bases for design of structures – Notations – General symbols' (1976, with Addenda, 1982 and 1983).

All symbols are defined in the text where they first appear. The principal symbols are given below. For brevity, subscripts are listed separately.

A cross-sectional area; amplitude; alteration
a distance; coefficient; acceleration
B breadth of steel flange; spacing of beams
b breadth of member; breadth of concrete in cross section of column
C torsional rigidity; constant; spectral constant
c cover to reinforcement; dimension in cross-sectional plane
D overall depth of section; flexural rigidity; outside diameter of steel tube; density
d clear depth of steel web; diameter; dimension in cross-sectional plane
E Young's modulus of elasticity
e eccentricity
F force; basic variable
f direct stress; natural frequency of vibration; strength of a material
G shear modulus of elasticity; permanent load
g acceleration due to gravity
H torsional rigidity of plate; transverse load on column; warping constant
h depth of member; depth of concrete in cross section of column
I second moment of area of cross section
i longitudinal rigidity per unit width of deck

J	torsional inertia
j	transverse rigidity per unit width of deck; number (in summations)
K	property of composite column; connector spacing per unit modulus; nondimensional coefficient; relative stiffness of member in frame; constant in equation for σ_r–N line; buckling coefficient
k	radius of gyration; connector modulus; constant
L	span of beam; length of shear surface; length of column; loaded length
ℓ	span of beam; effective length of column; transfer length for longitudinal shear force
M	bending moment
m	bending or torsional moment per unit width; nondimensional coefficient; exponent in equation for σ_r–N line
N	axial force; number of load cycles for fatigue failure; number of shear connectors; longitudinal force
n	number
P	strength of shear connector; prestressing force; wind load; point load; force
p	spacing of shear connectors
Q	longitudinal force; variable load
q	longitudinal shear per unit length
R	reaction at support; radius of curvature; ratio; resistance
r	radius of gyration of cross section; ratio; number in range 1 to n
S	plastic modulus of steel cross section; shape factor; function of actions
s	unit stress; spacing; stiffness coefficient; slip at steel-concrete interface; coordinate
T	tensile force; temperature; torque
t	wall thickness of tube; time; thickness of steel plate
u	displacement; constant; length of patch load
V	vertical shear force or resistance
v	wind speed; shear stress; velocity; nondimensional property of member
W	point load; total load on a span
w	thickness of steel web; load per unit length or area; displacement; width of crack
x	coordinate
y	coordinate; dimension
Z	section modulus
α	(in α_e) modular ratio; nondimensional coefficient; concrete contribution factor; torsional parameter
β	coefficient of linear thermal expansion; ratio of column end

moments; nondimensional coefficient; slenderness function
γ partial safety factor; constant
δ deflection
ε direct strain
η nondimensional coefficient; imperfection parameter
θ temperature; angle; flexural parameter
λ slenderness function
μ coefficient of friction; nondimensional coefficient
ν Poisson's ratio; nondimensional ratio
ξ coefficient for shear strength of concrete slab
ρ density; nondimensional coefficient; reinforcement ratio
$\sum$ summation
σ direct stress; strength (of a material)
τ shear stress
ϕ creep coefficient; curvature
χ nondimensional coordinate
ψ effective-breadth ratio; nondimensional coefficient
ω angular frequency

Subscripts

a   axial; due to HA loading; anchorage
b   bottom; due to HB loading; bond
c   concrete; circular; cylinder; connector; composite; compressive; creep
cr   crack; critical
cu   cube
D   drag (in coefficient C_D); design
d   difference; design; deck
E   Euler
e   effective; elastic
eq   equivalent
F   flange
f   (in γ_f) loading; flange
g   global; centroid; permanent load
H   limiting stress range
h   haunched; horizontal; hogging
i   initial (i.e. at transfer of prestress); interaction; hogging; imposed
k   characteristic
L   (in γ_{fL}) loading
LT   lateral-torsional
ℓ longitudinal; local
ℓi   limiting

M	(in γ_M) $\gamma_M = \gamma_m \gamma_{f3}$
m	(in γ_m) materials; mean; midspan
n	number (in frequency proportion factor)
o	outstand
p	plane of possible failure in longitudinal shear; prestress; plastic; plate; primary
q	variable load
R	resultant; reduced
r	steel in reinforcing bars; number of element; range
s	serviceability; steel; shear; stud connector; shrinkage; sagging; slab
t	tension; top; transverse; total
u	ultimate; uplift
v	shear; vertical
w	steel web; weld; weight
x	x-axis
y	yield
y	y-axis
0	at origin; initial

Chapter Six

Conception and Global Analysis

6.1 Introduction

The theme of this volume is the design of bridge superstructures in which the principal members rely on interaction between structural steel and reinforced or prestresssed concrete. The subjects treated include: behaviour of the structure during erection, in service, and at failure; the development of limit-state design methods from that behaviour; and examples of their use. Particular reference is made to the design philosophy and methods of the draft Eurocodes and the British Bridge Code.

The 'Bridge Code' is British Standard 5400, *Steel, concrete and composite bridges.*[1] It has ten Parts, first published between 1978 and 1983, which are revised and reissued periodically. The design of most bridges in the UK must also satisfy the requirements of the Department of Transport, which are published as Departmental Standards.[2] These can be revised more quickly than can Parts of the Code, and so are used to introduce new requirements and to resolve inconsistencies found in the Code. The Parts and Standards relevant to the second edition of this volume and current during its preparation are listed as references 1 and 2.

The Eurocodes are being prepared for the member states of the European Economic Community, in pursuance of the agreement reached in the Treaty of Rome (1957) to work towards the removal of barriers to trade between members of the Community. Eurocode 1, *Common unified rules for different types of construction and materials* (4th draft, July 1983), is a comprehensive statement of limit-state design philosophy in general terms, for use primarily by those drafting the other Eurocodes. Those relevant to composite bridge decks are as follows:

Eurocode 2, Concrete structures;
Eurocode 3, Steel structures;

Eurocode 4, Composite steel and concrete structures;
Eurocode 8, Structures in seismic zones.

The third drafts of Eurocode 2 (EC2)[3] and of EC3[4] were being studied and calibrated by the member states in 1984–85, and are expected to be revised during 1986 to take account of the national comments. Eurocode 4[5] was drafted in English in 1984, translated into other languages in 1985, and issued for comment by member states in 1986. Work on the first drafts of other Eurocodes is in progress.

The Eurocodes are intended to cover structures for both buildings and bridges, but the current recommendations for bridges are less comprehensive than the Bridge Code. For example, they exclude steel beams with longitudinal stiffeners, box girders and composite plates.

It is intended that, as a first stage of harmonisation, the Eurocodes will be approved for use in the member states as alternatives to the national codes. National load specifications will be used, with partial safety factors γ_f chosen by the individual countries. A consensus on values of materials factors (γ_m) is developing but, as yet, there is little progress towards the harmonisation of load specifications and factors.

To illustrate the many types of structure for which design methods are required, an outline is given in this chapter of existing practice in the design and construction of the superstructures of composite bridges. The relevant period is from the 1950s, when composite bridges first became economic in Great Britain, to the present time. However, one earlier development is worthy of mention.

Composite construction had been extensively adopted by the Tasmanian Public Works Department, Australia, as early as 1935.[6] Beam-and-slab highway bridges with spans of up to 26 m had been designed, for axle loads of up to 100 kN, and spans up to 18.6 m had been built. Deck slabs were about 80 mm thick, plus haunches, and spanned transversely between rolled-steel I-beams at spacings of between 1.8 and 2.7 m. Stresses were optimised by overpropping during construction, and for the longer spans, steel girders 0.85 m deep were fabricated from two I-beams 0.6 m and 0.25 m deep, so avoiding the use of plate girders. Shear connectors consisted of hooked bars of square section fillet welded to the steel flange and extending upwards at 45°. These designs were evolved by A. Burn and A. W. Knight, and further details are given in their papers in the Journal of the Institution of Engineers, Australia, February 1934 and March 1935.

Most existing composite bridges were designed for highway loading, with carriageways two or more lanes wide. Reference is made to many examples of such bridges in this chapter, to illustrate the main trends in

British practice and to draw attention to different lines of development in Continental Europe, North America and Australasia. Except where stated otherwise, it may be assumed that these bridges are permanent structures, with an *in situ* deck slab of normal-density concrete, composite with uncased steelwork, and are not prestressed, curved in plan, or skewed more than 20°. The emphasis is on conception and methods of analysis and design. Further information may be found in state-of-the-art surveys of composite construction.[7-9].

Some factors that influence the design of all highway-bridge super-structures are discussed in section 6.2 and illustrated by reference to the bridges for which data are given in tables 6.1 to 6.3, in the figures in the text, and in the plates. Notes on particular types of bridge and individual bridges are given in section 6.3. The applicability of various methods of global analysis to composite bridges is discussed in section 6.4.

6.2 Conception and design

There are significant and interesting differences between composite highway bridges, of similar span, built in different countries. Three possible reasons for these differences are now discussed.

(*1*) *Variations in design loadings*   In 1975 a survey was made[10] of the highway-bridge loadings specified in 18 countries. These were compared in terms of the maximum unfactored bending moment and shear force for simply-supported spans of up to 100 m, for both one lane and two lanes. The effects of the heaviest normal loading (the Class 60 loading of West Germany) were about double those of the lightest (the 1973 AASHO loading of the USA). The wide range of loadings specified thus reflects not so much the degree of development of the country concerned as its preferred mode of transport for heavy loads. In Great Britain, 1.2 times HA loading is specified in Part 2 of the Bridge Code for the serviceability limit state. This is heavier than all the other normal loadings studied, except that of West Germany.

Several countries require bridges on specified routes to be designed for abnormal loadings, but Great Britain is unusual in designing almost all its recent highway bridges for a vehicle as heavy as 1800 kN (45 units of HB loading). This causes bending moments and shear forces for one lane that exceed those for the loading specified in West Germany by between 30 and 65%; but when averaged over two lanes the effects of HB loading are slightly lower than those of the Class 60 loading. The use of this heavy HB loading has arisen partly because the railway loading gauge in Britain is smaller than those in most other industrialised countries, so that large and heavy objects normally have to be transported by road or sea.

(2) *Relative cost of labour and materials*   The higher ratio of cost of labour (including design time) to cost of materials in Canada and the USA explains some of the differences between practice in North America and in Europe.

(3) *Relative cost of steel and concrete*   The ratio of the costs of these materials has had little influence on the conception of composite structures of given span; but the ratios of overall costs of construction and maintenance do, of course, influence the range of spans within which composite construction is found to be competitive with concrete (for shorter spans) and steel (for longer spans). A movement in favour of concrete occurred in the early 1970s, following failures in several steel bridges between 1969 and 1971. In the UK this trend was reversed in the 1980s, when serious deterioration was found in some concrete bridges, due mainly to chloride attack or alkali/aggregate reaction. Between 1980 and 1984, contracts for eight road bridges, viaducts or jetties were won by contractors who submitted alternative designs using composite construction.[183] The breadths of deck ranged from 5 to 36 m, and the quantities of steel per unit gross area of deck from 113 to 191 kg/m². All of the multi-span decks were continuous, with spans ranging from 18 to 56 m. Permanent formwork (chapter 11) was used in six of the eight bridges.

It is clear that differences between the types of composite bridge used in various countries are due mainly to the specified loadings. These influence not just the thickness of a plate or slab, but the whole conception of the structure and its method of construction. For example, when the authors first noted that weights of steel per unit area of deck in Switzerland are usually less than half the typical values for Great Britain, for the same span, they concluded that Swiss highway loadings must be lighter. In fact, the Swiss loading per lane current in 1976 was about the same as the contemporary British HA loading for spans up to 70 m, and was 30 to 40% higher than the unfactored HA loading at 100 m span. The difference is due mainly to the widespread British use of HB loading. This may also explain why twin two- or three-lane superstructures, each supported on only two plate girders, are less common in Britain than in many European countries; for the effects of HB loading are relatively more onerous for such structures than (for example) for a six-lane deck of the type shown in plate 6.

The brief study shows that designers must be wary both of assuming that ideas from another country are necessarily economic in their own, and that their own experience is directly applicable when designing for an unfamiliar loading.

6.2.1 The deck slab

There are a few bridges where a composite plate has been used for the whole of the deck (sections 6.3.5 and 9.4), but most composite bridges have a reinforced or prestressed concrete deck slab. In highway bridges its minimum thickness is determined by the need to support wheel loads applied at any point, and in practice lies between 200 and 300 mm. Except in filler beam construction, this minimum thickness is almost always used in Great Britain, where it is considered that any additional strength needed for other purposes is better built into the steel structure, due to the higher strength:weight ratio of steel. This conclusion seems to be based on the assumption that for economy in construction, the deck slab should be of uniform thickness, except for quite narrow haunches to take up differences of level due to superelevation or camber. In France and Switzerland, by contrast, slabs of varying thickness are often used.

In many countries, the mean design traffic loading per unit area of deck is reduced as the loaded length increases. This does not affect the design of the deck slab, which must be capable everywhere of supporting the load specified when the loaded length is small. In the UK, there is little variation between bridges in the weights of the waterproofing layer and the surfacing, so that the design load per unit area for the deck slab is almost independent of its breadth and span if normal-density concrete is used. At the longer spans, it is worth considering the use of lightweight-aggregate concrete for the deck slab.

For British highway loadings, the economic spacing of lines of support for a continuous deck slab is between 2.5 and 3.5 metres. The many different ways in which support from steelwork has been provided at this or a closer spacing have been influenced by three interrelated factors: possible methods of erection, the cost of joints between intersecting steel members, and the problem of load distribution.

DIRECTION OF SPAN OF SLAB

As the top of a deck slab has to be almost flat and its thickness uniform, the top of the supporting steelwork has to lie in a plane, so that transverse members intersect longitudinal girders rather than rest on them. This makes the joints expensive. Their number is normally

Table 6.1 Examples of composite I-beam and plate-girder bridges

No.	Location and date	Max. span (m)		Breadth of deck (m)	Span: Depth Ratio	Main girders		Spacing of cross-members (m)	Notes and references
						No.	Spacing (m)		
1.1	Chapel St, Luton (1969)	17.5	S	18.3	18.4	12	1.6	none	Curved, 460 m radius
1.2	Vionnaz, Switzerland (1976)	32	C	10.4	17.0	2	5.6	6.4	Deck constructed by 'ripage' (p. 13)
1.3	Marton-le-Moor, Yorkshire (1963)	34	S	11.6	20.4	5	2.4	6.8	Plate 1. Cross frames of 90 × 90 mm angles[16]
1.4	S. Mimms, Herts M25/A1(M) interchange (1985)	37.2	C	35.9	24.6	10	3.5	6.0 to 9.0	Integral composite crossheads each on 5 bearings. Design checked for 305-tonne vehicle
1.5	Blacow, Lancashire (1968)	45	C	16	29	10	1.6	7.5	Partly Preflex. Fully encased. Part curved, 250 m radius[14a, 14f]
1.6	Conflans-Sainte-Honorine, France (1974)	90	C	20.2	—	2	8.8	5.0	Variable-depth girders. Longitudinal and transverse prestress. Plate shear connectors.[12] Plate 10, fig. 6.1

Notes: S, simply supported; C, continuous; CS, cantilever and suspended span

Table 6.2 Examples of composite closed-top and hybrid box-girder bridges

No.	Location and date	Max. span (m)		Breadth of deck (m)	Span: Depth Ratio	No. of boxes per deck	Max. spacing of webs (m)	Spacing of cross-members (m)	Notes and references
2.1	Moat St., Coventry (1964)	21	CS	19.1	19	10	1.4	—	Plate 2. Some spans curved. Propped construction[28]
2.2	Saltings Viaduct, S. Wales (1974)	31	C	23.6	20.2	4	3.8	—	Boxes 1.2 m square with no longitudinal stiffeners[36]
2.3	Poyle, Middlesex M25/M4 interchange (1985)	50.2	C	18.4	25.7	4	3.5	—	Integral composite crossheads. Precast concrete permanent formwork. Plates 11 and 12[183]
2.4	Glâne, Switzerland (1967)	71	C	11.2 (×2)	22.2	1	5.9	14.2	Radius 1060 m. Hybrid box, launched. Precast deck slabs with bitumen joints near supports[19]
2.5	Avonmouth, Bristol (1974)	73	C	40	25.3	2	18.5	3.0	Deck spans longitudinally. Longer spans have steel deck. Plate 5, fig. 9.10
2.6	Tay, Scotland (1966)	76	C	17.8	—	2	3.7	25.3	Variable-depth boxes in main span. Concrete diaphragms over piers[16, 24]
2.7	White Cart, Scotland (1968)	77	CS	29	—	2	9.9	3.0	Deck spans longitudinally. Radius 1270 m. Trapezoidal boxes, variable depth in main span[146]
2.8	Losendse, Switzerland (1977)	90	C	13.6	—	2	6.7	90	Boxes 0.9 m wide, 1.5 m to 3.0 m deep. Precast slabs. Longitudinal prestress
2.9	Sarine, Switzerland (1965)	106	C	11.0 (×2)	24	1	5.9	—	Plates 8 and 9. Hybrid box, launched. Precast slabs, prestressed both ways[16, 17, 19]
2.10	Veveyse, Switzerland (1968)	129	C	16.5 (×2)	24.8	1	6.3	—	Plate 7, fig. 6.2. Radius 907 m. Closed box, launched[151, 152]
2.11	Friarton, Scotland (1978)	174	C	10.4 (×2)	—	1	4.3	3.0	Lightweight-concrete deck slab, density 1680 kg/m^3, spans both ways

Table 6.3 Examples of composite open-top box-girder bridges

No.	Location and date	Max. span (m)		Breadth of deck (m)	Span: Depth Ratio	No. of boxes	Max. spacing of webs (m)	Cross-member spacing (m)	Notes and references
3.1	Genoa, Italy (1964)	25	C	7.9	18	1	3.1	—	Minimum radius 150 m. Length 4.5 km. Steel permanent formwork used across top of boxes[37]
3.2	Essen, W. Germany (1963)	27	C	3.9	35.3	1	1.8	—	Footbridge. Centre support lowered 0.55 m. Precast deck, HSFG bolts[153]
3.3	Mülheim, W. Germany	44	C	24	—	2	4.6	—	Slab post-tensioned both ways; variable-depth boxes[16]
3.4	Raith, Scotland (1967)	52	C	17.1	21	2	2.1[a]	6[b]	Plate 4. Internal supports lowered 0.7 m[16]
3.5	Richelieu, Canada (1965)	57	C	12.7	23	2	3.4	11[b]	No composite action in negative moment regions[16]
3.6	Weiberswoog, W. Germany (1966)	72	C	15.7	17.7	1	5.6	4	Figure 6.3. Radius 250 m. Slab post-tensioned both ways[16]

Notes: (a) Including secondary longitudinal girders; (b) Composite minimised by designing the slab to span one way only, so that the restriction on spacing imposed by the strength of the slab applies to the webs of either the main girders or the cross girders, but not both. This is evident from tables 6.1 to 6.3. For the shorter spans, the slab spans transversely, the number of main girders is determined by the breadth of the deck, and there are no cross-girders (e.g. bridges 1.1, 2.1, 3.1, and 3.2). At longer spans it becomes more important to reduce the number of main girders, so the spacing of their webs tends to increase (e.g. bridges 2.6 and 3.3). For decks of moderate breadth, secondary longitudinal girders are sometimes provided.[183] They are supported by cross-bracings that span between a pair of main girders (e.g. bridge 3.4, plate 4). If the breadth of the deck is large and the design imposed load is heavy, the design changes to a slab spanning longitudinally between composite cross-girders at not more than 3 m centres (e.g. bridges 2.5 and 2.7), and the spacing of the main girders can become very wide. Where there are widely-spaced cross-members (e.g. bridges 1.2 to 1.5, 2.6, and 3.5) they do not support the slab and are provided to improve the distribution properties of the deck and to form part of the lateral bracing of the main girders during erection.

LONG-SPAN DECK SLABS

For the traffic loadings specified in France, Switzerland and certain other countries, it has been found to be economical[11] to use deck slabs spanning up to about 10 m, with side cantilevers of up to 5 m. These are supported on only two main girders, as shown in plate 10. If the overall breadth of deck required exceeds about 20 m, as for the Veveyse bridge (plate 7) separate structures are built for each carriageway.

Details of a selection from the many bridges of this type are given in table 6.1 (bridges 1.2 and 1.6) and table 6.2 (bridges 2.4, 2.8, 2.9, and 2.10). The slabs are usually of variable depth, and those of longer span are prestressed transversely. The three quite different methods of construction in current use for such deck slabs[11] are now discussed.

(1) Cast in situ The casting of the deck at Conflans-Ste-Honorine (fig. 6.1) has been described.[12] For the two slightly narrower decks at Veveyse[13] the long span and deep valley made it necessary to launch the steelwork, and as the bridge is curved in plan, closed-top boxes were chosen. The problem of fit between the box and a precast deck slab, and the availability of the top plate as formwork, led naturally to the choice of an *in situ* slab, which was unusual in contemporary Swiss practice. The superelevation of the slab was provided by using an asymmetrical box (fig. 6.2) and a slab of variable depth. This ingenious solution enabled

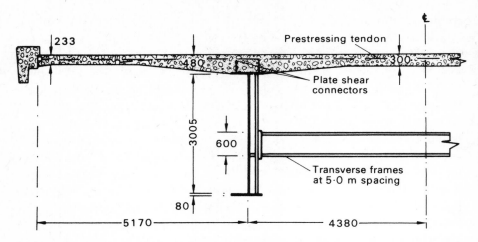

Fig. 6.1 Bridge at Conflans-Sainte-Honorine. Transverse section at centre of 90-m span

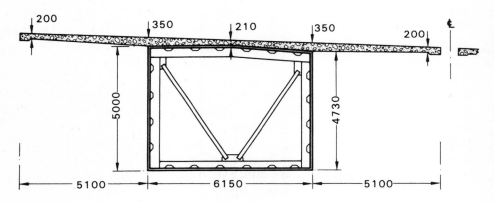

Fig. 6.2 Typical part cross section of Pont de Veveyse

transverse prestress at variable eccentricity, to suit the bending moment diagrams, to be provided using almost straight tendons. The prestressing force is 1.4 MN/m at the slab edge, increasing to 2.8 MN/m over the central 10.3 m of the breadth of the deck. This is sufficient to keep the slab in compression under dead load and up to 75% of the design imposed load. The formwork for the side cantilevers was supported on frames fixed to temporary lugs on the sides of the box girder (plate 7).

(2) Precast Particularly in long-span bridges, the casting of the deck slab *in situ* can be a slow process. Precast slabs offer the opportunity of rapid erection by a special crane (like a fork-lift truck) working from the slabs laid previously. The inherent problems are the bedding of the slabs

on the steelwork, the provision of shear connection, and the provision of continuity for vertical shear, local bending, and global tension or compression across the joints between the slabs.

In the 1960s, it was the usual practice to place the slabs on mortar beds[14d] and the method was used both with steel I-beams and open-top box girders.[15] The shear connectors projected through small holes in the slabs that were later filled with dry mortar. These uses of mortar were found to be costly and sensitive to bad weather. Other methods were tried, for example at the Weiberswoog Bridge (bridge 3.6 in table 6.3, and fig. 6.3). Here the shear connectors were attached to steel plates which were cast into the deck slabs. After erection, these plates rested on narrower plates which acted as the top flange of the steel girder during its erection, and slid along them when the slab was prestressed longitudinally. Finally, the two plates were welded together. This would appear to involve the difficult process of making overhead welds while working from temporary stagings under the deck. Another unusual feature is the use of outrigger struts at 4 m centres to support the edges of the deck slabs.

In the Sarine Bridge, near Fribourg (plates 8 and 9), 138 15-tonne precast slabs, each 11 m by 2 m, were placed directly on the top flanges of the steelwork, without bedding[16] at a rate of 20 per day. The shear connectors were large welded tees at 1 m spacing, which projected into holes in the slabs large enough for concrete, rather than mortar, to be used when filling them in.

Study of this and other designs suggests that precast slabs can be used without any form of flexible bedding in the following circumstances.

(i) The slabs should be supported on not more than two steel flanges, the lateral spacing of which should be several times the breadth of each slab. The fit between the slab and the flanges can never be perfect, but with this layout the slab has sufficient flexibility in

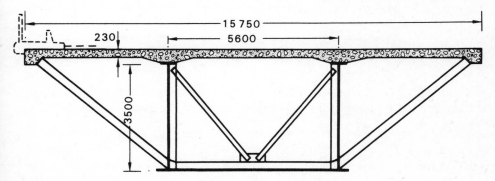

Fig. 6.3   Typical cross section of Weiberswoog Bridge, West Germany

torsion for the self-weight stresses in it, caused by non-uniformity of bearing on the steel flanges, to be small.

(ii) It should be assumed that imposed load is transferred from the slab to the steel flanges only through the *in situ* concrete placed in the holes for the shear connectors. These holes should therefore be large enough and sufficiently closely spaced for the local bending stresses in the slab due to imposed load to be small.

(iii) The location of the bridge and the detailing of the cantilever edges of the slabs should together be such that neither rain nor salt-laden spray can enter any gap between a deck slab and the steel flange. The risk of corrosion at this point is the most awkward of the problems created by this very economical method of construction. It can be reduced by injecting a suitable sealant into any gaps when the steel girders are repainted, or by the use of weathering steel.

The detailing of the transverse joints between the slabs on the Sarine Bridge involved a rather fragile projecting nib[17] and experience of its use led to a more robust design (fig. 6.4(a)) which was used in bridges at Glâne (table 6.2, bridge 2.4) and elsewhere. The cross section shows that

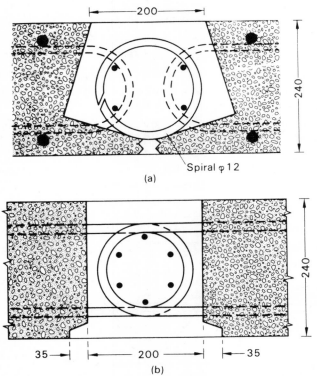

Fig. 6.4 Details of transverse joints between precast deck slabs

the longitudinal bars in the slabs do not overlap, but are both linked to a spiral bar of 12 mm diameter by threading four smaller bars through the joint along the whole width of the deck.

This type of joint requires an accurate match between the levels of the two projecting ribs at the deck soffit. To achieve this, great care must be taken over the stacking of the slabs after precasting, and in ensuring that any prestressing tendons follow exactly the same profile in each slab; otherwise creep deflections can cause differences of level of several centimetres between adjacent slabs, midway between the steel girders.

In 1974 it was noted[11] that precast slabs were becoming less common, due to problems with joints; but they are still used. The detail shown in fig. 6.4(b), which is less sensitive to small differences of level, was used in the Losendses Bridge (bridge 2.8 in table 6.2). After making the joints, the slab was prestressed longitudinally in negative moment regions, before the holes into which the shear connectors projected were concreted.

(3) *Sliding ribbon*   This method, known as 'ripage' in French, consists of casting a 20 to 30-m length of slab on the steelwork near one end of the bridge, jacking it longitudinally to clear the formwork, casting another length against its end, jacking again, and so on until the whole deck is complete. It is obviously suitable only for decks straight in plan or curved at a constant radius, but can be used for lengths of up to about 400 m, and decks of total area up to 20 000 m².[11] After completion of the sliding, the shear connectors are welded in groups of up to ten or twelve through large holes in the slab, at about 1.0 m centres, which are then filled with a dry concrete mix.

Sliding friction is reduced by the method shown in fig. 6.5. Where the slab is cast, the steel girders are provided with an additional top flange plate about 10 mm thick. During jacking, steel or cast iron 'skates' are

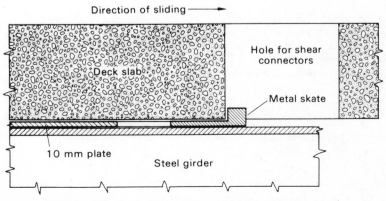

Direction of sliding ⟶

Hole for shear connectors

Deck slab

Metal skate

10 mm plate

Steel girder

Fig. 6.5   Detail of sliding ribbon method of construction

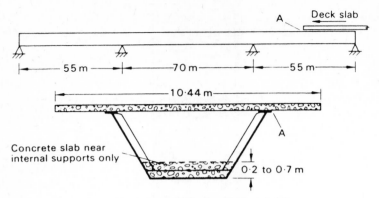

Fig. 6.6   Pont d'Illarsaz, Switzerland

placed in the shear-connector holes, at about 3 m intervals, as the slab leaves the end of the raised casting area, so that the whole of the slab except the last section cast is supported on the skates, which are lubricated with graphite. In this way the mean coefficient of friction can be reduced to about 0.08, and if the deck is on a slope care has to be taken to prevent it 'running away'. The skates are left in position, and are assumed to support the slab. The gap between the slab and the steel flange is sealed at the flange tips with rubber or mastic strips, and then filled with a fairly liquid mortar, which serves only to protect the steel from corrosion. The shear-connector holes are then filled with a non-shrinking mix of normal-aggregate concrete.

Care must be taken in design to make adequate provision for the moving point loads applied by the skates to the top flanges of the steelwork at a time when these are not restrained torsionally or laterally by the slab. In 1973, during the sliding of the deck slab for the Pont d'Illarsaz, near Aigle, Switzerland, there was a buckling failure of the compression flange of the open-top box at the point A shown in elevation and section in fig. 6.6.

6.2.2 *Load distribution and the provision of cross-members*

The subject of load distribution is introduced by reference to the simply-supported right bridge deck shown in plan in fig. 6.7. It consists of a deck slab ABCD, forming the compression flange of a number of composite beams, whose webs (which are the effective lines of support for the slab) are shown by the lines PQ, RS, etc. In principle, the deck structure has to be designed for two alternative loadings: uniformly distributed load, which causes the highest bending moments and shear forces averaged

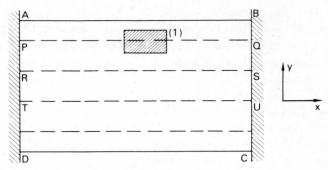

Fig. 6.7 Patch load on beam-and-slab deck

across the breadth of the deck; and a group of concentrated loads due to an exceptionally heavy vehicle (an 'abnormal load') which may be located anywhere on the deck.

If the flexural rigidity per unit width of the slab in the transverse y-direction is negligible in comparison with that of the composite beams in the x-direction, an abnormal load on area (1) in fig. 6.7 will not be distributed laterally, and beam PQ must be designed to carry almost the whole of its weight. As its position can vary, all the other beams must be equally strong, with the result that the total longitudinal flexural strength of the deck is much greater than that needed for the uniformly-distributed load. Another problem is that the deflection of beam PQ relative to those either side of it is likely to cause overstraining and severe cracking of the concrete slab.

The well-known 'statical' method of load distribution of BS153[18] assumed that, for point loads, the slab is simply-supported on the beams either side of the point considered. For British highway loadings, this forced designers either to use a thicker slab than necessary, or to reduce the spacing of longitudinal beams to about 1.5 m. It is only possible to use the optimum slab thickness with the more economic beam spacing of about 3 m when design is based on a more rational method of load distribution.

EFFECT OF WARPING RESTRAINT ON LOAD DISTRIBUTION

At first sight, one would expect statical distribution of load to give the correct answer for a deck slab supported on two girders only (e.g. as in fig. 6.1), but even here it is conservative, as it neglects the warping rigidity of the structure. Dubas[19] has given a simple but accurate method of hand analysis that agrees well with the results of loading tests[20] on bridges. It was found, for example, that for a bridge with a two-lane deck 13 m wide, warping rigidity reduced the proportion of the loading on one lane that was carried by one girder from the expected 75% to 64%.

IMPROVEMENT OF LOAD DISTRIBUTION

Kavanagh[21] has discussed distribution properties of bridge decks from the designer's point of view. Lateral distribution properties are particularly important for spans between 17 and 60 m. Below 17 m the deck slab alone usually provides sufficient lateral and torsional stiffness, and above 60 m the effect of the distributed dead load tends to predominate. Where the slab alone is inadequate, distribution can be improved in several ways. The choice between them may be influenced by the method of erection, because in many bridges cross-members are needed during construction to prevent lateral buckling of the steel girders before or during concreting of the slab. Open-top trapezoidal box girders are particularly flexible during erection. Data are available[22] on the influence of various types of bracing on their deformations and stresses.

In straight composite bridges, bracing at top-flange level serves no purpose once the deck is complete, so that temporary bracings can be used. In long-span or curved bridges, particularly if cantilever erection is used, plate girders may need to be connected by horizontal trusses at the levels of both flanges, so forming a hybrid box girder, to provide sufficient lateral and torsional stiffness during erection. These make a substantial contribution to load distribution. Their use in the construction of the Sarine bridge is shown in plate 8.

Lateral buckling of steelwork in plate-girder bridges is usually prevented by transverse trusses (fig. 6.2) or frames (fig. 6.1) in the vertical plane, at intervals along the span. Where the main girders are shallow, composite cross-girders are sometimes used, as in the River Stour bridge at Wimborne.[23]

Frames seem to be preferred when the main girders are widely spaced in relation to their depth; and also when the slab is cast *in situ*, because the cross-members then provide convenient supports for formwork and there are no diagonals to impede its movement along the bridge.

The influence of these members on load distribution is now considered. As deflection for a given load varies as the cube of the span, just a single stiff cross-connection at midspan between main girders greatly reduces relative deflection due to non-uniform load. There are diminishing returns as the number of cross-beams is increased, until their spacing becomes small enough to enable the slab to span longitudinally between them. Just two cross-bracings at the third-points of the spans of the Tay Bridge (bridge 2.6, table 6.2) transfer, from one box girder to the other, 30% of the difference in the live loadings on the girders.[24] There are four cross-members within the longest span of the Richelieu Bridge (bridge 3.5, table 6.3).

Where a large improvement in load distribution is required, it is necessary to increase the torsional rigidity of the main beams, either by cross-bracing plate girders together in pairs, as noted above, or by using box girders. Where the deck is so wide in relation to the span that several box girders would be needed, the torsional and transverse stiffnesses of the deck slab in turn become inadequate. The transverse rigidity can be significantly increased by using composite cross-girders, spaced closely enough to allow the slab to span longitudinally rather than transversely. The spacing of the webs of the main girders is then limited only by the flexural strength of the cross-girders. As noted earlier, it is economical to use the minimum number of main girders, both to simplify erection and because, for given overall dimensions, each steel plate is thicker, so increasing the buckling stress of stiffened panels of given size. This leads to cross sections of the type used for Tinsley Viaduct (plate 6) and Avonmouth Bridge (fig. 9.10) in which a deck 40.5 m wide is supported by only four longitudinal webs.

To sum up: in principle, the optimum degree of transverse load distribution is that which will ensure that each main girder need be made no stronger for abnormally heavy point loads than it need be for the alternative normal distributed traffic loading. But in practice other factors, such as curvature, skew, continuity and problems of erection, influence the choice of solution to the distribution problem, as is evident from the examples given.

6.2.3 *Prevention or control of cracking at internal supports*

In a continuous composite bridge deck, there is inevitably a short region over each internal support where high tensile strain occurs in the top flange of the steel girders. Due to the hogging curvature in this region, the notional strain at the level of the top surface of the slab is higher still, and is further increased by the primary and secondary effects of the shrinkage of concrete and by temperature effects when the slab is colder than the steel members.

It is generally agreed that wide cracks in the deck slab must be avoided, because they may contribute to rupturing of the waterproof membrane or the deck surfacing, and to corrosion of the slab reinforcement that is difficult both to detect and to repair. This is essentially a long-term problem, so inevitably there is little guidance from research on just how wide cracks can safely be, on what proportion of the design short-term loading should be included in the analysis for crack width, and on whether account should be taken of the even shorter-term wheel loadings in combination with the effects of overall loading. The consequence is that the preferred solutions to the problem vary widely from country to country. We find, for example, that the use of longitudinal prestress is common in Continental Europe, but rare in Great Britain and North America. The case for prestressing was clearly put by Roik[14d] and most of the objections to it were raised in the subsequent discussion.[14(p. 135-42)] There is also a study by Dubas[19] of possible solutions to the problem, with examples of their use in practice. These solutions are now considered in turn.

LONGITUDINAL PRESTRESSING BY TENDONS OR JACKING AT SUPPORTS

German standards as early as 1952 were based on the principle that, to ensure durability, it was better to prevent cracking of the concrete than to use a membrane to keep water out of cracks. This approach made prestressing essential, either by tendons or by altering the relative levels of supports during construction. Roik notes that the latter method is usually cheaper, and describes several procedures for causing permanent compressive *prestrain* in the slab. The objectors argue that because the corresponding stresses in the steelwork are far lower than in the tendons in conventional prestressed concrete, most of the initial *prestress* will, in time, disappear due to shrinkage and creep of the concrete. It is true that losses of prestress are high when jacking is used; but even with tendons, losses are likely to be much higher than in prestressed concrete structures, as is shown in the following example. The reason is that the flexural and axial stiffnesses of the steel section alone are not negligible in comparison with those of the composite section, particularly in long-span girders. For the limiting case of an infinitely stiff steel member it is easily shown that the loss of prestress due to creep in a concrete flange attached to it before prestressing is $100\phi/(1 + \phi)$ per cent, or 75% for a creep coefficient ϕ (ratio of creep strain to initial elastic strain) of 3.0.

In continuous members, prestress also causes changes in support reactions (secondary effects) that tend to increase losses. These effects can be eliminated by appropriate changes in the relative levels of supports, so that if a structure is prestressed by tendons, jacking of supports is often used as well.

The following simplified example is given to illustrate these points. It takes account only of the principal effects that are significant in practice and elastic behaviour is assumed.

A uniform composite propped cantilever of length L (fig. 6.8) is to be prestressed either by stressing a tendon at mid-depth of the slab to a force P, or by jacking up at the simple support, so increasing the reaction R, or by some combination of these procedures. The initial prestress (at time t_0) at the fixed end at the level of the tendon (point B) is to be f_0 in each case, f_0 being limited by the compressive strength of the concrete. Due to

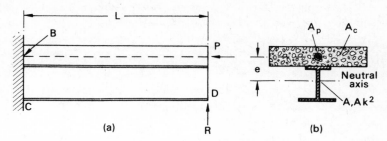

Fig. 6.8 Example to illustrate loss of prestress due to creep

creep of the concrete, this stress eventually falls to f_1 (at time t_1), so that the loss of prestress can be defined as:

$$\text{loss of prestress } (\%) = 100\left(1 - \frac{f_1}{f_0}\right) \tag{6.1}$$

Account is taken of the elastic shortening of the tendon, but not of relaxation or other losses of prestress, or the effects of shrinkage, self-weight of the structure or external loading. The effects of creep are allowed for by the 'effective modulus' method, so that only the initial and final situations need be analysed. The initial tension in the tendon is assumed to be $100f_0$ and comparison is also made of the relative areas of tendon needed to provide the initial prestress f_0 as these are roughly proportional to the cost of the prestressing operation.

The only properties of the composite section that enter into the analysis are shown in fig. 6.8(b). They are:

A_p the cross-sectional area of the tendon
e the eccentricity of the tendon
A_c the cross-sectional area of the concrete slab ('concrete' units)
A the transformed area of the uncracked composite section ('steel' units)
Ak^2 the second moment of area of the composite section.

All the parameters in the analysis except A_p, A_c, and the span L are functions of the modular ratio, α_e. This has two values, α_{e0} and α_{e1} corresponding to the initial and final situations, so the suffixes 0 and 1 are used generally.

Six methods of prestressing are considered:

(1) Slab separated from steel girder, prestressed by force P_0, allowed to creep until time t_1, and then fixed to the girder. Force R is zero throughout. This method is impracticable, as t_1 needs to be at least a

year, but it is obviously the 'best' method, and is included to set a standard for comparison with the others.

(2a) Slab separated from the girder, prestressed by force P_0, and at once fixed to the girder, which then restrains shortening of the slab due to creep.

(2b) As in (2a) except that the slab is attached to the girder before the tendons are stressed.

(3) A propped cantilever without prestressing tendons. The initial prestress is caused by jacking at end D at time t_0. The level of D is then kept constant, so that the initial force R_0 falls to R_1 due to creep.

(4a) As in (2a) except that the member is a propped cantilever. There is no change in the relative level of the supports, so that prestressing causes a downwards reaction R, which gradually increases due to creep.

(4b) As in (4a) except that the slab is attached to the girder before the tendons are stressed.

A set of numerical results from the analyses is given in table 6.4. They are calculated from the properties of the midspan cross-section of a typical plate girder, of depth 1.46 m, composite with a deck slab of thickness 220 mm and effective breadth 1.57 m, for the modular ratios $\alpha_{e0} = 7.4$, $\alpha_{e1} = 18.5$.

Two main conclusions can be drawn from these results:

(1) The greater the restraint offered by steelwork or redundant supports to the shortening and curvature of the slab caused by prestressing, the

Table 6.4   Examples of the influence of method of construction on loss of prestress due to creep

Method	Relative area of tendon	Reduction in force in tendon (%)	Loss of prestress (%)
(1) Cantilever, with slab free to slide relative to girder (impracticable)	1.0	9.4	9.4
(2) Cantilever, with shear connection between slab and girder effective:			
(a) Just after stressing tendon	1.0	6.6	24.5
(b) Before stressing tendon	1.2	6.4	25.2
(3) Propped cantilever without tendon, stressed by jacking at support	0	—	41.7
(4) Propped cantilever with tendon, with shear connection between slab and girder effective:			
(a) Just after stressing tendon	1.0	4.1	53.1
(b) Before stressing tendon	2.7	6.5	68.6

larger the area of tendon needed and the greater the loss of prestress in the concrete.

(2) Provided that the slab is not excessively restrained at the time of stressing tendons, the loss of prestress is less than that for prestressing by jacking at supports (e.g. methods (2a) and (2b) cf method (3)). However, jacking is often so much cheaper than prestressing that it is found to be economic in practice.

The figures given in table 6.5 for the losses of prestress at the Sarine Bridge (plates 8 and 9) show that the values found in the preceding example represent fairly well what may be expected in practice. At Sarine, both longitudinal prestress (1900 tonnes) and lowering of internal supports were used. The stresses given are for the top surface of the deck slab at an interior support (tension positive).

For a given arrangement of spans, the distance that the internal supports must be jacked down to create a given level of prestrain in the deck slab is roughly proportional to the span. The 27-m span footbridge at Essen (bridge 3.2 in table 6.3) was jacked down to 0.55 m, which is a simple operation, but the cost increases rapidly with the span and weight of the bridge. The much larger Sarine Bridge at Fribourg (spans 85, 106, 85 m) was jacked down 1.6 m.[17] The temporary supports at the internal piers are visible in plate 9.

The Raith Bridge (plate 4) is one of the few in Great Britain where internal supports have been jacked down, in this case by 0.7 m.

For a large prestressed bridge, the optimum sequence of casting lengths of deck, prestressing by tendons, and jacking at supports will depend on many factors that relate to that bridge alone, so its selection is too complex a matter to discuss in detail. But, four general comments on the subject can be made:

Table 6.5 Losses of prestress at Sarine Bridge

	Tendons (N/mm^2)	Jacking (N/mm^2)	Total (%)
Initial prestressing force	− 6.5	− 7.4	100
Relaxation of tendons	+ 0.7		5
Creep: primary (isostatic)	+ 1.4	+ 2.1	25
secondary (hyperstatic)	+ 0.4	+ 1.2	11
Shrinkage: primary		+ 0.7	5
secondary		+ 0.6	4
Prestress after losses		− 6.9	50

(1) It is usual to cast and to prestress lengths of slab over the internal supports before the rest of the deck is cast, whereas in a non-prestressed deck these regions are cast last.

(2) In a prestressed deck, tension due to dead load appears as a reduction of compression. There is an associated reduction in loss of prestress due to creep. With careful planning it is possible to arrange for the effective losses of dead-load prestress to be small. (For example, if it were practicable to provide prestress that exactly nullified the dead-load and shrinkage tensions in concrete *as they occurred*, the stress in the concrete would always be zero; there would be no creep, and so no loss of prestress due to creep.)

(3) Shrinkage cracking in young concrete can be eliminated by making quite small changes in the relative levels of supports, so allowing the concrete to develop sufficient strength to prevent it cracking later. This benefit from the use of jacking is not much affected by loss of prestress, for the reason discussed in (2) above.

(4) It is common for estimates based on preliminary designs to show a composite bridge to be cheaper than an alternative design in concrete only if it can be built without prestressing. This gives steelwork fabricators a strong incentive to develop alternative methods of controlling cracking. It was noted in 1972[8](p. 112) that the use of longitudinal prestress was declining. This trend has continued.

JOINTS IN THE SLAB ABOVE INTERNAL SUPPORTS
Some bridges have been constructed with continuous steelwork but with shear connectors in positive-moment regions only and a joint in the deck slab above each internal support. This is not a good solution, for satisfactory joints are expensive and it is doubtful whether the slab slides on the steelwork as is assumed.

CONTROL OF CRACK WIDTH BY LONGITUDINAL REINFORCEMENT
This method is widely used in Great Britain, and is gaining popularity in Continental Europe. It is cheaper than prestressing and allows composite action to be developed in negative-moment regions, unlike the two methods discussed above. The width of the cracks is minimised by an appropriate sequence of construction of the deck, and controlled by the use of closely-spaced longitudinal reinforcement, as discussed in section 8.3.3. The top surface of the slab is invariably protected by a waterproof membrane. It is, of course, not yet known how well these measures will protect the reinforcement from corrosion over the 120-year design life

specified for most highway bridges, but experience so far has been satisfactory.

PRACTICE IN NORTH AMERICA

In the United States and Canada, practice until the mid-1970s was to use simply-supported spans wherever possible[7(p.1111)] and to design continuous spans as composite in positive-moment regions only. Decks were neither prestressed, as in Germany and other Continental European countries, nor waterproofed, as in Great Britain. The reasons were that prestressing was not considered economically feasible[7(p.1094)] and that the welding of shear connectors to flanges in tension was thought to reduce their fatigue strength.

With this method of design, the tensile strain in the deck slab near a support is not necessarily less than it would be if this region were designed as composite, so crack-control reinforcement is needed. It was found[25] that the inevitable tension in this steel caused overloading and premature fatigue failure of connectors near points of contraflexure. Anchorage connectors are therefore now placed in this region[26] to develop the tension force in the longitudinal reinforcement.

6.2.4 The effects of creep and shrinkage

Methods of allowing for creep and shrinkage vary from country to country. In the worked example in section 6.2.3, the 'effective modulus' method was used for creep because it is generally agreed in Great Britain that the higher accuracy that can be obtained by using more complex methods is apparent rather than real, due to the inevitable errors in the assumed loadings, properties of materials and environmental conditions.

In West Germany, composite bridge decks have been designed in accordance with the German design rules for prestressed concrete, which require the effects of creep, creep recovery on unloading, and shrinkage to be analysed in great detail. The work has been simplified a little by the development of formulae and tables[27] that give equivalent modular ratios in terms of the load history, properties of the concrete and proportions of the composite cross-sections, but the methods are still more complex than those used in Great Britain.

This difference of approach stems from the different policies on prestressing. British designers allow cracking, and so cannot accurately calculate flexural rigidities. This causes uncertainties greater than those due to creep and shrinkage, so a detailed study of these effects is of no value, and in these matters experienced designers use judgement rather than analysis. In Germany, it is necessary to consider shrinkage and creep in detail, because of their effect on loss of prestress, and it is possible to do so because prestressing prevents loss of stiffness due to cracking.

The adoption of limit-state philosophy has made the effects of creep less significant in the design of non-prestressed structures at the ultimate limit state because the partial safety factors have been reduced for dead load and increased for imposed load. Except in prestressed or cable-stayed decks, it is found that most of the compressive stress in concrete is due to short-term loading.

6.2.5 Methods of construction

The longer the span of a bridge, the more essential it is to consider the proposed method of construction at an early stage of the design process. This is true for the construction of the deck slab (section 6.2.1), for decisions relating to the use of prestress (section 6.2.3) and for the erection of the main girders, which is now discussed.

For short-span bridges, the steel girders may be erected complete using a crane. For bridges of any span, where suitable foundations can be constructed not too far below the structure, falsework can be used and the steelwork erected in sections that are bolted or welded together on site. Here the engineer has to decide whether to leave the temporary supports in position while the deck slab is constructed. The implications, for design, of this choice between propped and unpropped construction are considered in detail in Volume 1, because both methods are in use for composite members in buildings. In bridges, propping may allow a lighter steel section to be used at midspan; but, the outcome is sensitive to any settlement of the props, as was found[28] from measurements taken on the Moat Street Flyover (bridge 2.1 in table 6.2) during the two years after its completion. The temporary supports are usually removed before negative moment regions of the deck are constructed, to reduce the longitudinal tensile strain and associated cracking in the deck slab. This is therefore a mixture of propped and unpropped construction, for which it is essential for the construction sequence to be agreed at the design stage. Ciolina has discussed this subject.[29]

Most bridges are constructed unpropped. The steelwork for medium spans may be erected by a purpose-designed gantry travelling on the spans already constructed, or by step-by-step cantilevering from supports, using cranes for the erection of short lengths of steelwork. This method has the disadvantage that many site joints are required.

Where the site allows assembly of a long length of steelwork behind one abutment, the launching of the complete steel structure is often the best method. It was used for three of the four Swiss bridges given in table 6.2. At Veveyse (plate 7), two box girders of the section shown in fig. 6.2, curved to 907 m radius in plan, and of total length 298 m, were each jacked in one piece across a deep valley, using a launching nose 20 m long. This is a remarkable achievement, for when the box was in cantilever over the longest span (129 m), the lateral displacement of its centreline, due to the curvature, reached 1.5 times its breadth, and almost all the vertical reaction was transmitted to the supporting pier through one web only. The two intermediate piers are about 50 m high and quite slender. Due to the use of roller bearings during jacking of the deck, the longitudinal forces were sufficiently low to be resisted by the piers without temporary bracing.

The design calculations are inevitably complex for composite bridges in which the deck is cast *in situ* in stages or where prestressing or jacking is used. The timing of the concreting is also important, for increments of wet concrete should be added to the deck only when concrete already placed is either very young or old enough to be able to resist the resulting stresses without local damage, which may otherwise occur near the shear connectors (section 15.2).

6.3 Examples of types of composite bridge

The principal classification is by type of longitudinal girder, and the examples given in tables 6.1 to 6.3 are in order of the longest span of the bridge considered (parallel to the main girders for skew bridges). The letters indicate whether this span is simply-supported (S), continuous (C) or cantilevers with a suspended span (CS). The large overlaps between the ranges of spans in the three tables are due to the many factors other than span that influence the conception of the structure. Where two carriageways are carried on separate bridges, details are given for one only. References to more detailed information on the bridges are given in the tables.

A number of design aids are available. The British Constructional Steelwork Association produced tables of the geometrical properties of composite sections in 1967[30] and a booklet on the influence of fabrication on design in 1985.[31] There is a worked example, in accordance with the Bridge Code, of a simply-supported span using Universal Beam sections,[49] the example of a continuous structure given in this book, and tables and graphs for the design of simply-supported beam-and-slab bridge decks.[32]

The three main types of bridge deck are now considered in turn.

6.3.1 Rolled steel I-beams

This straightforward form of construction is represented by bridge 1.1 in table 6.1. At the lower end of its span range (about 18 m) it gives way to structures in reinforced or prestressed concrete, and its maximum economic span of about 25 m is determined by the largest size of available rolled steel sections. Cross-girders are usually provided at supports, but elsewhere it is cheaper to rely on the concrete slab for load distribution than to provide permanent cross members, even though it may then be necessary to reduce the beam spacing to between 1.5 m and 2.5 m. Temporary bracing for the compression flanges during construction is less often required for I-beams than for plate girders.

A variant of this type is the Preflex deck[14h] in which the stiffness of the steel beam is increased by a factory process described in section 13.3. The system was used in the UK in the period 1960 to 1980, mainly where problems of clearance made a high span/depth ratio essential, and where a fully encased steel section was required to eliminate periodic painting, as for example in bridges over electrified railways.

Standard designs for composite motorway bridges of spans from 15 to 29 m were prepared in 1970[33], and cost comparisons between these and concrete structures showed that below about 22 m the difference in the cost of the deck was so small that in practice the choice would depend on other factors associated with the particular job. Plate girders were used for the 29-m span, where there was found to be a clear advantage in using composite construction. This choice has also been discussed by authors from the Department of the Environment.[34]

In multi-span bridges with spans in this range, both simply-supported and continuous construction have been used. The former has been the more common in North America, but both there and in the UK, local damage and excessive corrosion have occurred at the joints between spans. Continuous spans, now preferred,[183] give better riding quality and longer life, but are more complex to design, and can be more expensive.

Simple I-beam construction is suitable for skew decks, and a succession of straight girders has often been used for multi-span curved decks of large radius. For sharper curves, strength in torsion is needed, so box girders are used, as shown for example in plates 2 and 3.

6.3.2 Plate girders

This type of construction (plates 1 and 10) has been used for a wide range of spans, as shown by the details for bridges 1.2 to 1.6 given in table 6.1.

In comparison with rolled I-sections, the greater costs of fabrication and temporary bracing can be set against the more efficient use of material; for example by providing flanges of unequal size at midspan, and thicker webs near supports. Variable-depth girders are used for the longer spans. A comparison with box girders is given in section 6.3.3.

In these bridges, the stiffness of the concrete slab is insufficient to provide adequate load distribution, so that intermediate cross-frames are provided.

Blacow Bridge (bridge 1.5) is of interest for its high span/depth ratio (29 on the skew span) and the use of curved fully-encased plate girders in conjunction with straight preflexed girders. For this highly skewed deck, Sawko found[14a] that it was structurally more efficient for the internal cross-members to be parallel to the abutments than at right angles to the main beams. This leads to more complex joints with the main beams, a problem solved in this bridge[14f] by using reinforced concrete cross-members.

The last example in table 6.1 represents developments in large plate-girder bridges in France that are quite different from British practice. Another bridge of the same type was built at Pontoise, and a third, with spans of 90 and 50 m, was completed at Cergy in 1972. There are accounts in French[12] of the development of the shear connection for the Pontoise bridge and of the designs for Cergy and Conflans-Sainte-Honorine.

COMPOSITE CROSSHEADS

One of the most common uses of continuous composite beams is in motorway viaducts and interchanges. The breadth of the deck is often similar to a typical span (e.g. bridge 1.4 in table 6.1). It can then be economic to use, for each span, ten or more longitudinal plate girders, or up to six box girders. To reduce the number of bearings required, and also the size and cost of the piers, composite plate-girder crossheads are often provided, supported on bearings that each lie between a pair of main girders. The crossheads are usually of similar depth to the main girders, and require substantial site joints, which are usually bolted, as at Poyle interchange (table 6.2 and plates 11 and 12).

6.3.3 Box girders

Box-girder decks have been subdivided into those where the top plate of the box is initially of steel, later forming part of a composite slab (table 6.2 and plates 3, 6, 7, 11, and 12), and those where the composite box is formed by an open-top steel trough (sometimes known as a 'wash-tub' member) and a concrete slab (table 6.3, figs 6.3 and 6.6, and plate 4). Both types have been used for spans ranging from 20 to 75 m, with an apparent preference for closed boxes in Great Britain and open boxes elsewhere in Europe. For spans above about 75 m, the problems of erection are such that closed or hybrid boxes are normally used.

Box girders have several advantages over plate girders:

(1) An inherently high torsional rigidity, which makes them an automatic choice for sharply-curved spans.

(2) A neater appearance, because all stiffeners can be placed within the box.

(3) Less potential corrosion and so easier maintenance.

(4) A shallower depth for a given span, because flanges of large area can be provided easily and stiffened sufficiently for high stresses to be used. (To illustrate the importance of shallow depth, Elliott states[14j] that for a motorway overbridge having embankment approaches and 18 m spans, an increase of 180 mm in the depth of the superstructure increases the overall cost of the works (including earthmoving) by about 10%.)

(5) A better aerodynamic shape leading to lower lateral loading from wind.

(6) They are better suited to prefabrication in large sections, and are more easily handled and erected. If transported by road and rail, their cross section is limited to about 3.5 m square; but large boxes have on occasions been sealed and floated into position. The growing demand for protective systems to be applied under controlled conditions has also encouraged the use of prefabrication.

FAILURES IN STEEL BOX GIRDERS

The above advantages led to an increase in the use of box construction that outran the growth of knowledge of their structural behaviour, with the inevitable result that there were several failures in steel bridges during the period 1969 to 1971. There followed a phase of intensive research into the problems so revealed, leading to new design rules[1,35] in respect of flatness of plates, construction tolerances, residual stresses, and local stresses in diaphragms and near bearings. Many existing structures had to be strengthened to meet the new standards, and design calculations became more complex.

The immediate effect of these changes was to make clients and designers reluctant to choose steel box-girder construction, and to cause fabricators to quote artificially high prices for this type of work. Thus the cost advantage of steel or composite box girders over plate girders and, for shorter spans, over concrete structures was reduced.

The failures arose mainly from inadequate attention to the details of design or erection of steelwork, and did not cast any doubt on the fundamental soundness of steel and composite box girders. As designers and fabricators became more familiar with the new rules, and then with BS 5400, steel construction recovered its competitive position.[183] The Saltings Viaduct[36] (fig. 9.2) is an example of a post-Merrison box structure, the conception of which allowed the design, fabrication, and erection of the steelwork to be kept simple.

Due initially to the concentration of effort on steel structures, and then to a general reduction in funding for research, very little research was done in the period 1970 to 1985 on subjects such as the behaviour of composite crossheads, of composite plates in closed-top box girders, or of shear connectors subjected to biaxial loading, even though better understanding of these subjects should lead to economy in design.

CLOSELY SPACED SMALL BOXES
These have been used in several short-span bridges curved in plan, effectively in substitution for the I-beams that would be used for a similar straight bridge. They require little stiffening and have an elegant appearance (e.g. plate 2).

OPEN-TOP AND CLOSED-TOP BOXES
Both of these types of construction have been widely used. The open-top box makes more use of the torsional and flexural stiffnesses of the concrete deck slab, and needs fewer shear connectors because in closed boxes they are provided across the whole width of the top plate (plate 6) to resist local loads and, in some bridges, to stabilise the plate. During fabrication, access for welding is better and less steelwork is needed in the final structure.

The advantages of the closed-top box are mainly in construction and erection. The open-top box has negligible torsional stiffness and inadequate lateral restraint for its top flanges until the deck slab is complete, so extensive temporary bracing is needed.[22] The closed box provides permanent formwork for part of the deck slab, and its interior is less liable to corrosion. These last two problems have been solved in open-top boxes by using corrugated metal decking[37] or a thin steel membrane[38] as permanent formwork.

The precast deck slabs used for the Sarine Bridge at Fribourg, Switzerland (plates 8 and 9) are supported by a pair of plate girders interconnected by two tubular trusses in the planes of their flanges, forming a partly plated and partly latticed structure, here called a 'hybrid box'.

RECTANGULAR AND TRAPEZOIDAL BOXES

The spacing of webs at the top of a box is related to the design of the concrete deck and the method used to provide load distribution, as discussed earlier. The need to stabilise the top flange during erection has little influence on the shape of box chosen but in continuous decks the bottom flange can be more highly stressed near supports if it is made as narrow and thick as possible. This also reduces the loss of effective cross section due to shear lag, and leads to the use of trapezoidal boxes, as for example in plate 3.

Fabrication becomes complicated if the box depth is varied as well, so rectangular boxes are normally used for the longest spans (e.g., bridges 2.5 and 2.6). The White Cart Viaduct[14e] is an exception, probably because the advantage of using constant-depth trapezoidal boxes in twenty approach spans outweighs the extra cost due to this shape in the three variable-depth main spans.

CELLULAR BOXES

This term is used for wide shallow boxes with internal webs. They are not often used in composite structures. A typical cross section of Wardley Moss Bridge (plate 3) is shown in fig. 6.9. Where there is sufficient transverse stiffening to prevent cross-sectional distortion of individual cells, as here, the internal webs receive very little torsional shear stress, and effectively resist vertical shear only.

6.3.4 Tied arches

There are two tied-arch bridges in Great Britain with steel arches and a composite deck, each with a single span of about 100 m. In the Scotswood Bridge (1968),[39] which is 22.2 m wide, the slab spans transversely between composite stringers. These are supported on non-composite

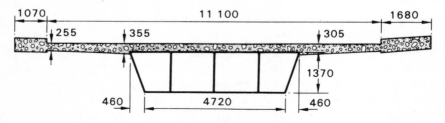

Fig. 6.9 Typical cross section of Wardley Moss Bridge

cross-girders that span between the tie-members of the two arches. In Bonar Bridge (1974),[40] the trusses are only 12.1 m apart, and the slab spans longitudinally between composite cross-girders. In both bridges the deck slab is shear-connected to the main structure at the ends of the arch, so that its longitudinal reinforcement resists part of the tie force due to imposed load.

6.3.5 Suspension bridges

In the lower part of the range of spans for which suspension bridges are normally used, it has been found economic in France to construct the deck from a thin composite plate, rather than to use steel battle deck. This type of construction, known as a 'Robinson slab'[41] was used for the deck of the Tancarville bridge, near Le Havre, which was completed in 1959. Its main span of 608 m was then the longest suspended span in Europe.[42,43]

The deck consists of a steel plate, 10 mm thick, composite with 95 mm of normal-density concrete, which is covered by 50 mm of bituminous surfacing. The composite plate spans transversely between longitudinal stringers at 2 m centres, which are themselves supported by cross-girders at 10.67 m spacing, which span 16 m between the stiffening trusses. Details of the composite plate and results of tests on a prototype are given in section 9.4.1.

Other bridges with composite-plate decks are listed in section 9.4.

6.4 Global analysis

The initial conception of a bridge superstructure emerges from the interaction between the client's requirements, the site, and the designer's experience. The proposed sizes of members are checked by rough calculations of the type described by Hayward[44] for decks carried on two girders only, by Wex[45f] for the preliminary design of box girders, and by Tung and Fountain[46] for curved members. The design loadings for the structure at the various limit states can then be found.

The next stage, known as 'global analysis', is the calculation of stress resultants in critical regions throughout the structure for the various loading cases and limit states. This work becomes impracticable unless the principle of superposition can be used. This implies the use of linear-elastic analysis.

Inelastic behaviour is acceptable at the ultimate limit state, but appropriate methods of analysis are available only for compact cross sections and a few special situations, such as the shedding of longitudinal stress from steel webs to flanges, which are discussed later. For global analysis, elastic theory is almost always used, for want of anything better. The chief exception is the simply-supported beam-and-slab deck, for which a combination of yield-line theory for the slab and plastic theory for the beams has been found to be practicable for the structures studied.[47,48] There is no known reason why the method could not be used for spans up to about 25 m where the slab is composite with five or more closely spaced steel beams of compact section. Ultimate strengths would be found to be significantly higher than those given by current methods based on elastic global analysis. In a model set of calculations for a structure of this type,[49] the design of the outer main beams was found to be governed by the serviceability limit state, so the use of plastic theory would not necessarily lead to lighter or cheaper structures.

The method is not allowed by BS 5400 or Eurocode 4. The objections appear to be that the applicability of yield-line theory to regions of slabs that also resist significant in-plane forces and transverse shear is not proven, and that current methods for assuring the stability of steel bottom flanges may not be valid in regions of extensive inelastic behaviour. Therefore, only elastic global analysis is now considered.

Nearly all of the extensive literature on bridge-deck analysis is either not specific as to material, or else considers steel structures or concrete structures. Most of the work on concrete decks and on steel bridge structures can be applied to composite bridge structures, provided that care is taken to consider their special features, given below.

(1) *Cracking of concrete*   For concrete box-girder bridges, 'It is generally accepted that for design purposes the distribution of internal forces, moments and displacements . . . can be based on an elastic analysis of an uncracked homogeneous concrete system'.[45c] This is also true for composite bridges, but for composite members the difference between the reductions of flexural stiffness in positive-moment and in negative-moment regions due to cracking is greater than in all-concrete structures, so 'uncracked' analysis of continuous members is less accurate.

The extent of cracking is strongly influenced by locked-in stresses due to shrinkage, and also makes uncertain the prediction of the effective breadth of a concrete flange. The coincidence in composite structures of these uncertainties with the need for accurate evaluation of stresses in steelwork, to avoid buckling and fatigue failures, is the principal difficulty that is peculiar to composite structures.

(2) *Shrinkage of concrete*   In comparison with an all-concrete member, axial shortening due to shrinkage is less, but changes of curvature are

greater, as the shrinkage occurs in one flange of the member only. In continuous members, the secondary effects of shrinkage (section 8.7.2) may influence the design of the steelwork or the planning of the sequence of construction.

(3) *Temperature effects* These may be more severe than in non-composite structures, because bare steel both conducts and radiates heat more rapidly than concrete.

(4) *Slip at shear connectors* Although shear connectors are ineffective until there is some slip, experience has shown that when they are provided over the full length of composite members, slip can be ignored in global analyses.

(5) *Method of construction* Changes in the effective cross sections of members during construction are likely to occur in long-span bridges in any material, but the widespread use of unpropped construction forces the designer to consider construction stresses in almost all composite bridges.

6.4.1 Scope of analytical methods

The principal methods that are in use for global analysis, and their applicability to the various types of composite bridge, are discussed in the following pages. Extensive bibliographies of this subject are given in references 7, 8 and 50 and further details of the methods are available.[51, 52]

Most composite bridge decks fall into one of two groups: those which can be idealised as an orthotropic plate or grillage, and those in which the longitudinal members consist of a small number of closed boxes. Those in the first group include beam-and-slab and multi-box decks (plates 1 and 2), and are the simpler to analyse. For uniform loading, elementary beam theory gives useful results, but distribution analysis by an orthotropic plate or grillage method is needed for more complex loadings.

The second group are known as spine box-girder bridges (plates 4 to 9). In general, the effects of loading on such structures are best evaluated by dividing the load into uniform, torsional and distortional components. Simple beam theory can be used for the first, and the Bredt theory[53] for thin-walled hollow sections gives the primary shear stresses due to torsion. These are accompanied by longitudinal torsional warping stresses which reach their maximum values at internal supports, and at cross sections subjected to eccentric point loads.

The distortional loading causes the cross section of the box to change shape by bending of its walls in the transverse plane. The associated transverse bending and shear stresses are usually acceptable in the relatively thick walls of concrete boxes, but in steel and composite boxes, diaphragms or cross-frames may be required to control the distortion. There are also longitudinal distortional warping stresses. In concrete boxes these can be several times as large as the torsional warping stresses[54] and in composite boxes their relative size depends on the spacing of the cross-frames.

Two other types of structural action that occur in both plate-type and box-type structures, are warping of cross sections in bending due to shear lag in flanges with high ratios of breadth to span, and local bending and shear in deck slabs in the vicinity of point loads.

Shear lag is usually neglected in global analysis, but must be considered when calculating stresses, as it frequently causes the longitudinal stress at a flange/web intersection to exceed the mean stress in the flange by 20% or so.[45f] It is allowed for in calculations based on the elementary theory of bending by using an effective flange breadth less than the actual breadth. The values given in the Bridge Code are based on a study of steel box girders[55] and will be discussed later (section 8.4.1).

Calculation of local bending stresses in plates is simplified by the use of influence surfaces.[56] Those for simply-supported plates tend to give high values in box girders as they do not consider the true boundary conditions of the plate panels.[54 (p. 259)]

Finally, the local effects of reactions at bearings cause complex states of stress in the support diaphragms and stress concentrations in the adjacent webs. This is a particular problem with steel diaphragms[57a] which suggests that concrete or composite members could be used more often in this location, where extra weight is easily carried. Kristek[63] has studied the influence of such diaphragms on longitudinal warping stresses.

These modes of structural action have been clearly explained by Horne[187] and in a paper[54] on concrete box-girder bridges, which lists 13 available analytical methods and is based on a survey[58] of about 500 references. Of the methods discussed below, the folded-plate, finite strip and finite element methods are capable of giving comprehensive analyses. There is some disagreement on the applicability of grillage analysis to box structures with fewer than five cells[54(discussion), 59] and it is not suitable for studies of warping stresses, shear lag, or local effects in decks.

There are methods, more specialised than the four just mentioned, that are useful in the study of particular aspects of box behaviour. The analogy with a beam on an elastic foundation[60] simplifies the calculation of transverse bending and distortional warping stresses, and can be used in conjunction with Heilig's method[61] for torsional warping stresses. Knittel[62] gives algebraic expressions for torsional and distortional stresses, neglecting warping effects and shear lag, and applies them to a four-cell trapezoidal box. A more general algebraic method[63] for tapered box girders of deformable cross section considers distortional effects in detail, and gives a useful rule for the spacing of diaphragms in single-cell boxes.

6.4.2 Orthotropic plate methods

The elastic properties of an orthotropic plate are defined by two flexural rigidities D_x and D_y and a plate torsional rigidity H. The governing equation relating deflection w to load p acting normal to the plane of the plate is:

$$D_x \frac{\partial^4 w}{\partial x^4} + 2H \frac{\partial^4 w}{\partial x^2 \partial y^2} + D_y \frac{\partial^4 w}{\partial y^4} = p(x, y) \qquad (6.2)$$

Design charts for decks[64] that can be idealised as orthotropic plates have been derived from series solutions of this equation by Morice and Little and by Cusens and Pama.[50] They give deflections and longitudinal and transverse moments due to a point load, and so provide a rapid method of distribution analysis. Their applicability is limited to simply-supported decks of skew not exceeding 20° whose elastic properties can be represented solely by length, breadth, and the three quantities D_x, D_y, and H.

In composite structures, they can be used[34] for beam-and-slab decks with not less than five equally spaced longitudinal members of uniform section, and no transverse members other than the concrete slab and diaphragms over the supports: but in practice the method has been superseded by grillage analysis.

FOLDED-PLATE ANALYSIS

Direct solution of equation (6.2) in cartesian or polar co-ordinates by harmonic or finite difference methods is straightforward by computer, and this enables assemblages of orthotropic plates to be analysed by the matrix stiffness method. Folded-plate analysis is applicable to box-girder decks[45c, 65] of uniform cross section. Its use for a deck slab composite with three open-top steel boxes (fig. 6.10) led to the simplified rules[45d] for load distribution in decks of this type given in the 1969 AASHO specifications.[26] For these small boxes, in which distortional and

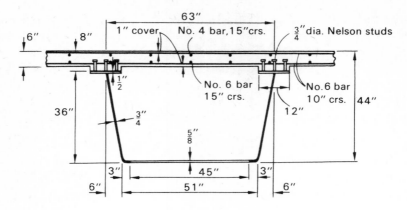

Typical girder cross-section

Bridge span-80ft.

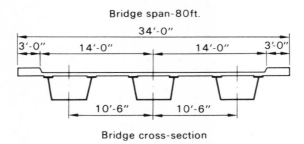

Bridge cross-section

Fig. 6.10   Cross sections of prototype bridge analysed by the folded-plate method

warping stresses were shown to be negligible, these simple rules are also applicable to continuous spans; but this may not be generally true, due to the warping restraint provided at internal supports by adjacent spans.

The folded-plate method based on Fourier solutions of equation (6.2) is not applicable to skew decks[45c] due to coupling between the harmonics, and is normally limited to assemblages of rectangular orthotropic plates. These may extend over several spans but must be simply-supported at the extreme ends, with rigid diaphragms over the end supports.[50] Its advantages are that it can give a complete and accurate solution in much less computer time than is needed for finite element methods, and can accept a wide variety of types of loading and both displacement and force boundary conditions.

Developments in folded-plate methods for bridges were reviewed by Scordelis[66] in 1975.

FINITE DIFFERENCE ANALYSIS OF ORTHOTROPIC PLATES

For boundary conditions such as column supports that rule out harmonic analysis, equation (6.2) can be solved by finite differences.[50] This method is not recommended for skew or tapering decks. It was used by Buragohain[67] for the analysis of deck slabs curved in plan and stiffened by both circumferential and radial beams eccentric to the slab.

6.4.3 Grillage analysis

In orthotropic-plate analysis the deck structure is 'smoothed' across its length and breadth, and so treated as a continuum. Grillage analysis is the converse in that the structure is represented by a plane grillage of discrete but interconnected beams. Almost any arrangement in plan is possible, so skew, curved, tapering or irregular decks can be analysed. But, the usual layout is sets of parallel beams in two directions. This will now be discussed, assuming the plane of the grillage to be horizontal.

In a simple form of grillage analysis, each beam is allotted a torsional stiffness and a flexural stiffness in the vertical plane. Vertical loads are applied only at the intersections of the beams. The matrix stiffness method of analysis is used to find the rotations about two horizontal axes and the vertical displacement at these nodes, and hence the bending and torsional moments and vertical shear forces in the beams at each intersection.

CHOICE OF GRILLAGE

Care must be taken to select an appropriate idealisation for a continuous structure, and also in deducing the stress resultants in it from the results of the grillage analysis. Detailed recommendations for concrete bridge decks are available.[59,68] Those in reference 68 are based on a comparison of test results with those given by grillage analysis for 53 bridge decks. The principal conclusions relevant to the design of composite structures are as follows:

(1) Grillage analysis is the most suitable method for simple beam-and-slab decks.

(2) Each longitudinal web should be represented by a grillage beam (so that a simple box becomes two beams) and there should be at least five and preferably nine such beams.

(3) Each transverse diaphragm or cross-frame should be represented by a transverse beam. If there are none within the span, or if their spacing exceeds the mean spacing of the longitudinal beams by more than 50%, additional transverse beams representing the slab alone should be added, and there should be at least five in all.

(4) The transverse grillage members should extend to the edge of the (real) slab and their ends should be attached to longitudinal grillage beams, even if the real slab has no significant edge stiffening. The reason is that some grillage programs can only consider loads that act on areas bounded on all sides by grillage members.

(5) For skew decks, the grillage cross-beams should be parallel to the real cross-members, if any. The stiffness and position of the real supports should be accurately modelled in the grillage.

Real cross-members can with advantage be skewed for skews up to about 20° but for high skews, they should be orthogonal. For decks without cross-members, skew grillages are preferred for skews up to about 30° because the preparation of input data for an orthogonal grillage is more complex. For highly skewed decks, the cross-members (if any) and the grillage beams should be orthogonal, to avoid the high torsional moments inherent in skew layouts. Transverse reinforcement in the slab should be parallel to the transverse grillage beams.

COMPUTER PROGRAMS
Not all of the many programs available include shear flexibility of beam elements, which substantially improves the representation at nominal extra cost.[59] Some programs can accept continuous or curved or non-uniform members, or random supports, or allow loads to be applied to points other than joints. Sawko[14a] has discussed the errors to be expected when a curved deck is simulated by straight beam elements, and gives details of grillage analyses of two such decks, one with ten concrete-encased plate girders and the other with eight box girders.

APPLICABILITY TO BOX STRUCTURES
West[68] gives no specific guidance on the applicability of grillage methods to composite box-girder decks, but advises against their use for concrete box or cellular structures where the area of the void exceeds 60% of the total area of each box cell. The reason is that the simple grillage method takes no account of the reduction of load distribution in the transverse direction due to distortional bending of the box walls. This restriction is relevant to concrete boxes without internal diaphragms or stiffeners.

Where the boxes are of steel, the need to allow for distortional bending in the grillage analysis can be assessed by estimating the average distortional stiffness per unit length of each box, and calculating the wall thickness of an equivalent uncracked but unreinforced concrete box of uniform thickness and equal stiffness, taking account of the modular ratio. If the voids ratio of this box exceeds about 60%, distortion effects

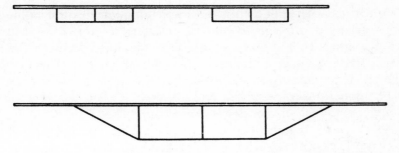

Fig. 6.11 Cross sections of decks analysed by the grillage method

should be allowed for in the analysis by giving the lengths of transverse grillage beams between the box webs the appropriate flexibility in vertical shear.

Grillage analysis has been used for concrete spine-beam decks with as few as two cells[54 (p. 257)] and guidelines for its use in multi-cell decks are available.[59] It neglects torsional warping stresses and the corresponding increase in effective torsional rigidity. Sawko notes[14a] that this error should be unimportant in composite box-girder decks, except for those with concrete end diaphragms (as for example in the Tay Bridge[24]) since these have significant out-of-plane stiffness.

The grillage method was used[45e] for the Saltings Viaduct[36] in which a deck 23 m wide was supported over seventeen 31 m spans on four widely spaced box girders 1.2 m square, without cross-girders. It has been used for the analysis of decks with cross sections of the types shown in fig. 6.11, and the results are similar to those given by finite strip analysis.[59] With proper modelling, the method can be used for any multi-box deck. It is not suitable for single-box decks unless the box is wide and has at least four cells.[54]

6.4.4 *Stiffness of grillage members*

Much of the guidance given by West[68] and by Cusens and Pama[50] on the evaluation of stiffness parameters for grillage beams is applicable to composite structures, if areas of structural steel are transformed to concrete on a modular ratio basis.

Shear lag can be neglected in global analysis, except as noted below, because its effects have little effect on the relative stiffness of members. It can be significant in partially concreted spans, for which the effective breadth of an isolated section of deck slab can be found from the tables given in codes, by using its length/breadth ratio in place of the span/breadth ratio for the whole span.

For regions where concrete is in compression, the flexural stiffness of a slab is strictly that of the uncracked reinforced section, but it is usually accurate enough (and inevitable before the reinforcement has been designed) to use the uncracked unreinforced section. The neglect of reinforcement cancels out to some extent the neglect of shear lag.

The reductions of stiffness due to cracking of concrete increase throughout the life of a bridge, so they cannot be allowed for precisely. The extent of the region where concrete is in tension is different for every load case. The concrete may not be cracked, and when it is, there remains the effect of tension stiffening. For these reasons, it is common to assume in global analyses that all concrete is uncracked and unreinforced.

If the longitudinal tensile stress in the slab near an internal support is found to be high, this is likely to be so for all the main girders. The effect of cracking on transverse load distribution will be negligible; but the effect on the longitudinal distribution of moments may be significant, and is considered in chapters 8 and 9.

Where composite cantilever cross-girders and an edge-stiffening member are used, allowance for cracking in the cantilever slab may lead to more accurate assessment of the action effects in the edge member, and lower design moments for the cantilevers.

LONGITUDINAL GRILLAGE MEMBERS

Flexural rigidities of longitudinal members depend on the breadth of concrete slab that is assumed to be associated with each steel web. For the deck shown in fig. 6.12(a), arbitrary division of the deck midway between webs would result in the neutral axis of the beam with web EH being higher than that of the beam with web FG. More rigorous analysis of concrete decks shows the neutral axes of all the longitudinal members to be almost at the same level across the width of the deck.[59] This suggests that the breadths of deck allocated to the webs EH and FG should be more nearly equal, as shown in fig. 6.12 by the lines A, B and C.

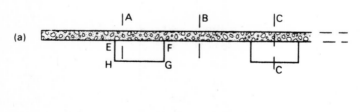

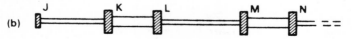

Fig. 6.12   Idealisation of multi-box deck for grillage analysis

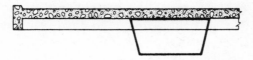

Fig. 6.13   Stiffening member at edge of cantilever deck

This should give a better reproduction of the longitudinal shear flows in the slab.

In a deck with a large edge cantilever, fig. 6.13, it is not obvious what breadth of slab should be associated with an edge-stiffening member when considering loading in its vicinity. Results from a grillage analysis can be misleading, and should be checked against results of an analysis of this specific problem.[45a]

TRANSVERSE GRILLAGE MEMBERS IN BEAM AND SLAB DECKS

The stiffnesses of a member that represents only a certain length of deck slab can be based on the whole of the slab replaced.

The stiffnesses of a member that is aligned with diaphragms or cross-frames should take account of the steelwork, which may be composite with the slab or connected to it only via the main girders, and of an associated effective breadth of slab. For T-beams, this breadth can be assumed to be one-quarter of the span of the transverse member, or one-sixth for cantilevers, as given in BS 5400: Part 3.

The flexural and shear stiffnesses can then be found by analysis of a transverse slice of the structure. In bracings of the type shown in fig. 6.14(a), any member AB is provided mainly for erection and can subsequently be ignored, as it will be close to the flexural neutral axis for the composite cross-frame. It can be assumed initially that the shear connection is rigid in the transverse plane, that the vertical stiffeners act with an effective breadth of steel web to form T or cruciform sections,

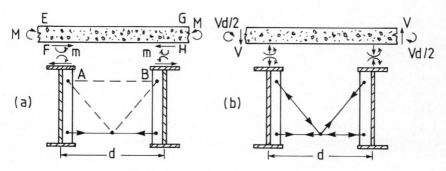

Fig. 6.14   Treatment of cross-bracing in global analysis

and that the bracing members are pin-jointed. The relative rotations between cross sections EF and GH due to transverse moments M can be calculated and hence, so can the mean flexural stiffness for length d of the grillage member. Similarly, the shear stiffness can be found from the force system shown in fig. 6.14(b).

If the results of a grillage analysis using these stiffnesses are such that the vertical tensile force at one side of the shear connection appears to be excessive, then account could be taken of the reduction in stiffness due to the axial flexibility of stud connectors; or the design could be modified. For example, member AB could be made composite, which would reduce the local moments m, or a more flexible bracing, such as a frame instead of a truss, could be used.

TRANSVERSE GRILLAGE MEMBERS IN BOX-GIRDER DECKS

If there are no composite cross-members, the grillage members consist of lengths of slab alone, such as JK and LM in fig. 6.12, alternating with lengths (KL, MN, etc.) that have the transverse stiffnesses of the composite box. The effective breadths of both the slab and the box are equal to the spacing of the grillage members, and the stiffnesses can be found by analysing a typical transverse slice in bending and shear. The modelling of the distortional stiffness of a box member is discussed by Hambly.[52]

TORSIONAL STIFFNESSES

Values of torsional stiffness for grillage members representing beam-and-slab decks can be found as described by West.[68] Methods for composite box girders are discussed in section 9.2.

6.4.5 *Interpretation of the output of a grillage analysis*

A typical output gives values of the vertical shear, bending moment and torsional moment for each grillage member on both sides of each joint in the grillage, and also the external reactions at each support. Care is needed in deducing design values for the real members from these results.

The bending and torsional moments will, in general, show a discontinuity at each joint (fig. 6.15). For an orthogonal grillage, each change in bending moment is equal to the change in torsional moment at that joint in the member at right angles to the one considered; and, similarly, the change in torsional moment equals the change in bending moment in the perpendicular member.

The design bending and torsional moments at each joint can be taken as the mean of the values on each side of the joint, for the member

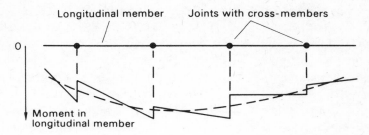

Fig. 6.15   Typical bending or torsional moments given by a grillage analysis

considered (dashed line in fig. 6.15).

The vertical shear forces in the output are equal to the slopes of the 'saw-tooth' line in fig. 6.15. They are larger than the shears due to bending alone, as they include the shear forces due to torsion, and they should be used for the design of members.

The results for vertical shear force are likely to be too low in members that meet at a joint which is an external support. To explain this, we consider the grillage shown in fig. 6.16, in which A and B are external supports. The computer program replaces a load W by the statically equivalent loads W_a, W_b, W_c and W_d at the four nearby nodes A, B, C and D. Loads W_a and W_b increase the external reaction, but do not enter into the grillage analysis, so that the sum of the calculated shear forces at A in members AB, AD, and AE is too low by an amount W_a; similarly at B. The error can easily be allowed for by comparing each computed external reaction with the sum of the shear forces in the members meeting at that point.

In skew bridges without transverse composite members, the directions of the reinforcing bars in the deck slab may not correspond with the directions of the grillage members. The reinforcement can be designed by the method of Wood and Armer[69] if at each grillage joint the bending

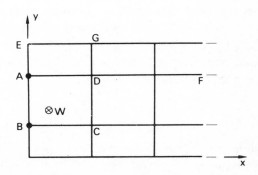

Fig. 6.16   Plan of grillage members near external supports A and B

and torsional moments m_x, m_y and m_{xy} ($=m_{yx}$) per unit width of slab are known. The method of finding them is now explained with reference to point D (fig. 6.16) assuming that ADF represents a composite girder of flange width b_x and GDC represents a strip of deck slab of width b_y.

As explained above, the moments M_x, M_y, M_{xy} and M_{yx} are found by averaging the computed values. The composite girder resists bending moment M_x and torsional moment M_{xy}, and from analysis of its cross section (using its effective breadth) the direct stresses in the slab in the x-direction and the torsional moment per unit width of the flange m_{xy} can be found. It will be conservative to assume that this exists over the whole of the actual width of the flange.

In the y-direction the bending moment m_y in the deck slab is M_y/b_y, and the torsional moment m_{yx} is M_{yx}/b_y. This will in general be different from m_{xy}. If the difference is small (e.g. less than 20%), the design of the reinforcement should be based on the mean value $(m_{xy} + m_{yx})/2$. For larger differences, the designer should satisfy himself that he understands the reason, and should consider designing for the higher of the two values.

6.4.6 Finite element analysis

This is the most versatile of the matrix stiffness methods of elastic analysis, and can in principle solve almost any problem of global analysis of a bridge deck. Its main disadvantage is its cost, which is so much higher than for other methods that, in practice, it is mainly used as a research tool and for checking methods that require less computer time. Other problems are that judgement and experience are needed in selecting an appropriate pattern of elements and interpreting results. These can be misleading in regions of steep stress gradient, because the conditions of static equilibrium are not necessarily satisfied. Manual preparation of computer input and interpretation of output takes up so much time that procedures for mesh generation and the plotting of trajectories of principal moment or stress are now incorporated into programs.

For these reasons the method has little application in routine design work, and is not summarised here.

6.4.7 Finite strip analysis

The finite strip method combines certain features of the folded-plate and the finite element methods. Its advantages and limitations are similar to those of the folded-plate method, except that it is free from the restriction that a single displacement function must be found that is valid over the whole area of each plate. Cusens and Pama show[50] that there is negligible difference between the results given by the two methods for longitudinal stresses in a three-cell, two-span continuous box girder, and note that the method requires a computer time that is usually between 2% and 40% of that for a comparable finite element solution. It is more suitable for box girders than for beam-and-slab structures, where grillage analysis can provide a cheaper solution.

The method is essentially a finite element analysis with a small number of elements, each of which consists of an orthotropic-plate strip running the length of the structure. The displacement function for the out-of-plane deformation of each strip is a harmonic series in the longitudinal x-direction and a polynomial in the transverse y-direction; for example, the third-degree function (the simplest practicable form) for a plate of length L is:

$$w(x, y) = (a_1 + a_2 y + a_3 y^2 + a_4 y^3) \sum_{n=1}^{\infty} \sin(n\pi x/L)$$

The boundary conditions of simple support at the ends of each strip are automatically satisfied, and the values of the coefficients a_1 to a_4 are chosen to provide continuity of slope and deflection at the longitudinal edges of the strips. The amplitudes of these slopes and deflections are the basic unknowns in the analysis, and are calculated by the matrix force method in the usual way. A similar procedure is used for in-plane deformations of each strip.

The auxiliary nodal line technique[45b, 50] is an improved method of solution which can be used for composite box-girder decks with one or several cells of uniform longitudinal cross section.

Transverse or longitudinal ribs eccentric to the plate or slab can be simulated if their spacing is small enough to allow their influence to be smoothed over the span in the longitudinal direction and over the width of a strip in the transverse direction, and programs for these methods have been written for both straight and uniformly curved decks.[45]

As harmonic analysis is used in the longitudinal direction only, point loads acting within strips have to be replaced by their statical equivalents acting on the boundaries between strips, so the type of loading to be considered may influence the location of these boundaries. It should also be remembered that for a given level of accuracy many more harmonics must be used in an analysis for point loads than for distributed loads.

Plate 1   Marton-le-Moor Bridge, Yorkshire

Plate 2   Moat Street flyover, Coventry

Plate 3   Wardley Moss Bridge (courtesy Lancashire Sub-unit, North Western Road Construction Unit)

Plate 4   Raith Bridge, Scotland, during construction

Plate 5   Cantilever cross-girders for an approach span, Avonmouth Bridge (courtesy Freeman, Fox and Partners, consulting engineers)

Plate 6   Tinsley two-level viaduct, Sheffield

Plate 7   Bridge at Veveyse, Switzerland. Launching of the upstream box and construction of the concrete deck slab (courtesy Ateliers de Constructions Mécaniques de Vevey, S.A.)

Plate 8 Pont de la Madeleine over the River Sarine, Fribourg, Switzerland, after launching of the steelwork (courtesy Ateliers de Constructions Mécaniques de Vevey, S.A.)

Plate 9　Sarine Bridge, showing jacking of the superstructure at the internal supports during erection of the precast deck slabs (courtesy Ateliers de Constructions Mécaniques de Vevey, S.A.)

Plate 10　Bridge at Conflans-Ste-Honorine, France. The toothed-plate shear connectors are visible at the top right

Plate 11 Box girders and plate-girder crosshead for motorway interchange at Poyle, near London, built in 1984–85 (courtesy Cass, Hayward and Partners, consulting engineers)

Plate 12 'Omnia' planks used as permanent formwork spanning 3.5 m between box girders at Poyle interchange (courtesy Cass, Hayward and Partners, consulting engineers)

Chapter Seven

Design Philosophy, Materials and Loading

7.1 Limit-state design philosophy for bridges

The principal terms used in limit-state design philosophy were introduced in chapter 1 (Volume 1). These are the *partial safety factors* for loads and load effects (γ_f) and for materials (γ_m), the *characteristic* and *design* values of loads and materials' strengths, and the *serviceability* and *ultimate limit states*.

In the Bridge Code[1] 'characteristic' has been retained in definitions of the strengths of concrete, reinforcement and tendons, but has elsewhere been replaced by 'nominal', the meaning of which is defined in BS 5400: Part 1. The *nominal strength* of a material is the specified strength, to be confirmed by tests on samples of the material used in the structure. Where the number of tests is sufficient, the nominal strength is taken as that below which not more than 5% of the results may be expected to fall. It is then the same as the *characteristic strength*, as used for example in BS 8110.[185] It differs from the characteristic strength when it is confirmed by an established procedure of batch or sample testing that does not give sufficient results to enable the 'five per cent' level to be determined, as is the usual practice with rolled structural steelwork.

The *nominal loads* specified for bridges are those which may be expected to occur not more than once or twice within the design life of the bridge, which in Great Britain is usually taken as 120 years. Where adequate statistical evidence is available, as for wind load, the nominal value is that which has a probability of 63% of being exceeded during this period.

The philosophy and terminology of BS 5400 and of the Eurocodes were parallel developments from earlier work by the European Committee for Concrete. The treatment in BS 5400 dates from about 1976, whereas current versions of Eurocodes 2, 3 and 4[3-5] were drafted between 1981 and 1984. Their presentation, symbols and terminology have been influenced by work since 1981 of the Eurocodes Coordinating Committee, and by the need for unambiguous expression in all of the relevant languages. In consequence, there are inconsistencies between BS 5400 and the Eurocodes, as explained below.

7.1.1 Design philosophy and notation of Eurocodes 2, 3 and 4

Three types of *basic variables* are defined:

(1) *Actions*, which include loads, imposed deformations (e.g. due to settlement) and certain consequences of prestress and of shrinkage of concrete.

(2) *Strengths*, and some other properties of materials.

(3) *Geometrical data*, such as initial bow of columns and tolerances in fabricated steelwork.

Representative values of variables, F_{rep}, include *characteristic* and *nominal* values, defined above. These are multiplied by coefficients γ_f that take account of deviations and uncertainties related to them. The *action effects* are functions S of the actions. They are multiplied by coefficients γ_{Sd} to give *design action effects* S_d:

$$S_d = \gamma_{Sd} . S(\gamma_f F_{rep}) \tag{7.1}$$

where γ_{Sd} allows for the uncertainties in the model for the limit-state condition used in determining the action effect.

Similarly, the *design resistance* R_d of a member or cross section is given by:

$$R_d = (1/\gamma_{Rd}) . R(f_k/\gamma_m) \tag{7.2}$$

where f_k is a characteristic strength, γ_m allows for deviations relating to f_k, and γ_{Rd} allows for uncertainties in the model for the limit-state condition related to the assessment of resistance capacity (function R) and of geometrical parameters.

A general condition for satisfactory design for the ultimate limit state is:

$$R_d \geqslant S_d \tag{7.3}$$

and the comparison may be scalar, vectorial (e.g. when a moment/shear interaction diagram is used), or more complex.

In the Eurocodes, equations (7.1) and (7.2) can be simplified where appropriate to:

$$S_d = S(\gamma_F F_{rep}) \tag{7.4}$$

and:

$$R_d = R(f_k/\gamma_M) \tag{7.5}$$

where:

$$\gamma_F = \gamma_{Sd}\gamma_f \qquad \gamma_M = \gamma_{Rd}\gamma_m \tag{7.6}$$

Representative permanent and variable loads are denoted by G and Q, respectively, and γ_F is written as γ_G, γ_Q, etc., as appropriate. Several types of *load combination* are defined; only the *fundamental* (also named 'rare' or 'infrequent') combination is now considered. Its design action effects are given by

$$S_d = S\left[\gamma_G G + \gamma_{Q1} Q_1 + \sum_{i>1} (\gamma_{Qi}\psi_{0i}Q_i) \right] \tag{7.7}$$

where:

Q_1 is the *basic variable action*
Q_i is an *accompanying variable action*
ψ_{0i} (<1) is the *combination coefficient* for Q_i.

7.1.2 Design philosophy and notation for BS 5400

It is relevant to BS 5400 that from equations (7.1) and (7.2), equation (7.3) can be written:

$$R(f_k/\gamma_m) \geqslant \gamma_d S(\gamma_f F_{rep}) \tag{7.8}$$

where:

$$\gamma_d = \gamma_{Rd}\gamma_{Sd} \tag{7.9}$$

For reinforced concrete, the left-hand side of equation (7.8) becomes:

$$R(f_{cu}/\gamma_c, f_{ry}/\gamma_r)$$

where subscripts c and r refer to concrete and to reinforcement.

For structural steel, the evidence from tests and the nature of the function R have caused the expression $R(f_k/\gamma_m)$ to be changed in Part 3 to $(1/\gamma_m)R(\sigma_y)$ where σ_y is the nominal yield strength of the steel, and γ_m has values ranging from 1.05 to 1.30, depending on the type of resistance considered. These are based on statistical analyses of test data[150] and are chosen to give a calculated probability of failure of approximately 6×10^{-5}.

The coefficient γ_d of the Eurocodes is essentially γ_{f3} of BS 5400, known as the 'gap factor', which takes account of 'inaccurate assessment of the effects of loading, unforeseen stress distribution in the structure, variation in dimensional accuracy achieved in construction, and the importance of the limit state being considered' (Part 3[1]). For composite bridges, its values are 1.0 at the serviceability limit state and 1.1 at the ultimate limit state. Most of the uncertainty represented by the value 1.1 relates to action effects in concrete bridges and to resistance of members in steel bridges. Consequently, in BS 5400, γ_{f3} is applied to resistances in Part 3 (steel bridges), implying that $\gamma_{Rd} = 1.1$, $\gamma_{Sd} = 1.0$; and to action effects in Part 4 (concrete bridges), implying that $\gamma_{Rd} = 1.0$, $\gamma_{Sd} = 1.1$.

In Part 5 (composite bridges) as amended by the Department of Transport[2], it is stated that the relation to be satisfied is that of Part 4, which is:

$$R(f_k/\gamma_m) \geqslant \gamma_{f3}S(\gamma_{fL}Q_k) \qquad (7.10)$$

where γ_{fL} and Q_k are equivalent to the γ_f and F_{rep}, respectively, of the Eurocodes. This is misleading, because the relation actually used in Part 3 for composite members (e.g. in the note to clause 9.9.1.2) at the ultimate limit state is different and more complex:

$$(1/\gamma_{f3}\gamma_{ms})R(\sigma_y, f_{cu}\gamma_{ms}/\gamma_{mc}) \geqslant S(\gamma_{fL}, Q_k) \qquad (7.11)$$

where:

γ_{ms} is γ_m for steel, which ranges from 1.05 to 1.30
γ_{mc} is γ_m for concrete, which is 1.50.

For example, the resistance of a composite beam of compact section in sagging bending, for which $\gamma_{ms} = 1.05$, is first calculated assuming that the stress in the concrete flange is $0.6 \times 1.05/1.5 = 0.42f_{cu}$ (compared with $0.40f_{cu}$ for a concrete T-beam), and is then divided by 1.155 ($= 1.1 \times 1.05$).

For beams in hogging bending, for which γ_{ms} is usually 1.20, Part 3 gives no guidance on whether account need be taken of the small difference between γ_{ms} and γ_m for reinforcing bars, which is 1.15. If it is ignored, as the authors recommend, and all the concrete is cracked in tension, the left-hand side of equation (7.11) becomes:

$$(1/\gamma_{f3}\gamma_{ms})R(\sigma_y, f_{ry})$$

For calculations at the ultimate limit state for deck slabs, shear connection and transverse reinforcement, it is convenient to calculate resistances following Part 4 or Part 5 of the Bridge Code (that is, excluding γ_{f3}) and to apply γ_{f3} to the effects of the factored loads, as stated in equation (7.10). This is the method used generally in this book. The alternative is to divide the resistances by γ_{f3}. This is sometimes simpler; for example, when design is governed by the minimum value of a loading, rather than the maximum.

For loads, the values of γ_{fL} given in BS 5400: Part 2 correspond in principle, but not in detail, to the format of equation (7.7). For example, the reduction of γ_{fL} for HA loading from 1.5 for combination 1 to 1.25 for combination 2 is equivalent to using $\psi_0 = 1.25/1.5 = 0.83$ when HA load accompanies wind load.

7.1.3 Load combinations and factors for highway bridges

Combination 1 (table 7.1) consists of all permanent loads (dead, superimposed dead, and the effects of shrinkage of concrete) plus primary live loads for all types of bridge and secondary live loads for railway bridges only. Primary live loads are the vertical loads due to the *mass* of traffic, and are treated as static loads. Secondary live loads arise from the *movement* of traffic; for example, centrifugal, longitudinal, skidding, and collision forces in highway bridges, and centrifugal, longitudinal, lurching, and nosing forces in railway bridges.

Table 7.1   Load factors γ_{fL} for load combination 1

Limit state:	Service-ability	Ultimate	Ratio ULS/SLS, including $\gamma_{f3} = 1.1$
Dead load:			
steel structural elements	1.0	1.05 or 1.1*	1.155 or 1.21*
concrete structural elements	1.0	1.15 or 1.2*	1.265 or 1.32*
load that reduces the effect considered*	1.0	1.0	1.0
Superimposed dead load: surfacing	1.2	1.75	1.604
load other than surfacing	1.0	1.2	1.32
load that reduces the effect considered*	1.0	1.0	1.0
Shrinkage of concrete	1.0	1.2	1.32
Highway loading:			
type HA	1.2	1.5	1.375
type HB (and HA in same lane)	1.1	1.3	1.3
Footway loading	1.0	1.5	1.65

* See section 7.3.1

Combination 2 consists of the appropriate combination 1 loads plus the effects of wind. Higher values of γ_{fL} for wind are specified (table 7.2) for bridges without live load than for loaded bridges, as the combination of full live load with maximum wind is highly improbable.

Combination 3 consists of the combination 1 loads plus the effects of temperature range and temperature difference. The factors γ_{fL} for highway load (table 7.2) are slightly reduced, to take account of the lower probability that extreme highway and temperature loadings will coin-

Table 7.2   Load factors γ_{fL} for load combinations 2 and 3

Limit state:	Service-ability	Ultimate	Ratio ULS/SLS, including $\gamma_{f3} = 1.1$
Combinations 2 and 3			
Dead and superimposed dead loads			
Shrinkage of concrete	} as for combination 1		
Temporary materials, plant and equipment			
during erection	1.0	1.15	1.265
Highway loading:			
type HA	1.0	1.25	1.375
type HB (and HA in same lane)	1.0	1.1	1.21
Footway loading	1.0	1.25	1.375
Combination 2 only			
Wind:			
with dead plus superimposed dead			
load only	1.0	1.4	1.54
Wind:			
with dead, superimposed dead, and			
appropriate live or erection loads	1.0	1.1	1.21
Wind:			
that reduces the effect considered	1.0	1.0	1.0
Combination 3 only			
Temperature (restraint to movement, except			
frictional)	1.0	1.3	1.43
Effects of temperature difference	0.8	1.0	1.375

cide. In 1984, it was concluded from a study of this subject[70] that the design highway loadings for combination 3 could be reduced to 0.5 HA for the serviceability limit state and 0.6 HA for the ultimate limit state, and that there was no case for including HB loading, except perhaps for bridges on routes to major ports.

The strongest winds never occur at the same time as extremes of temperature, so that no combination includes both types of load.

Temporary loads that can occur only during erection are considered in combination with wind and with temperature, but not in combinations 1, 4, or 5.

Combination 4, which does not apply to railway bridges, consists of the permanent loads, the secondary live loads, and the appropriate primary live loads associated with them.

Combination 5 consists of the permanent loads and the loads due to friction at bearings, which include frictional restraint of movement due to temperature effects. In Appendices X and Y no calculations for combinations 4 and 5 are given, as they have negligible influence on the design of the structures considered.

7.1.4 The limit states

ULTIMATE LIMIT STATES

An ultimate limit state is reached when there is failure or loss of static equilibrium of the structure or an element of it. For example, there may be loss of load-bearing capacity of a member due to the strength of its material being exceeded, or due to buckling, brittle fracture, or fatigue; or there may be overall instability caused, for example, by aerodynamic behaviour or transformation into a mechanism; or a cantilever may fail due to lifting of the far end of its anchor span from its bearing.

SERVICEABILITY LIMIT STATES

These are conditions 'beyond which a loss of utility or cause for public concern may be expected and remedial action required'.[1] Examples are:

(1) excessive deformation of all or part of the structure, affecting its appearance, clearances, or drainage; or causing damage to non-structural components such as surfacing
(2) localised fatigue damage
(3) excessive vibration, causing discomfort or alarm to users
(4) excessive local crushing, spalling, slipping, cracking or yielding, affecting the appearance, use, or durability of the structure.

The limits to calculated stresses in steel that are specified in BS 5400 are not intended to ensure that yield never occurs.[150] The residual stresses caused by rolling, flame cutting, or welding invariably result in local yielding well before working loads are reached. The specified stress limits are intended to ensure that when yielding occurs, it leads to shakedown (subsequent elastic behaviour) rather than to a gradual increase in crack widths in concrete (which would accelerate corrosion) and in deflections. Excessive yielding could also reduce fatigue life, as stress resultants would eventually differ from those given by elastic analysis and used in the design.

Highway bridges (but not railway bridges for high-speed trains) turn out to be so stiff for other reasons that limitation of deflection need rarely be considered in design. Vibration is found to be a design problem only in some footbridges (section 8.8.2) and in long-span suspended structures. Cable-stayed bridges, which sometimes have composite decks, can suffer from aerodynamic instability, which is treated in the literature on long-span steel structures.

Wide cracks in concrete lead to corrosion of reinforcement and deterioration of concrete. The need to avoid them influences the detailing of reinforcement (section 8.3.3) and the planning of the sequence of construction of deck slabs.

FATIGUE

The ultimate limit state may be reached if a critical fatigue failure occurs, but fatigue damage of a minor nature is a form of unserviceability. Because fatigue failures may be classified as either limit state, depending on the circumstances, fatigue checks are, in practice, done as a separate investigation, using a philosophy different from those for the serviceability and ultimate limit states.

The fatigue lives (N) for standard types of welded detail are given in Part 10 of the Bridge Code in terms of the applied range of stress (σ_r). These are based on extensive testing, and are used without modification by a partial safety factor. Only the fatigue behaviour of shear connectors is considered in this book. Their relevant properties are given in Appendix B.

The only loadings that have much influence on the fatigue behaviour of a composite deck for a highway bridge are those due to commercial and abnormal vehicles. Accurate prediction of the number, sizes, weights and spacings of these over a 120-year period is obviously not possible, and the variation from bridge to bridge is likely to be greater than for maximum static loads. The design highway loadings and associated numbers of repetitions (the *load spectra*) given in Part 10 of the Bridge Code are based[57b, 71] on observations of traffic and on the Motor Vehicles (Construction and Use) Regulations, and are used without modification by a load factor. Those used in this book are given in Chapter 10.

In relating these loadings to the specified fatigue lives of welded details, Miner's cumulative damage rule is normally used. This is the best method that is sufficiently simple for everyday use, but it is not very accurate, and so introduces a further uncertainty into fatigue calculations.

The margin of safety provided in design for fatigue is essentially included in the σ_r–N data and in the load spectra. It is accepted that fatigue failures will occur in some bridges during their design lives, and it is expected that the great majority of these will be of a minor nature.

7.2 Properties of materials and shear connectors

In the Bridge Code, properties of materials are specified as *nominal* values for some materials and as *characteristic* values for others, for reasons explained in section 7.1.

7.2.1 Structural steel and reinforcement

Structural steels are required by Part 6 of the Bridge Code[1] to comply with BS 4360,[72] unless otherwise specified. Only the two most widely used steels, Grade 43 and Grade 50, are used in examples in this book. The nominal yield strengths to BS 4360 depend on the thickness t of the plate or section, as shown in table 7.3. Account is taken[150] in the design methods of BS 5400 of the variation of σ_y with t, so that in Part 3, σ_y is taken as the value for $t = 16$ mm, irrespective of the actual thickness of the plate or section.

Table 7.3 Nominal yield strengths of structural steel

Thickness t (mm)	Grades 43A, B and C		Grades 50B, C and D
	Plate (N/mm²)	Rolled sections (N/mm²)	Plate and rolled sections (N/mm²)
$t \leqslant 16$	275	255	355
$16 < t \leqslant 25$	265	245	345
$25 < t \leqslant 40$	265	240	345
$40 < t \leqslant 63$	255	230	340

Design against fatigue and brittle fracture of structural steel in composite structures is as in steel structures, and so is not discussed in this book. The fatigue strength of shear connectors is treated in chapter 10 and Appendix B.

Reinforcement for concrete is required by Part 7 of the Bridge Code to comply with the relevant British Standards, BS 4449[73] and BS 4461[74] in particular. Hot-rolled high-yield bars, for which the specified characteristic strength, f_y, is 460 N/mm², are used in the examples in this book.

The elastic modulus for structural steel, E_s, is given in Part 3 of the Bridge Code as 205 kN/mm², and in Eurocode 3[4] as 210 kN/mm². Both Part 4 and Eurocode 2 give the elastic modulus for reinforcing steel as 200 kN/mm², but Eurocode 4 also allows it to be taken as equal to that for structural steel. This greatly simplifies calculations of transformed steel sections for composite members. The error is negligible because the area of structural steel in a composite section is always much greater than the area of reinforcement.

7.2.2 Concrete

Characteristic compressive strengths of concretes are given in both Part 4 and Part 7. Those that may be specified for reinforced concrete work are Grades 25, 30, 40 and 50, where the grade number is the characteristic cube strength in N/mm² at age 28 days. The associated values of the short-term modulus of elasticity of normal-density concrete and the corresponding modular ratios (assuming $E_s = 205 \text{ kN/mm}^2$) are given in table 7.4.

Table 7.4 Modulus of elasticity of concrete for short-term loading

Characteristic cube strength f_{cu} of concrete at the age considered (N/mm²)	Modulus of elasticity of concrete, E_c; mean and typical range (kN/mm²)	Modular ratio
25	26 (22 to 30)	7.9
30	28 (23 to 33)	7.3
40	31 (26 to 36)	6.6
50	34 (28 to 40)	6.0

In Eurocodes 2 and 4, the grade number refers to the characteristic cylinder strength, and provision is made for the use of Grades 20 to 50 in composite structures. The conversion factor from cube to cylinder strength depends on the grade of concrete, type of aggregate, and other variables, and values ranging from 0.67 to 0.85 have been used. The value used in the drafting of Eurocode 4, 0.80, is recommended for general use, in absence of better information.

Provision is made in Parts 4 and 5 of the Bridge Code for the use of lightweight concretes of density between 1400 and 2300 kg/m³. The short-term modulus for concrete of density D_c (kg/m³) is taken as the value in table 7.4 multiplied by $(D_c/2300)^2$. The treatment in Eurocode 4 is similar.

An assumption has to be made about the type of aggregate to be used, as well as the density, before the structure is designed. The coefficients of thermal expansion of structural steel, reinforcement, and many types of concrete may all be taken as 12×10^{-6} per degree Celsius. Limestone, lightweight and certain granite aggregates can give concretes with a coefficient of 9×10^{-6} per degree Celsius or less. The effects of differential thermal expansion should be considered in design if such an aggregate is used.

Creep is allowed for in the Bridge Code by the use of an effective modulus for calculations relating to permanent loading. It is stated in Parts 4 and 5 that a value half of that given in table 7.4 is appropriate. This implies that the creep strain ε_c equals the initial elastic strain ε_e (i.e. that the creep coefficient $\phi = \varepsilon_c/\varepsilon_e = 1.0$). The value 1.0 is in fact appropriate to creep after a year or two: values of ϕ for creep after several decades usually lie between 2 and 4.

The use of the short-term modulus for variable load and an effective modulus for permanent load is unnecessarily complex for global analysis, which is why it is stated in Part 3 that the 'long term' effective modulus (i.e. $\phi = 1.0$) can be used for all loads. In unpropped construction, only a small proportion of the total design stress in concrete may be due to permanent loads, but the error is partly offset by the fact that those loads cause more creep than is implied by using $\phi = 1.0$.

Creep has more influence on the results of elastic analyses for stresses in cross-sections than it has on global analyses, so that it is necessary in slender sections to calculate live-load and dead-load stresses separately. It is usual in Great Britain to assume $\phi = 1$ for the latter, but higher values are used in some other countries.

7.2.3 Static strength of shear connectors

Methods of shear connection are discussed and illustrated in Volume 1. The only types of connector for which static strengths are given in Part 5 of the Bridge Code are headed studs, bars with hoops, channels and friction grip bolts. The derivation of these strengths is described below and, for friction grip bolts, in section 7.2.4. Other types of connector, such as horseshoes (in Preflex beams) and bent-up reinforcement bars have been used for bridges, but are now rarely used in Great Britain. Design of connectors for repeated loading is described in Chapter 10.

No provision has been made for the use of epoxy resins for shear connection even though it is well established that with careful surface preparation and material control, a connection can be made with a shear strength exceeding that of the concrete. The reasons are: that this type of connection fails at a very low slip, and so has less ability to redistribute load than the mechanical connectors now in use;[75] that it attaches the steel to the underside of the slab only and so cannot comply with current rules for resistance to forces tending to cause uplift (section 8.5.3); and that there is inadequate information on its durability and fatigue strength, in relation to use in bridge decks. There is some evidence[7] of rusting of the top surface of the steel member due to penetration of the epoxy layer by water from the concrete above.

Typical nominal static strengths for stud, bar and channel connectors from Part 5 of the Bridge Code are given in table 7.5 and fig. 7.1, in terms of the characteristic cube strength of the concrete surrounding the connectors. They were derived by Menzies[76] from the results of push-out tests at the Building Research Establishment, at Imperial College, London and in the USA.

The British Standard method for finding the compressive strength of concrete (BS 1881) requires cubes to be stored under water until they are tested, and so Menzies reported his own results in terms of concrete strengths found from water-cured cubes. These strengths were on

Table 7.5 Nominal static strengths of shear connectors

Type of connector		Nominal static strengths in kN per connector for concrete strengths f_{cu} (N/mm²)			
		20	30	40	50
Headed studs					
Diameter (mm)	Overall height (mm)				
25	100	139	154	168	183
19	100	90	100	109	119
19	75	78	87	96	105
Bars with hoops					
50 mm × 40 mm × 200 mm bar		697	830	963	1096
Channels					
127 mm × 64 mm × 14.90 kg/m 150 mm long		351	397	419	442

average 20% higher than those of similar cubes moist cured in air alongside the push-out specimens. This latter method had been used in the work from which the design strengths in CP 117: Part 2[77] were deduced, and for the concrete cylinders to which the American results were related. These other results were adjusted to allow for the different methods of testing concrete, and mean static strengths were obtained for each type of connector by regression analyses on concrete strength. These are the 'nominal strengths' in the Bridge Code.

The values for typical sizes of bar, channel, and stud are compared in fig. 7.1 with the 'ultimate capacities' given in CP 117: Part 2. For stud

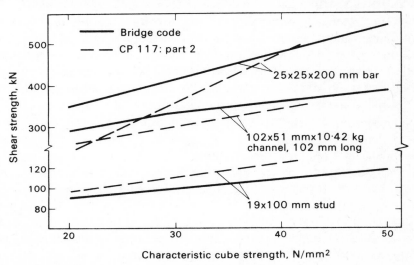

Fig. 7.1 Static strengths of typical shear connectors

connectors, the new strengths are about 10% less than those in CP 117.

The strengths of channel connectors were found to depend on the direction of loading. In CP 117: Part 2 it is recommended that the maximum shear force should be as shown in fig. 7.2, as it had previously been found in tests on specimens of moderate cube strength that channels were stronger when loaded in this direction. Menzies found[76] that for cube strengths above about 30 N/mm², channels were stronger when loaded in the other direction. The new nominal strengths are based on the lower of the strengths found for loading in the two directions, and therefore have a slight kink at about 30 N/mm². They are about 10% lower than strengths given by a formula based on tests in the USA[78] and up to 13% higher than those given in CP117: Part 2.

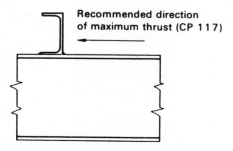

Fig. 7.2   Direction of loading for channel connector

The strengths of bar connectors will be discussed in terms of the ratio of the mean bearing stress at maximum load on the vertical face of the bar to the cube strength of the concrete. The strengths given in CP 117: Part 2 corresponds to a ratio of 2.34. Six tests by Graf (1950) gave ratios from 2.35 to 3.2 for cube strengths from 38 to 45 N/mm². The nominal strengths in the Bridge Code are given by the regression line for mean strength from 16 tests by Menzies. The ratio of bearing stress to cube strength is 3.49 at a cube strength of 20 N/mm², falling to 2.19 at 50 N/mm², so the specified strengths are substantially higher than in CP 117 for weak concrete, and about the same for strong concrete.

STEELS USED FOR SHEAR CONNECTORS

The strengths given in table 7.5 for bar and channel connectors are for Grade 43 steel (nominal yield stress, 255 N/mm²). For studs, the situation is more complex, though it is known that their push-out strength in strong concrete depends more on the ultimate tensile strength (UTS) of the stud material than on its yield strength.[7] (In weak concrete, it depends mainly on the strength of the concrete.)

The shanks of studs are normally drawn down from material of larger diameter, and so may not have uniform properties through the cross section. Most studs are too short for a standard tensile test to be done on the whole cross section because the specified gauge length for the determination of elongation is five times the diameter of the specimen. The results of tests on coupons cut from shanks of studs may depend on the diameter of the coupon, which has not been standardised. Further, the steel properties that most influence the strength and ductility of studs in beams are those of the heat-affected zone near the weld.

The specification for studs in the Bridge Code is limited to the material from which the studs are drawn. From Part 6[1]: 'Steel for headed stud type shear connectors shall have a minimum yield stress of 385 N/mm^2 and a minimum tensile strength of 495 N/mm^2'. The same figures are given in Part 5, with the addition of a minimum elongation of 18%. This value is believed to be based on the actual properties of steels used for early studs, rather than on analysis of the minimum required (which probably depends on the extent to which partial shear connection is exploited in the design of beams for buildings). The minimum elongation specified in draft Eurocode 4[5] is 15%, based on practice in West Germany.

Confusion can arise when these values are compared with the results of tests on as-drawn studs, because drawing increases the UTS and reduces elongation (in one set of tests, for example, from 19% to 13%). A survey has been made of the results of 108 tensile tests on finished studs done between 1971 and 1983, and taken from four sources. The lowest UTS was 488 N/mm^2; all the others lay between 500 N/mm^2 and 680 N/mm^2. Elongations, after correction to the standard ratio of gauge length to coupon diameter, ranged from 10.6% to 17%; three were below 12% and 93 below 15%.

It is clear that quality control for stud connectors (to BS 5400) has to be based mainly on test certificates for the parent material. Results of tests on studs should show the same or a higher UTS, but the elongation cannot be expected to reach 18%. Protection against excessive embrittlement of the material during manufacture of studs is given by the procedure trials and site tests described in section 15.3.

In this situation, test data from different countries have to be interpreted with care. For example, in pre-1965 tests in the USA, studs with tensile strengths up to 517 N/mm^2 were used; more recently, the tensile strengths have been close to the specified minimum value, 414 N/mm^2.

LOAD–SLIP CURVES
The mean load–slip curves obtained by Menzies[76] are compared in fig.

7.3 with results of tests in the USA and by Stark in the Netherlands. The studs tested by Ollgaard *et al.*[79] appear to be more flexible partly because their mean ultimate strengths were higher, as they were tested using four studs per slab, rather than two. The greater flexibility shown in the results reported by Stark may be due to the smaller diameter of the studs. These variables, and also the size and shape of the test slabs, the details of reinforcement near the connectors, and the size of the weld collars, have been found[80] to have a marked influence on the mode of failure of stud connectors, and on the maximum load and corresponding slip.

Slip reduces composite action and increases flexural stresses and deflections, so the stiffness of a shear connection should be high at the serviceability limit state. An appropriate measure of stiffness is the slip at about 55% of the nominal shear strength of the connector. Figure 7.3 shows that the slip at this load of stud connectors tested in different ways varied by a factor of about three and also that there was little difference between the three types of connector tested in similar ways by Menzies. His results show that in well-reinforced slabs, 25×25 mm bar connectors are about 1.6 times as stiff as 19-mm studs.

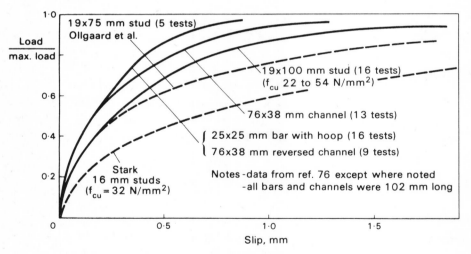

Fig. 7.3   Load–slip curves for stud, channel, and bar connectors

CONNECTORS IN LIGHTWEIGHT-AGGREGATE CONCRETE

Menzies[76] did push-out tests using two types of lightweight aggregate; expanded clay and shale (Leca) and sintered pulverised fuel ash (Lytag). The ratio of the strength of the stud connectors to their strength in normal-density concrete was about 0.6 for Leca concrete of density 1230 kg/m^3 and 0.85 for Lytag concrete of density 1590 kg/m^3. This 15% reduction in strength agreed well with results of tests in the USA in the

1960s using concretes of density between 1490 and 1730 kg/m³. This work led to the rule in Part 5 of the Bridge Code that for lightweight concretes of density exceeding 1400 kg/m³, the static strength of connectors should be taken as 15% less than the value specified for normal-density concrete of the same strength.

The load–slip curves for studs given by Menzies show that the ratio of the stiffness at working load of stud connectors in lightweight concrete to their stiffness in normal-density concrete is slightly less than the ratio of ultimate strengths. Figure 7.4 is typical of similar results found at the University of Missouri[81] both for 22-mm studs and for 100-mm channel connectors. In the figure, there is a negligible difference between the curve for normal-density concrete scaled down 15% (to allow for the reduction in design strength), and the lower of the two curves for lightweight concrete of similar strength.

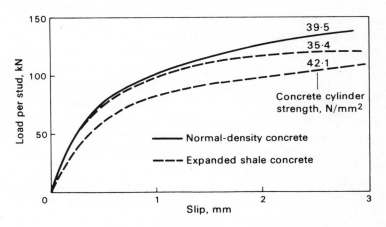

Fig. 7.4 Load–slip curves for 22.2-mm studs

RESISTANCE TO SEPARATION

The tendency for the slab of a composite member to separate from the steel girder is noted in Volume 1. In the standard push-out test, separation is prevented to some extent by the way in which the slabs are supported. Teraszkiewicz devised two types of push-out test[82] in which separation was restrained only by the connectors themselves. He found the shear strengths of 100 × 19 mm headed studs to be 20% to 35% less than the value given by standard tests, and that the reduction increased to 45% when the heads of the studs were machined to the same diameter as the shank, even though the separation in all the tests was always less than half the slip.

In bridge decks, certain patterns of loading tend to cause separation, so the shear connectors must prevent uplift as well as slip. This is assumed to be done by the underside of the head of a stud; the lower surface of the top of the hoop, for a bar connector; and the lower face of the top flange of a channel connector. This 'surface that resists separation' must be embedded in the concrete deck slab by a minimum distance, given by empirical rules in Part 5 of the Bridge Code. These rules effectively define the minimum height of connectors. They are based on rather limited evidence from tests. Buttry, for example, reports[83] a series of 57 push-out tests in which there were four failures by the studs pulling out of the slab; two of 19-mm studs 50 mm high, and two of 16-mm studs 63 mm high. These must be near the critical lengths, because 12.5-mm studs 50 mm high sheared off, and 19-mm studs 75 mm high caused local failure of the concrete slab. But variables such as bottom slab reinforcement, haunch shape, and global stress in the surrounding concrete are obviously relevant, so the rules in the Bridge Code are necessarily conservative.

The only relevant tests on beams known to the authors are nine by Schlaginhaufen[84] on simply-supported uncased beams of 4.5 m span under two-point loading. In four tests, the loads were hung from the steelwork. Headed studs were used in two beams, and unheaded studs in seven, and in four of these the shanks were threaded. Failure loads ranged from 85 to 90% of those calculated for full interaction. About half of the loss of strength was attributed to the absence of heads; the remainder was due to the use of partial shear connection (about 60%) in all beams.

It may be found in design that to satisfy the rules on embedment and cover, it is necessary to use connectors longer than those described in Part 5. The variation of the strength of studs with their height is now discussed.

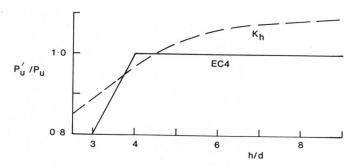

Fig. 7.5 Effect of h/d ratio on shear strength of stud connectors

Let P'_u be the static shear strength of a stud with ratio h/d of overall height to shank diameter and let P_u be the strength of a stud with the same diameter and $h/d = 4.2$. The variation of P'_u/P_u with h/d, based on regression analyses of many tests,[80] is shown by the curve K_h in fig. 7.5. Eurocode 4[5] gives the rule shown by the lines labelled 'EC4', with a lower limit on h/d of 3.0 (to prevent pull-out failure) but no upper limit. This rule gives the strength of a 19×75 mm stud as 0.99 times that of a 19×100 mm stud. The ratios of the values given in the Bridge Code range from 0.85 to 0.88, depending on concrete strength. Apart from the 19×75 mm size, all studs specified in the Bridge Code have h/d ratios of 4.0 or above. The strengths for studs of height exceeding 100 mm may be taken as those given for 100-mm studs, which is consistent with the Eurocode. Standard stud welding equipment can accept studs of 150 mm in length without modification, and some manufacturers provide equipment for welding stud anchors up to 500 mm in length.

No information is available on the shear strength of channel connectors of heights exceeding 127 mm. For bar connectors, an increase in the height of the hoop is most unlikely to have any effect on the shear strength, but no test data on this subject are available.

COMBINED SHEAR AND TENSION
There is no simple method of calculating the small tensile forces that commonly occur in shear connectors, due for example to the mode of failure shown in fig. 8.23. They need not be considered in design, because similar forces occur in the push-out tests from which design shear strengths of connectors are deduced.

A method is given in Part 5 of the Bridge Code for taking account of the much larger tensile forces that can occur in certain types of bridge deck. These are referred to as 'significant calculable direct tensions', and may arise in the global response of a box-girder deck to certain types of loading or, when the slab acts as part of a U-frame to resist lateral buckling, from transverse bending moment at a corner of the frame.

Where a stud connector is subjected both to shear Q and to tension due to uplift T_u, the equivalent shear Q_{max} used in checking the connector is taken as:

$$Q_{max} = [Q^2 + (T_u^2/3)]^{1/2} \qquad (7.12)$$

This rule is now compared with relevant research, omitting all partial safety factors and assuming that the design of the shear connection is such that Q_{max} is equal to the design ultimate shear strength, P_u. Equation (7.12) is plotted as curve AB on fig. 7.6.

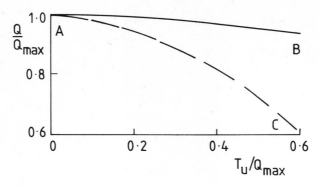

Fig. 7.6 Equivalent shear for connectors in combined shear and tension

Hughes concluded[86] from 176 pull-out tests on studs and other types of anchor that for depths of embedment of $9d$ or more (where d is the diameter of the stud shank) failure always occurred in the shank rather than the concrete. McMackin *et al.* report[87] about 60 tests on 19-mm and 22-mm studs of lengths 100, 178 and 203 mm, set in plain concrete blocks (cylinder strengths 34–37 N/mm²) and loaded in combined tension and shear. They conclude that the pull-out strength T_o of a stud of height nd and head diameter $1.5d$, made of steel with ultimate tensile strength f_u and set in normal-density concrete of cube strength f_{cu} is given by:

$$T_o/d^2 = 1.19n(n + 1.5)(f_{cu})^{1/2} \leqslant 0.785f_u$$

Putting $f_u = 500$ N/mm² and $f_{cu} = 30$ N/mm², this gives the minimum anchorage to develop the tensile strength of the shank as $7.1d$, and so is consistent with Hughes' results. It gives the pull-out strength of a 19×100 mm stud ($n = 5.3$) in Grade 30 concrete as 85 kN. For these studs, the Code gives $P_u = 100$ kN, so $T_o/P_u = 0.85$, which is well below the value 1.73 implied by equation (7.12).

The interaction curve that gave the best fit to these test results was:

$$(T_u/T_o)^{5/3} + (Q/P_u)^{5/3} = 1 \qquad (7.13)$$

Assuming that $T_o = 0.85P_u$ and that $P_u = Q_{max}$, equation (7.13) becomes curve AC on fig. 7.6, which is much lower than curve AB.

Equation (7.12) implies that the relation between shear strength and tensile strength for a stud is that between yield of its shank in shear and in tension, as given by the von Mises yield criterion. This is inaccurate for three reasons: resistance is determined more by the ultimate tensile strength than the yield strength; the shear resistance is augmented by the weld collar, but the tensile resistance is not; and tensile failure may occur in the concrete rather than the shank.

In practice, embedment lengths are less than $7.1d$, so concrete failure would occur if slabs were unreinforced. The increase in resistance to pulling out due to the reinforcement in slabs has not been quantified, particularly where the slab soffit is in tension. Fortunately, pull-out resistance is not relevant to practice, because design ratios of tension to shear are low.

In Eurocode 4, the effects of tension may be neglected when the design tensile force T_u is less than $0.1P_u$; higher tensions are considered to be outside its scope.

It is concluded that in situations where T_u exceeds, say, $0.2P_u$, it would be prudent to use curve AC for design rather than curve AB. At low tensions, the Eurocode method is obviously accurate enough.

DESIGN PHILOSOPHY FOR SHEAR CONNECTION

The static strengths of shear connectors given in Part 5 of the Bridge Code are called 'nominal' rather than 'characteristic' because they are mean values, not those corresponding to a 5% probability of failure. Mean values are used for the following reasons.

(1) Ultimate-strength design of the shear connection in a composite beam is based on the assumption that individual connectors possess sufficient plasticity to redistribute load between them until all those in a shear span fail as a group. The characteristic strength should therefore be that corresponding to a 5% probability of failure of a group of connectors in a beam. This strength cannot be determined, as very few failures of shear connection in beams have been reported.

(2) There are not even enough reliable data for the calculation of 5% values for failure in a push-out test. As explained above, the test data are influenced by the method of curing the concrete cubes or cylinders, and in many reports the method of curing is not given. There have been changes in the specification for shear-connector materials, and variations in the reinforcement in the test slabs and the number of connectors per slab. The size of the weld collar depends on the welding system used. Even for nominally identical specimens, the range of the results of three push-out tests often exceeds 10% of the mean value.

(3) If sufficient tests were done to establish characteristic strengths for the most widely used types of connector, similar work would be needed for any new type of connector, or to find the strength of an existing type of connector in a haunch of novel shape. Instead of push-out tests in groups of three, groups of about 40 would be needed.

The use of strengths based on mean results of push-out tests was first proposed in 1965, in CP 117: Part 1,[88] and has been satisfactory for these reasons: connectors used in practice have load–slip curves that allow substantial redistribution of load before failure, so that the strength of a group depends more on its mean strength than on the weakest member; and in CP 117 design was based on 80% of the push-out strength for reasons given in Chapter 3. This use of '80% design' has effectively been continued in the Bridge Code, as explained in section 7.2.6.

Another reason often quoted is that push-out tests underestimate the strength of connectors in positive-moment regions of beams but the evidence on this is inconclusive. Davies[89] gives load–slip curves for studs in push-out tests and half scale T-beams that show the studs to be stiffer, but not stronger, in the beams. A method was developed at the University of Missouri for deducing loads on studs in beams from strain readings on the steel joist. The load–slip curves so obtained lie well above the curves from push-out tests in some beams[90a] and well below in others.[90b] Further research on this subject is needed.

SHEAR CONNECTORS IN HAUNCHED SLABS

As explained above, the design load for a bar connector is based on the assumption that the concrete bearing against its vertical face can resist a compressive stress of several times its cube strength. Other types of connector also impose very high local stresses on the surrounding concrete, which can only be resisted if lateral restraint is available to prevent bursting of the sides of the haunches.[91] The design loads given in Part 5 of the Bridge Code can be used for connectors in haunched slabs only if the sides of the haunch lie outside a line drawn at 45° from the outside edge of the connector. This edge must have a concrete cover of not less than 50 mm (40 mm in buildings). These empirical rules together define the minimum size of a haunch and are illustrated in fig. 8.24. Together with the rules for reinforcement of haunches (section 8.6), they provide more guidance on haunch design than was given in CP 117: Part 2. But, as there are so many relevant variables, their backing from tests to failure is limited. Haunches fail suddenly, and so should be designed with caution.

No provision is made in Part 5 for the use of helical reinforcement around studs, which has been shown[91] to improve the behaviour of narrow haunches under static loading, probably because no comparable tests under fatigue loading were done.

PUSH-OUT TESTS FOR NON-STANDARD CONNECTORS AND HAUNCHES

If it is intended to use in design a type of connector or shape of haunch that does not comply with the requirements of Part 5 of the Bridge Code, the static strength of the shear connection should be established by push-out tests.

Apart from metrication of dimensions, the specification for the push-out test is as in CP 117: Part 2, and the nominal strength is taken as the lowest of three results, after correction for variation in the strength of the concrete. If the three test specimens are a random sample from the whole population of connectors of the type tested, there is an 87.5% probability that the nominal strength so found will be below the mean strength of the population, if the strength is normally distributed.

If a type of shear connector without resistance to uplift (section 8.5.3) is used, separate connectors to prevent uplift must be provided as well. It is required in Part 5 that these should be sufficiently strong and stiff to prevent the separation in a push-out test from exceeding half the slip.

As with most tests involving concrete, the push-out test should be carried out exactly as specified, and the concrete strength should be taken as that of water-cured cubes (section 7.2.3). Most of the variables that can influence results are given in Chapter 2 (Volume 1). Results obtained in other countries require skilful interpretation. For example, push-out specimens used in the USA, Netherlands and some other countries may have studs set at two levels, and give the mean strength of eight studs. This is usually higher than the mean strength of four studs at one level, given by the British test, probably because there is less axial tension in the studs. Variations in the way the slabs are supported[14b] and reinforced also affect the results.

7.2.4 Use of friction-grip bolts as shear connectors

Precast concrete slabs have been successfully used for composite bridge decks in several European countries. Sometimes shear connectors project through holes in the slab, which are later filled in (plate 9), but in other designs the slabs are firmly bolted to the steelwork, and shear connection is by friction.

The criteria for limit-state design using friction-grip bolts were assumed to be that there should be no slip at the serviceability loading, and no failure of the bolts or the surrounding concrete at the design ultimate loads. The clause on this subject in Part 5 of the Bridge Code is based mainly on tests by Marshall et al.[92] on five composite beams, in which general grade bolts to BS 3139: Part 1 (metric version, BS 4395:

Part 1[93]) were used. A design method is therefore given only for bolts of this grade.

SERVICEABILITY LIMIT STATE

Both Marshall and Sattler[27] agree that in design for working loads, the coefficient of friction at first slip, μ, may safely be taken as 0.45 for precast slabs on steel beams provided a uniform bearing is achieved. The increase to 0.50 for *in situ* concrete is based on Marshall's work. If the method of tightening the bolts complies with BS 4604: Part 1[94] the nominal initial tensile force in the bolt may be taken as the proof load. The designer must then estimate the long-term losses of force in the bolt (by methods similar to those used for post-tensioning forces in pre-stressed concrete) to obtain the net tensile force in the bolt, T. The nominal frictional resistance at the interface, μT per bolt, is divided by partial safety factors for both μ and T. These were both provisionally assumed to be 1.1, giving the design frictional resistance as $\mu T/1.2$ per bolt.

In a beam designed for unpropped construction and HA loading (for which γ_f at serviceability is 1.2), the overall factor of safety against slip for HA loading is thus 1.2×1.2, or 1.44. This is consistent with existing practice for steel/steel connections, for which BS 4604 gives a load factor of 1.4 with μ taken as 0.45, so the provisional factors 1.1 for μ and T were accepted.

ULTIMATE LIMIT STATE

Marshall tested bolts up to 16 mm diameter in slabs with cube strengths ranging from 36 to 63 N/mm² and concluded that the ultimate shear strength of the bolts could be taken as 600 N/mm². For the Bridge Code, this was provisionally reduced to $500/\gamma_m$ N/mm², with γ_m taken as 1.15 at the ultimate limit state. Using the factors $\gamma_{fL} = 1.5$ and $\gamma_{f_3} = 1.1$ for HA load, it can easily be shown that shear failure of the bolt at the ultimate limit state never governs design, so no design method is given.

The provision of local reinforcement and bearing plates, to prevent premature failure of the concrete slab at the bolts, is particularly important for bolts exceeding 16 mm in diameter.

DRAFT EUROCODE 4

The design method for friction-grip shear connection at the serviceability limit state is similar to that of BS 5400, except that:

(1) the coefficient of friction may be taken as 0.5 for all types of slab, provided that the steel flange satisfies certain conditions;

(2) the recommended factor γ_m is 1.0 not 1.2;

(3) the reduction in the initial tension in the bolt due to creep and shrinkage should either be found from long-term tests, or should be assumed to be not less than 40%.

Two design methods are given for the ultimate limit state.

(1) The shear resistance is assumed to be due to friction alone. The method is as above, except that the recommended γ_m is 1.25.

(2) The shear resistance is assumed to be developed by the bolts alone in shear and bearing. The design method is as given in Eurocode 4 for stud connectors, except that 'careful evaluation of the effects of slip should be made'.

7.2.5 Use of ribbed steel flanges for shear connection

Push-out and beam tests by Gogoi during 1962–63 on members with a ribbed or indented flange at the steel/concrete interface led to the conclusions[14b] that holding-down devices must be provided to mobilise the shear strength and that, if sufficient devices are provided to prevent the slab riding over the deformations until the ultimate moment is reached, then these devices tend in themselves to be an adequate shear connection.

More encouraging results were obtained from static and fatigue tests in Belgium[95,96] on plates with a diagonal pattern of ribs 2 mm high and 6 mm wide, larger than those used by Gogoi. It was concluded that use of such plates for top flanges of plate girders would enable the number of shear connectors to be halved.

Ribbed rolled steel sections have been developed in Japan, and their effectiveness has been established by tests on simple and continuous beams with spans up to 8.4 m.[191]

7.2.6 Partial safety factors for materials, γ_m

When the mean stiffness of a material throughout a member is relevant, as in global analysis or the calculation of deflections, γ_m is taken as 1.0. When a limiting stress or strain is relevant, as for crushing of concrete or yield of reinforcement, account must be taken of the variability of the material and the importance of the limit state. This leads to the use in several codes of $\gamma_m = 1.5$ for concrete and $\gamma_m = 1.15$ for reinforcement at the ultimate limit state.

For structural steel at the ultimate limit state, the treatment of γ_m in

Part 3 of the Bridge Code is explained in section 7.1.2. For composite members, a distinction has to be made between bending resistance governed by the strength of a steel component in compression, for which $\gamma_m = 1.20$, and other resistances, such as that of a beam in sagging bending, for which $\gamma_m = 1.05$. The higher of these values takes account of the greater scatter of the results of tests in which buckling occurs. For the resistance of the steelwork in shear, which benefits from tension-field action, $\gamma_m = 1.05$.

For serviceability limit states, γ_m is usually 1.0, either because mean properties are relevant, or because the consequences of the limit state being reached are less adverse than for ultimate limit states. An exception is that in Part 4, $\gamma_m = 1.33$ for concrete in a cross section subjected to uniform compressive stress. This is probably because there is less scope for redistribution of stress at points of local weakness than there is for triangular stress distributions, for which $\gamma_m = 1.0$.

In the Bridge Code, γ_m factors appear in three different forms.

(1) For concrete, they are incorporated within specified values for design stresses. For example, the two compressive stresses referred to above are given as $0.38f_{cu}$ and $0.50f_{cu}$, respectively.

(2) For reinforcement, the characteristic yield strength is preceded in formulae by the factor 0.87, which is $1/1.15$.

(3) For structural steel, the symbol γ_m is included in formulae for resistance, rather than applied to the nominal yield strength. This is because the variability that it represents was determined from tests on components, not on samples of the steel.

SHEAR CONNECTORS

The lack of test data and other difficulties (section 7.2.3) that led to the use of mean values for the strengths of shear connectors also make it impossible to determine values of γ_m on a statistical basis.

At the ultimate limit state, the failure of some connectors (e.g. bars and large studs) occurs mainly in the surrounding concrete, whereas small-diameter studs may shear off without much damage to the concrete. Thus γ_m for connectors should lie between the values for concrete and structural steel. It should also allow for the difference, if any, between the strengths of connectors in beams and in push-out tests, and should vary with the type and size of connector, strength and density of concrete, and perhaps even the local arrangement of reinforcement! This would be far too complicated, and in any event the data are not available. In developing the Bridge Code, a single value of γ_m was deduced as follows.

Let us consider HA traffic loading (as connectors resist mainly live loads) and studs (the most common type of connector). It was assumed that at the ultimate limit state the load on a connector (calculated by elastic or ultimate strength theory, as appropriate) should be limited to $0.8 P_u/\gamma_m$, where P_u is the nominal static strength, and that the number of connectors should be as given by CP 117: Part 2. That code allowed a static load of $0.4 P_o$ per stud for HA loading, and gave a design strength P_o about 10% higher than P_u. So for a given girder, the number of studs is proportional to $(HA)/0.4 P_o$. For the new code to give the same number:

$$\frac{HA}{0.4(1.1 P_u)} = \frac{1.1 \times 1.5 (HA)}{0.8 P_u/\gamma_m}$$

whence:

$$\gamma_m = 1.102 \qquad (7.14)$$

This result is consistent with the other criteria above, and so γ_m was taken as 1.10 at the ultimate limit state, and the design resistance per stud as $0.8 P_u/1.1$, or $0.73 P_u$.

At the serviceability limit state it is appropriate to take γ_m as 1.0, as for other materials. The design resistance per stud can be taken as $k P_u$, where k is given by a similar comparison to that above. γ_{fL} for HA loading is now 1.2, so that:

$$\frac{HA}{0.4(1.1 P_u)} = \frac{1.2(HA)}{k P_u} \qquad (7.15)$$

which gives $k = 0.53$, which is rounded up to 0.55.

These design resistances for shear connectors, given in Part 5 (1979) were modified by a subsequent Department of Transport Standard[2] to:

$0.714 P_u$ at ULS, presented as P_u/γ_m with $\gamma_m = 1.4$
$0.54 P_u$ at SLS, presented as P_u/γ_m with $\gamma_m = 1.85$

These expressions do not include γ_{f3}, because the 1979 edition of Part 5 was drafted on the assumption that γ_{f3} would be applied to loads. When this is not done (i.e. for use in conjunction with Part 3), the resistance at ULS given above must be divided by 1.1, so that:

design resistance of shear connectors at ULS $= P_u/\gamma_m \gamma_{f3}$
with $\gamma_m \gamma_{f3} = 1.54$

7.3 Loading

The specified live loading for a bridge is the same whether the bridge is of steel, concrete or composite construction, but there are substantial differences between the loadings in use in different countries. In a comparative study[10] of highway loadings in 18 countries, it was found, for example, that for two-lane decks with spans between 10 and 100 m, the maximum bending moments and shear forces due to the AASHO loading (USA) were about half of those for the West German loading. The British HA loading (in 1975, since increased) gave forces about 80% of those for West Germany, and 45 units of HB loading (a vehicle weighing 1800 kN) was one of the heaviest 'abnormal vehicle' loadings in use. Some road bridges in Great Britain are designed for even heavier vehicles, of up to 3 MN.

In BS 5400: Part 2[1] the HA and HB loadings are retained in a slightly modified form. An account of the changes is available.[57b] A full discussion of these loadings and the rules for their application would be outside the scope of this book, but problems that cause particular difficulty and details relevant to the design of a composite bridge deck with one two-lane carriageway are now discussed. Certain types of highway loading that do not influence the worked examples given here have been omitted. They are: restraint at bearings; differential settlement; and braking, centrifugal, skidding, and accidental loads. All the specific loads given in this section are nominal values.

MAXIMUM AND MINIMUM LOADS

The load factors γ_{fL} from Part 2 of the Bridge Code are summarised in tables 7.1 and 7.2. Most of these, of course, refer to the maximum values to be considered in design, but careful assessment of minimum values is equally important, for three reasons:

(1) In continuous construction, maximum stresses occur in a span when the adjacent spans carry minimum loads.

(2) Where two loads are balanced against each other, the margin of safety against overturning is least when one load is a maximum and the other a minimum.

(3) The fatigue life of welds is a function of range of stress rather than maximum stress, so that minimum loads have to be considered.

7.3.1 Dead load

The partial safety factors at the ultimate limit state for the weight of the structure itself are much lower than before, so particular care must be taken to assess dead load accurately. For example, the assumed density for the concrete and assumed percentage addition for weight of bolts in steelwork should be verified. Where this is done, Part 2 of the Bridge Code permits γ_{fL} at the ultimate limit state to be taken as 1.05 for steel and 1.15 for concrete. If minor approximations are made in the assessment of dead load, then the values of γ_{fL} should be appropriately increased.

MINIMUM DEAD LOAD

At first sight it seems logical to reduce γ_{fL} to 1.0 or even 0.9 for those parts of the structure where the dead load reduces the effect being considered. In complex structures this would make global analysis excessively complicated. It is also an over-conservative assumption. If, for example, the weight of a deck slab is underestimated, the error is far more likely to occur in all spans than in alternate spans.

The factors for maximum load at the ultimate limit state are therefore normally applied to all parts of the dead load, irrespective of whether they have an adverse or relieving effect on the member considered. An exception is made in Part 2 of the Bridge Code, for members where the total effect considered becomes more adverse when γ_{fL} for dead load *throughout the whole structure* is reduced. At the ultimate limit state, γ_{fL} for dead load should then be taken as 1.0. Although not stated in Part 4 of the Bridge Code, it would be logical also to reduce the factor γ_{f3} for dead load from 1.1 to 1.0 when designing reinforced concrete.

In design according to Part 3, it is clear that γ_{f3} should not be reduced to 1.0 in this situation because it is applied to resistances, not to loads. For simplicity, it is suggested that this practice be followed also in the design of composite members to Part 5.

SUPERIMPOSED DEAD LOAD

This consists of the weight of all materials, finishes, services, deck furniture, etc., which are not structural elements. The high load factors for surfacing (1.2 and 1.75) result from experience, which has shown that its weight tends to increase during the life of a bridge. Minimum superimposed dead load is normally treated in the same way as minimum dead load, as explained above.

CONSTRUCTION AND ERECTION LOADS

Calculations for the serviceability limit state involving erection loads are required only to the extent that may be necessary to satisfy the requirement in Part 2 of the Bridge Code that 'nothing shall be done during erection that will cause damage to the permanent structure or will alter its response in service from that considered in design'.

At the ultimate limit state, erection loads are not included in combination 1, as this is essentially a check on exceptionally high traffic loads. They are included in combinations 2 and 3 (with $\gamma_{fL} = 1.15$) together with 'appropriate' live loads; also, when the period during which the erection loads will be applied is known to be short, the wind and temperature effects allowed for are reduced below the 120-year values. Erection loads are unlikely to govern the design of the short-span bridges studied in Appendices X and Y, so no calculations of this type are given.

In composite bridges, the formwork for the deck slab is usually supported by the steelwork alone, and is not removed until after the deck becomes composite. The stresses induced in the structure by these operations may be small enough to neglect, but in principle they are a form of permanent prestressing, and if considered at all they should, in the authors' opinion, be included in all five load combinations, in the same way as other permanent effects. This is easily done by adding the weight of the formwork (referred to as 'construction load' in the worked examples) to the weight of the wet concrete that is imposed on the steelwork, and allowing for the removal of the formwork by deducting its weight from the superimposed dead load that is later applied to the composite structure. Attention must be paid to the correct use of partial safety factors in these calculations.

7.3.2 Highway loading

HIGHWAY LOADING TYPE HA

This loading represents a train of vehicles on a single traffic lane, and consists of *either* (1) a uniformly distributed lane load and (2) a knife-edge load *or* (3) a point load representing a heavy wheel load. The details are as follows.

(1) Lane loading (HAU)   30 kN per linear metre of lane for loaded lengths up to 30 m. For longer lengths, lane loading is given in the first (1978) edition of BS 5400: Part 2[1] as:

$$W = 151L^{-0.475} \text{ but not less than } 9.0 \qquad (7.16)$$

where:

W is the load per metre of lane in kN, spread uniformly over the width of the lane, and

L is the total loaded length in metres, for the lane considered.

For bridges in the UK, this loading for $L > 30$ m was superseded by a curve given in the *Interim Revised Loading Specification* of the Department of Transport (August 1982, revised November 1982).[186] The reduction in W with increasing loaded length was made less steep. For example, at $L = 100$ m the revised curve gave $W = 20.8$ kN/m, replacing 16.9 kN/m; and at $L = 250$ m it gave $W = 19.0$ kN/m, replacing 11.0 kN/m.

The changes were based on evidence that the ratio of heavily loaded lorries to other vehicles in typical streams of traffic was occasionally much higher than had been assumed in the derivation of equation (7.16).

(*2*) *Knife-edge loading* (*HAK*) A uniform line load of total weight 120 kN extending across the width of the lane.

(*3*) *Wheel load* (*HAW*) A load of 100 kN placed anywhere on the carriageway with a contact area of either a circle of diameter 0.34 m or a square of side 0.30 m, corresponding to a bearing pressure of 1.1 N/mm².

For static loading on a two-lane single carriageway, full HA load is applied to one or both lanes, and over the lengths that give the extreme values of the stress resultant being considered.

Hard shoulders or marginal strips at the level of the traffic lanes are treated as 'carriageway'. Verges that are separated from the carriageway by kerbs or fences are assumed to be unloaded for the overall design of the structure, but must be capable of supporting any four wheels of 25 units of HB loading.

The loaded length L for the member under consideration is normally the total base length of the adverse area or areas of the relevant influence line. More detailed requirements for continuous and multi-level superstructures are given in Part 2 of the Bridge Code.

The knife-edge component of loading does not represent a single heavy axle, but is a device to ensure that HA loading correctly represents both the vertical shears and the longitudinal moments that may occur in practice. It is located, for design purposes, along a line perpendicular to the traffic lane or, in skew decks, in line with the bearings, and in the longitudinal positions that give extreme values of the effect being considered.

Examples of the application of HA load are given in Appendix X.

HIGHWAY LOADING TYPE HB

This represents a single heavy vehicle defined in terms of a number of 'units'. One has a nominal weight of 40 kN, consisting of sixteen equal wheel loads of 2.5 kN. The wheels are arranged on four axles as shown in fig. 7.7.

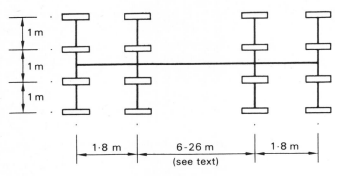

Fig. 7.7   Spacing of axles and wheels of HB vehicle

The spacing of the inner axles is intended to be that which gives the most severe effect in the situation being considered, and must be taken as the most adverse of the dimensions 6, 11, 16, 21 and 26 m. The corresponding overall lengths of vehicle are 10, 15, 20, 25 and 30 m.

In Great Britain, most highway bridges are designed for 45 units of HB loading. Each wheel load is then 112.5 kN. The assumed bearing pressure is 1.1 N/mm², giving a contact area of diameter 0.3 m, if circular, or 0.32 m if square. Type HB loading is assumed to act either in one traffic lane or straddling two lanes, in place of HA loading (including knife-edge load) over a length 50 m greater than the overall length of the HB vehicle. The vehicle is placed centrally within this length, in the most severe lateral position, but with at least 250 mm between a kerb and the centre of the nearest wheel. The remainder of the lane carries HA loading as described above, with the loaded length determined as if the HB vehicle had not replaced part of the HA loading. When part of the HA loading is applied in conjunction with HB loading in this way, the load factors γ_{fL} are taken as those for HB loading (tables 7.1 and 7.2), not the higher values specified for HA loading when it is applied on its own.

DISPERSAL OF WHEEL LOADS

The contact areas specified for the HA and HB wheel loads are at road surface level. These loads may be assumed to spread out at a slope of one horizontal to two vertical through asphalt or similar surfacing, and at a slope of 45° through a structural concrete slab, down to the level of the neutral axis. The loaded area for the design of the deck slab is thus significantly larger than the contact area.

FATIGUE LOADING

The load spectra, and other data given in Part 10 of the Bridge Code, that are relevant to a highway bridge are summarised in Chapter 10.

FOOTWAY LOADING

On footbridges, for loaded lengths up to 30 m, this consists of a uniformly distributed live load of 5 kN/m². The minimum loading is zero.

7.3.3 Wind loading

The maximum loads from wind occur during gusts. The nominal gust speed v_c for a given site is a function of location, design return period, height above site ground level, and horizontal loaded length of bridge deck considered. Values for sites in Great Britain can be assessed from data given in Part 2 of the Bridge Code. A reduced wind speed can be used in erection calculations, to allow for the much shorter period during which the relevant structure is at risk.

The statistical backing for wind loads is sounder than it is for imposed loads, because they are based on extensive meteorological observations of phenomena that are more predictable than traffic is over a 120 year period, and an extensive study of forces on structures due to wind.

The basic transverse wind load P_t, normally taken as horizontal, on a structure or element of net exposed area Ld is given by

$$P_t = \tfrac{1}{2}\rho v_c^2 L d C_D \qquad (7.17)$$

where:

ρ is the density of air, taken as 1.226 kg/m^3

L is the length of the structure or element

d and C_D are an effective depth and a drag coefficient respectively, defined in Part 2 of the Bridge Code, that depend on the type of structure and parapets, the ratio of its overall width to its overall depth, and on the presence or absence of live load.

Detailed rules are also given for determining the basic longitudinal wind load P_ℓ and vertical wind load P_v and for the combinations and application of these loads.

These rules take account of the combined effects of pressures above and below ambient. Wind is assumed to act on the whole structure or not at all, so values of γ_{fL} are given (table 7.2) for wind on those parts of the structure where wind load reduces the effect being considered, and are to be used with minimum gust speeds given in Part 2 of the Bridge Code.

7.3.4 Effect of temperature

Non-uniform distributions of temperature in a composite bridge deck are caused by direct sunlight, wind, and the differences between the thermal capacity and conductivity of steel and concrete. These very complex distributions have to be simplified for design purposes. The thickness of a typical concrete deck varies little over the length and breadth of a bridge, and the cross-sectional shape of most steel members is constant along their length. In practice, only temperature distributions through the depth of typical cross sections of the main structural members need be considered. Those given in Part 2 of the Bridge Code for composite decks with surfacing of thickness 100 mm are reproduced in fig. 7.8. They are referred to as *temperature differences*, as the word 'gradient' implies a linear distribution.

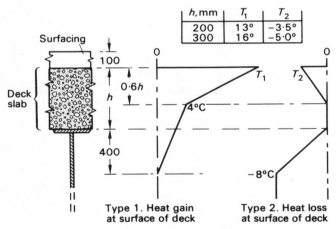

h, mm	T_1	T_2
200	13°	−3·5°
300	16°	−5·0°

Fig. 7.8 Specified temperature distributions through the depth of a composite girder

Type 1 represents a rapid rise in the temperature of the top of the deck slab, due to direct sunlight, while the steel is shaded. Distributions of type 2 occur when the ambient temperature is falling, because thin steel plates cool more quickly than the thick concrete slab. Both types of temperature difference cause longitudinal bending and shear stresses in all types of composite member.

Both of these temperature distributions may cause longitudinal stress at the level of the neutral axis of a typical composite girder, resulting in a change in the overall length of any member free to expand. This change is likely to be much less than the variations of length due to the temperature changes discussed below, and can usually be neglected in practice.

Each temperature distribution can be considered as being super-imposed on a certain overall, or uniform, temperature of the bridge. This is referred to as the *effective bridge temperature*, to distinguish it from the ambient shade temperature, which may be different. Variation in the effective bridge temperature is referred to as *temperature change*. Maximum and minimum effective bridge temperatures are specified in Part 2 of the Bridge Code, and the difference between them is the *temperature range*.

Temperature change causes change of length in all members, and longitudinal stress in any member not free to expand or contract. If the coefficient of thermal expansion of the concrete deck (β_c) differs signifi-cantly from that of the structural steelwork (β_s), (e.g. if limestone aggregate is used), temperature change causes thermal stress in com-posite members even where there is no external restraint.

To calculate these stresses it is necessary to define a *datum temperature*, at which they are assumed to be zero. In theory, this is the temperature of the bridge when restraint of thermal movement first became effective during its construction. In composite structures where β_c and β_s differ, the shear connectors cause some restraint of thermal movement as soon as each section of deck is cast, so that the datum temperature may not be uniform over the bridge, but usually this complication can be neglected and an average value can be assumed for design purposes.

A clear distinction can be drawn between the principal effects of temperature difference and temperature change. The first causes change of curvature; the second causes change of length, and of curvature also when $\beta_c \neq \beta_s$. Most bridge decks are designed to be free to expand longitudinally, but change of curvature is restrained in continuous members. So, there are two types of thermal stress: primary stresses (P), which occur in all members where there is temperature difference; and secondary stresses (S), which result from changes in the external reactions of continuous members, caused either by temperature dif-ference, or by temperature change in members with $\beta_c \neq \beta_s$. These conclusions are summarised in table 7.6.

EXAMPLE: CALCULATION OF NOMINAL TEMPERATURES

The use of Part 2 of the Bridge Code to derive nominal values of

Table 7.6 Thermal stresses and changes of length for longitudinally unrestrained members

Type of composite member	Temperature change		Temperature difference	
	Thermal stress	Change of length	Thermal stress	Change of length
Statically determinate, with β for concrete and steel:				
the same	None	Yes	P	Negligible
different	P	Yes	P	Negligible
Statically indeterminate, with β:				
the same	None	Yes	P and S	Negligible
different	P and S	Yes	P and S	Negligible

temperature difference and effective bridge temperature is now illustrated by an example. We consider a composite beam-and-slab bridge with a 220-mm deck slab, and 80 mm of surfacing, including waterproofing. It is assumed to be located near Birmingham, England, and to be designed for a life of 120 years.

The temperature differences are specified in fig. 7.8 in terms of T_1 and T_2. When $h = 220$ mm, interpolation gives $T_1 = +13.6°C$, $T_2 = -3.8°C$. These temperatures are sensitive to the thickness of surfacing, assumed in fig. 7.8 to be 100 mm. Corrections for other thicknesses are given in Appendix E of Part 2. For 80 mm, they are $+2.0°$ and $-0.3°$, so the nominal temperature differences are as in fig. 7.8, with $T_1 = +15.6°$, $T_2 = -4.1°C$.

Isotherms of minimum and maximum shade temperatures in Great Britain with a 120-year return period are given in Part 2. The values at the site of the bridge are $-20°$ and $+35.5°C$ at sea level, and become $-20.5°$ and $+34.5°C$ after correction for the altitude of 100 m. Nominal effective bridge temperatures are tabulated in terms of shade temperature, for four types of bridge superstructure. For a composite deck with 100 mm of surfacing they are $-17°$ and $+38.5°C$. Correction for thickness of surfacing gives the final values, $-17.6°$ and $+39.3°C$, so the nominal effective temperature range is $56.9°C$. This range is surprisingly large, and corresponds to a change in length of 68 mm for a bridge 100 m long.

COMBINATION OF TEMPERATURE DIFFERENCE AND TEMPERATURE CHANGE
A temperature difference of type 1 (fig. 7.8) can occur at any time between spring and autumn. In Part 2 of the Bridge Code it is assumed that for composite bridges, it can co-exist with effective bridge temperatures at or above 15°C.

It is considered that differences of type 2 can occur at any time of year. These are assumed to co-exist with any effective bridge temperature between the minimum and 4°C below the maximum.

For the bridge girder considered above, fig. 7.9 shows the combined temperature distributions through the depth of the member that are given by these rules, rounded off to the nearest degree. The extreme effective bridge temperatures are labelled L and H, and the combinations of these with temperature differences of type 1 or 2 are H1, L2, etc. The combinations M1 and M2 include intermediate effective bridge temperatures.

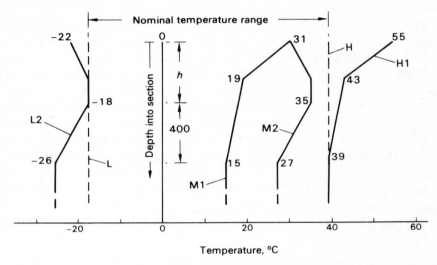

Fig. 7.9   Combination of temperature change and temperature difference

Temperature distributions are assumed to exist along the whole length of a bridge superstructure, not just along lengths where their influence on the effect considered is adverse.

CALCULATION OF PRIMARY EFFECTS OF TEMPERATURE

Changes in the overall length of a bridge deck must be calculated, for use in the design of bearings and expansion joints. Usually, the effects of temperature difference can be neglected, and the calculation is straightforward.

The calculation of temperature stresses is more complex. In longitudinally unrestrained decks with equal coefficients of thermal expansion, β_c and β_s, only two temperature distributions need be considered (the two types of temperature difference). Most composite decks are of this type. For these, it is convenient in calculation to take the datum temperature for both distributions as that of the lower part of the steel web (i.e. to add 8°C to the type 2 distribution shown in fig. 7.8), so that 'free' changes of length due to both temperature and shrinkage occur only in the slab and top 400 mm of the steel section. This is done in the worked example in Appendix X.

In structures where $\beta_c \neq \beta_s$, consideration must be given to the primary stresses and curvatures caused by all six of the distributions corresponding to those shown in fig. 7.9. This can be done by calculating stresses due to a uniform rise of temperature of 1°C, scaling them to give results for other temperature changes, and then combining them with the two sets of stresses due to temperature difference.

As an example, this was done for a typical cross-section of a composite plate girder. The unhaunched concrete flange (assumed to be uncracked) was 1.57 m wide and 0.22 m thick. The steel girder was 1.46 m deep, with a 12-mm web, 300×20 mm top flange, and 500×40 mm bottom flange. The datum temperature was taken as 10°C. The coefficients of thermal expansion were $\beta_s = 12 \times 10^{-6}$ and $\beta_c = 8 \times 10^{-6}$ per degree Celsius. (Part 2 of the Bridge Code, as amended by the Department of Transport, gives values for β_c ranging from 9.0×10^{-6} to 13.5×10^{-6} per degree Celsius, depending on the type of aggregate. These include an allowance for restraint from reinforcement.)

The maximum longitudinal stresses found at five levels in the cross section and the corresponding temperature distributions (as labelled in fig. 7.9) are given in table 7.7. It so happens that all six distributions contribute to the maximum stresses.

Table 7.7   Nominal primary temperature stresses in a composite girder with a slab of lime-stone-aggregate concrete

Level in cross-section	Maximum tension		Maximum compression	
	Stress (N/mm²)	Governed by	Stress (N/mm²)	Governed by
Top of deck slab	+1.6	M2	−2.6	M1
Bottom of deck slab	+2.1	H1	−1.8	L2
Steel top flange	+15	L	−15	H
0.4 m below top flange	+24	L2	−10	H
Steel bottom flange	+2	H	−5	L2

In practical design, these stresses can usually be ignored. They are either very low, as in the steel bottom flange; or are tensions in uncracked concrete, which would be relieved by cracking; or are stresses in steel near to the neutral axis, which would not govern design of the cross section. The only stress that is significant, in comparison with the design strength available, is the 2.6 N/mm² compression at the top of the slab. This is unlikely to cause load combination 3 to govern, rather than load combination 1, unless the composite section has to resist significant dead-load sagging bending moment. This subject is considered further in section 8.7 and in Appendices X and Y.

DESIGN EFFECTS OF TEMPERATURE

Stress resultants, stresses, or displacements due to the specified temperatures are nominal values, and are multiplied by the values of γ_{fL} in table 7.2 to give design values for the two limit states for use in load combination 3.

7.3.5 Effects of shrinkage of concrete

This subject is introduced briefly in chapter 3 (Volume 1). In composite bridges, shrinkage of the concrete slab can cause compressive stresses in steelwork (which need careful evaluation where the steel members are slender), redistribution of bending moments in continuous structures, and additional deflection of beams. Shrinkage also influences crack widths in concrete.

Most shrinkage occurs when the concrete is young, so that shrinkage stresses are overestimated unless allowance is made for the effects of creep. Detailed information is available on both shrinkage and creep, and quite accurate methods of calculation exist, but they are too tedious for routine use in any except the largest structures. Designers therefore require information at several levels of complexity and accuracy.

For large structures, the effects of shrinkage and creep can be analysed from first principles using data given in Appendix C of Part 4 of the Bridge Code, which gives values of the creep coefficient ϕ for many situations. This coefficient is the ratio of long-term creep strain to initial elastic strain due to a constant compressive stress, f_{cc}, say. The total long-term strain is therefore:

$$\varepsilon_{c\infty} = (1 + \phi)f_{cc}/E_c$$

where E_c is the short-term modulus of elasticity for concrete, so that the effective modulus for constant stress is:

$$E'_c = f_{cc}/\varepsilon_{c\infty} = E_c/(1 + \phi) \tag{7.18}$$

This overestimates creep due to shrinkage because the latter is time-dependent. The practice in several European countries, due to Fritz[97] is to use instead the relation:

$$E'_c = E_c/(1 + \psi\phi)$$

and to assume $\psi = 0.5$ for calculations of long-term stresses due to shrinkage. Values of ψ that take account also of the type of composite cross-section have been computed by Haensel.[98]

The method given in Part 5 is simpler, in that free shrinkage is assumed to depend only on mean ambient humidity. This assumption is reasonable only for concrete mixes where the cement and water contents lie within certain ranges, given in Part 5. Three environments are defined, and values of free shrinkage and a reduction factor for creep (ϕ_c) are given for each. The reduction factor is defined by:

$$E'_c = \phi_c E_c \qquad (7.19)$$

To allow for creep, the short-term modular ratio is therefore divided by ϕ_c. Comparison with equation (7.18) shows that:

$$\phi_c = 1/(1 + \phi)$$

The values of shrinkage strain and ϕ_c given in Part 5 are plotted in terms of ϕ as points 1, 2 and 3 in fig. 7.10. They are based mainly on the more detailed figures given in the rules for prestressed concrete current in 1975 in West Germany, which are also shown in fig. 7.10. A disadvantage of this method is that the modular ratio for shrinkage calculations may differ from those used for dead and imposed load, so that a separate set of properties of transformed composite cross sections has to be calculated.

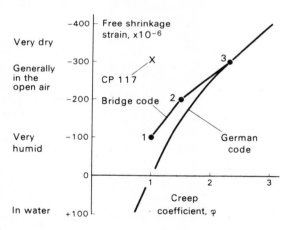

Fig. 7.10   Shrinkage strains and creep coefficients

A simpler method, which is always conservative, is to assume $\phi_c = 0.5$ for all environments, so that the effective modulus is the same as that used for permanent load. This method is quite accurate for humid environments, but can overestimate shrinkage stresses for dry environments by up to 50%.

A shrinkage strain of ε_{cs} in the concrete of a slab with reinforcement ratio ρ is equivalent to a fall in the temperature of the slab, relative to that of the steel beam, of $\varepsilon_{cs}/[\beta_c(1 + \rho\alpha_e)]$, where β_c is the coefficient of thermal expansion of the concrete and α_e is the modular ratio. Primary shrinkage stresses and curvatures can therefore be calculated by the methods used for temperature effects, discussed above; the only difference is in the use of an effective modulus to allow for creep. Relevant equations are given in Appendix A.

Shrinkage effects are normally considered at the serviceability limit state for compact members, and at the ultimate limit state for non-compact members. The partial safety factors γ_{fL}, given in Part 5, are 1.0 and 1.2, respectively, for use with all load combinations. The specified shrinkage, or a more accurate value if available, is taken as both the maximum and the minimum value. No more accurate treatment would be justified unless account were also taken of sequential casting, cracking of concrete, and other variables, which is impracticable.

Where shrinkage has a relieving effect, the treatment of partial safety factors should be similar to the procedure adopted for relieving effects of dead load (section 7.3.1).

7.4 Global safety factors

It is rarely possible to do full-scale tests of bridge components, and so designers rely heavily on accumulated experience of the behaviour of bridges in service. The partial safety factors given in the Bridge Code are based partly on this experience, so that structures designed to this Code should not be substantially lighter or heavier than those given by earlier methods. This has been checked by design studies on four existing bridges. A simpler and more approximate check is to compare values of 'global safety factors'.

As an example, we consider a simply-supported composite girder carrying HA loading only. The load factor on first yield of steel given by design to BS 153[18] was about 1.7. Assuming that the member is slender, the corresponding safety factor for design for the ultimate limit state (combination 1) by the Bridge Code is the product $\gamma_{fL} \cdot \gamma_{f3} \cdot \gamma_m$, which is $1.5 \times 1.1 \times 1.1 = 1.815$. For dead load which is half concrete and half steel, the global safety factor is similarly found to be $\frac{1}{2}(1.05 + 1.15) \times 1.1 \times 1.1 = 1.331$; so the new safety margins are higher for HA loading and lower for dead load. These simple examples of course neglect the influence of method of construction on the total stress in the steel member.

For HB loading, BS 153 allowed 25% overstress, so that the effective safety factor is 1.7/1.25, or 1.36. The new value is $1.3 \times 1.1 \times 1.1 = 1.573$, again showing the higher margin now adopted for live loads.

FUTURE DEVELOPMENTS

Codes of practice can only make use of experience and the results of research that are so well established as to be beyond question or doubt. At any time, there is also a body of incomplete evidence, that points the way to further inquiry, and suggests where there may be scope for improving design procedures. This evidence is relevant also to the prediction of overload capacity for existing bridges.

It was concluded from such an assessment for beam-and-slab composite superstructures[195] that their capacity is more likely to be limited by criteria of serviceability or repeated loading than by their ultimate resistance, even though the ultimate limit state governed most aspects of their design. This suggests that for these bridges, the design methods of BS 5400 for the ultimate limit state are more conservative, in relation to expected behaviour, than those for the serviceability limit state.

Chapter Eight

Design of Superstructure for Beam-and-Slab Bridges

8.1 Introduction

Computers can be programmed to calculate stresses in a bridge deck of given materials and dimensions, but not to select the materials or decide the dimensions. Once the decision has been made to design a composite structure of steel and concrete, the choice of type or quality of these materials is fairly straightforward. For the longer spans, possible savings of weight and erection cost with the use of lightweight-aggregate concrete or high-strength steel must be set against the additional costs per tonne of material. The expected recurrent cost of maintenance will be compared with the extra capital cost of using weathering steel; and so on. But these matters are not specific to composite bridges, and will not be discussed further.

The subject of this and the following chapters is the determination of the detailed dimensions of the deck structure. The choice between rolled sections, plate girders, or box girders, and the number of girders and their arrangement will be based on the designer's experience. The choice of detailed dimensions is essentially a matter of trial and error, and the errors decrease rapidly as experience grows.

There is much interaction of one decision with another, particularly for the longer spans. The initial choices of dimensions and the calculations for checking them must be done in a rational sequence that minimises the amount of work to be repeated when the size of a member turns out to be wrong. There are optimum sequences for the various types of composite bridge deck. Those presented here are quicker than those likely to be followed by an engineer unused to limit-state design but they include many checks that turn out not to govern, some of which would be omitted by an experienced designer.

THE WORKED EXAMPLES

The design methods for superstructures that are explained in chapters 8 to 10 are illustrated by two worked examples (Appendices X and Y): a three-span highway bridge, and a footbridge. These include the principal calculations for the composite structure, in accordance with Parts 2, 3 and 5 of the Bridge Code, and some calculations to Part 4, for the deck slab. There are no complete calculations for the non-composite behaviour of the structural steelwork, because such work would involve the design of stiffened plating and the discussion of subjects such as welding, buckling and plate imperfections, that are outside the scope of this volume.

For similar reasons, fatigue calculations are given for the shear connectors only, and are in accordance with Part 10 (Fatigue).

Much of the detailed arithmetic has been omitted from the worked examples. The comments are fuller than might be expected in practice, and further discussion is given in the main text. To save space and repetition of calculations the 'trial and error' part of the work has usually been omitted, only the final dimensions and related calculations being given.

8.1.1 Which limit state governs?

The requirements of the Bridge Code are such that the members of a composite structure should be designed initially for the ultimate limit state, assuming full shear connection. Erection requirements usually ensure that the top flanges of steel members are wide enough for the design of the shear connection, which is often governed by the serviceability limit state, to be straightforward.

Checks on stresses in steelwork at the serviceability limit state are required by Part 3 to be made whenever design for the ultimate limit state is such that inelastic behaviour is assumed, or local effects (e.g. slip at friction-grip bolts, local bending stresses in stiffened flanges) are neglected. There are two of these situations of particular relevance to composite members.

(1) For all composite cross-sections designed as compact, stresses in the smaller flange have to be checked at the serviceability limit state, treating the member as non-compact. The 'smaller' flange is not defined. By implication, it is the one for which elastic analysis gives the higher bending stress; that is, the one for which the extreme fibre is further from the elastic neutral axis for the transformed cross section. For unpropped construction, the smaller flange may thus be the top flange when checking the steel section alone, and the bottom flange when checking the completed composite section.

(2) In longitudinally-stiffened beams, the distributions of bending stress at the ultimate limit state are first found by elastic theory. If these cause yield of the tension flange, but not buckling or yield of the compression flange, redistribution of stress from the tension flange is allowed by Part 3, and the section can be designed using elastic-plastic theory. When this is done, the section has to be checked at the serviceability limit state.

Checks on stresses in reinforced concrete at the serviceability limit state are required by the Bridge Code to be made whenever the methods used for the ultimate limit state give insufficient indication of the behaviour to be expected in service. Linear-elastic analysis is used. The situations of relevance to composite members are as follows.

(1) Where precast concrete deck slabs are used with *in situ* concrete topping.

(2) Where effects of temperature, creep and shrinkage are not considered at the ultimate limit state. This applies to all composite sections designed as compact.

(3) Where the combined effects of global and local loading are not considered at the ultimate limit state.

(4) In hogging moment regions, where specified limits apply to the width of cracks in the deck slab under the combined effects of global and local loading.

Fatigue is unlikely to influence design of the deck slab, but may govern the shear connection in midspan regions. When it does, there are no consequences for other components, so the check can be made late in the design process.

The design of the deck of a highway bridge is likely to be governed mainly by load combination 1, but in establishing and checking the main dimensions it seems to be necessary to consider both the serviceability and ultimate limit states in parallel, as the following example shows. This is based on the authors' experience in preparing worked examples and trial designs in accordance with the Bridge Code.

8.1.2 Sequence of design calculations

For a beam-and-slab simply-supported deck, it has been found convenient to do the first set of design calculations in the following order. (For brevity the serviceability and ultimate limit states and load combinations 1 and 3 are referred to as S-1, S-3, U-1 and U-3.)

(1) Initial member sizes, based on experience, other designs and rough calculations.

(2) Dead and imposed loadings on an edge girder for U-1 using static distribution for the HB vehicle at the edge of the deck; hence longitudinal bending and vertical shear stresses, and check on size of edge girder.

(3) Analysis of deck slab for wheel loads for U-1 and S-1.

(4) Distribution analyses for the deck for HB vehicle (U-1) at midspan at centre of deck and at edge of deck; hence transverse moments in deck slab for S-1 and U-1 and distribution coefficients for general use.

(5) Check on thickness and concrete strength for deck slab from results of (3) and (4). First design of slab reinforcement.

(6) Calculation of shrinkage stresses for S-1 or U-1, as appropriate.

(7) Accurate calculation of envelopes of bending moment and vertical shear for the steel and composite sections in an outer girder for U-1, and check that an inner girder is less heavily loaded. Check on longitudinal bending stresses in concrete and steel (U-1) from results of (3) and (6).

(8) Maximum longitudinal shears for S-1 and U-1 found from results of (6) and (7).

(9) Design of transverse reinforcement (U-1) and of shear connection for static loading (S-1).

(10) Calculation of bending stresses and longitudinal shears due to temperature. Results added to scaled values from (7) and (8) for check on bending stresses and transverse reinforcement (U-3) and shear connection (S-3).

(11) If plastic analysis of sections has been used at U-1, check longitudinal stresses at midspan (S-1 and S-3).

(12) Check on transverse bending stresses in deck slab (S-1).

(13) Check on design of shear connection for repeated loading.

8.1.3 Elastic properties of cross sections of members

The properties required are areas of transformed cross sections, neutral-axis depths, second moments of area, section moduli, shear factors ($A\bar{y}/I$), and torsional rigidities. Their calculation for composite members is straightforward, and well suited to a desk-top computer. Their values depend on the modular ratio, the effective breadth of the slab, and on whether the concrete is cracked, as well as on the geometry of the section. At first sight, it seems that about fifteen sets of values are needed for each section. It will now be shown that simplification is possible.

In beams for buildings, the elastic neutral axis often lies within the concrete slab; in bridges this is so uncommon that the case need not be considered. The slab can be assumed to be in one of two conditions: uncracked and unreinforced (U), or cracked and reinforced (C). The U properties, used in global analysis, are needed for all cross sections; the C properties only for regions of hogging bending.

All U properties have to be determined for two modular ratios, α_e, (for permanent and for variable loads) but modular ratio has no influence on C properties or on those of the steel section alone (S). A third modular ratio is appropriate for the analysis of shrinkage effects (section 7.3.5), but the error in using the value for permanent loads is usually negligible.

When shear lag is significant, one cannot avoid using the full flange breadth (F), a mean effective breadth (M) and the effective breadth for the section considered (E). As table 8.1 shows, one then needs six sets of properties for sagging regions and four for hogging regions, plus one for the steel section.

For many composite cross sections, effective-breadth ratios exceed 0.95. The error in neglecting shear lag is then negligible. The F, M and E properties become identical. One then needs only two sets of properties for sagging regions and three for hogging regions (MU with two modular ratios, and MC).

In the rare event that significant longitudinal torsional moments are applied to composite rolled sections or plate girders, it may be necessary to consider longitudinal stresses due to restraint of warping. The Vlasov theory for thin-walled elastic beams is applicable at the serviceability limit state, and simplified expressions for the torsional properties, bimoments, and stresses are available.[192]

Table 8.1 Properties of a composite section required in calculations

Breath of concrete flange:	Concrete in tension assumed to be:	
	Uncracked, unreinforced	Cracked, reinforced
Full (F)	For global analysis (FU); two values of α_e	—
Mean effective breadth for span (M)	For longitudinal shear and for deflections (MU); two values of α_e	For deflections (MC); no α_e
Effective, at section considered (E)	For check on tensile stress in slab at internal support (EU); two values of α_e	For longitudinal bending stresses (EC); no α_e

8.2 Distributions of bending moment and vertical shear

It was considered to be outside the scope of Part 5 of the Bridge Code to give detailed guidance on methods of global analysis for complete deck structures (discussed in section 6.4), but methods are given for allowing for the effects of cracking of the deck slab near internal supports of continuous girders. Loss of stiffness due to cracking is more significant than in continuous reinforced concrete girders, in which its effects are partly offset by cracking in the midspan regions. As explained in Volume 1, it is found that in continuous composite beams for buildings the bending moment at an internal support at the serviceability limit state may be from 15 to 30% lower than that given by an elastic analysis in which no account is taken of cracking. In internal spans, the midspan moments are therefore up to 30% higher than predicted. Errors of this order are unacceptable in bridge girders because they may lead to premature fatigue failures or buckling of slender steelwork.

Even though computers are widely used in bridge design, it was thought to be essential to provide a method suitable for hand calculation, for use for small-span bridges and preliminary designs and because a hand method is usually easier to understand. For simple structures, the method should be applicable before areas of slab reinforcement are known, and should not require the analysis of continuous members of non-uniform cross section, because influence lines for such members are not generally available.

The subject is now introduced by a review of the method given in CP117: Part 2. The new method is then explained and related to the computer studies by which it was checked. Its application to global analysis of a complete deck, rather than an isolated girder, raises questions of interpretation not covered by the Bridge Code, some of which are considered below.

8.2.1 Serviceability limit state

Clause 12 of CP 117: Part 2[77] 'Continuity of composite section over supports', allows the longitudinal moments and shears to be calculated initially by assuming uncracked cross sections, and using effective flange breadths that are constant over each span, but may vary from span to span. From the bending moments so obtained, the maximum tensile stress in the uncracked concrete flange over each internal support (f_{tc}) is calculated *for each particular case of loading*. The effects of cracking on the distribution of longitudinal moments are neglected at each support where f_{tc} does not exceed the lesser of $0.14f_{cu}$ and 5.86 N/mm². Where f_{tc} exceeds this limit, the stiffening effect of the concrete at that support has to be neglected (i.e. the slab is assumed to be fully cracked), over the length of the negative moment region *for that case of loading only*, and the analysis is repeated; or else the two adjacent midspan moments *for the case of loading in question* are increased by 15%, without reduction in the corresponding hogging (negative) moment. Shear lag is considered in calculating both the bending moments and the stress f_{tc}.

The first alternative is rational, provided the reinforcement in the cracked slab is included in the second analysis (which is not clear in CP 117), but is laborious, because the length of girder where the stiffness is reduced is different for every load case considered. The second alternative introduces a step-function into the methods of analysis, illustrated in fig. 8.1, which leads to an illogical discontinuity in design. Where the slab is 'cracked' at two adjacent supports, the total increase in bending moment in the intervening midspan region can reach 30%. This agrees with the maximum value quoted above for buildings but computer studies for the new Code found it to be over-conservative for most bridge girders.

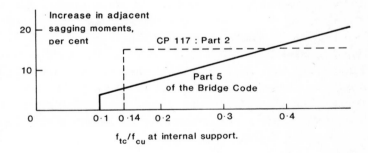

Fig. 8.1 Approximate methods of allowing for cracking

Although the method in the Bridge Code appears to be similar to that in CP 117, there are important changes. The phrases in italics (above) have all been replaced by 'due to the maximum design hogging (negative) moment', in the belief that if cracking occurs due to this maximum loading, it remains effective in reducing stiffness for all lesser loadings.

In continuous girders where adjacent spans do not differ appreciably in length, redistribution of moments due to cracking is roughly proportional to f_{tc}. The 'simple' method of allowing for cracking has therefore been changed from '15% increase' to an increase of $40f_{tc}/f_{cu}$ per cent in the maximum design sagging moment in each of the two midspan regions adjacent to an internal support where the calculated 'uncracked' stress due to the maximum design hogging bending moment is f_{tc}, provided that f_{tc} exceeds $0.1f_{cu}$. As shown in fig. 8.1, this does not eliminate the 'step' in the CP 117 method, but it reduces it from 15% to 4%.

In bridges where adjacent spans do 'differ appreciably in length', the simple method becomes increasingly inaccurate. It is not possible to give general guidance on what is an 'appreciable difference'. The highest ratio of adjacent spans in the bridges shown in fig. 8.2 is 1.57 (in bridge 3). In the analyses to be described below, the results for this bridge conformed least well to the general pattern (fig. 8.3). However, results from bridge 1, with equal spans, were also atypical. It may be that the ratio of the cracked to the uncracked flexural rigidity (k) is the more significant variable, as these two bridges had the lowest values.

For large bridges, or where the 'simple' method is thought to be too conservative because moments at internal supports are not reduced, the alternative method of allowing for cracking should be used. This is to repeat the analysis neglecting the stiffening effect of concrete at internal supports, as in CP 117: Part 2, except that the length to be taken as cracked is now a fixed 15% of the span on each side of each support where f_{tc} exceeds $0.1f_{cu}$.

Both in the initial 'uncracked' analysis and in any re-analysis, the longitudinal moments may now be found neglecting shear lag, because Moffatt and Dowling found[55] that it had little effect on the distribution of longitudinal moments.

These are major simplifications, for they enable published influence lines[99] for members of uniform section to be used for the initial analysis of composite decks where the steel members are parallel and have little or no variation of section from span to span. For 'cracked' analyses, the assumption that changes of section occur only at 15% of the span from each end makes feasible the use of moment distribution (for which relevant formulae were given in the first edition of this book), and also of influence lines, which could be prepared for members with various ratios of cracked to uncracked flexural rigidity.

CHECK ON METHODS OF ALLOWING FOR CRACKING

These new methods were checked by means of elastic analyses by computer of six continuous girders in which allowance was made for cracking adjacent to the internal supports, over various proportions of

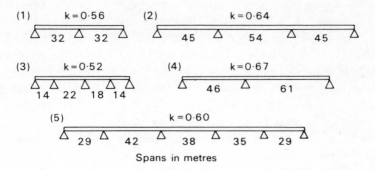

Fig. 8.2 Spans of bridges used in study of cracking

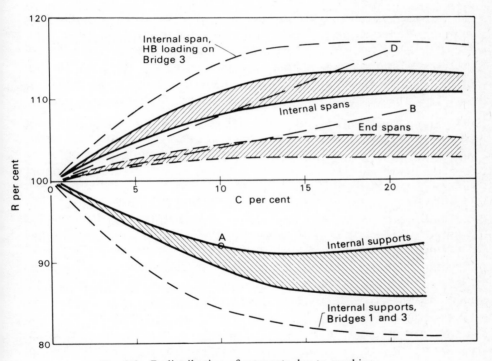

Fig. 8.3 Redistribution of moments due to cracking

the spans. The data for five of these were taken from real bridges, the spans of which are shown to scale in fig. 8.2. The values of k, the ratios of the cracked to the uncracked flexural rigidity in the negative moment regions, ranged between 0.52 and 0.67. Maximum bending moments were calculated for both HA and HB loading, at internal supports and in midspan regions in both internal and end spans. The results are given (fig. 8.3) in terms of R, the ratio of the maximum bending moment to the maximum value found in the 'uncracked' analysis, and C, the proportion of each span assumed to be cracked. It was not practicable to simulate the 'partially cracked' situation in this work; the slab was assumed to be either 'uncracked', or to have a stiffness corresponding to its reinforcement only. The values of C therefore do not define positions of points of contraflexure, but the lengths of regions where the negative moments were large enough for the stiffness of the concrete to be negligible, and the term 'cracked' is used below with this meaning.

All the results lay within the cross-hatched areas, except that the changes of bending moment were unusually large in two regions in bridge (3) and in one region in bridge (1), as shown. The following conclusions can be drawn from the results.

(1) Provided that 10% or more of the span is cracked, as is likely in practice, the reduction in support moments due to cracking will exceed 8% (point A in fig. 8.3). The assumption that the difference between an 'uncracked' and a 'cracked' analysis is equivalent to 10% redistribution of the 'uncracked' support moments (made in the design methods for slender beams at the ultimate limit state) is reasonably accurate.

(2) Bending moments found in a 'cracked' analysis assuming $C = 0.15$ (the method given in Part 5) will be correct to within about 5% if in fact any proportion of the span between about 8 and 25% is cracked.

(3) The errors in an 'uncracked' analysis increase steadily with the extent of cracking, so the rule for increasing sagging moments to allow for cracking (fig. 8.1) is of the correct type. It is only possible to examine the accuracy of the proposed increases by making arbitrary assumptions. The rule implies that when f_{tc}/f_{cu} just reaches 0.1 at an internal support, cracking increases midspan moments in an end span by 4%. If it is assumed that the proportion of the span cracked is then 10%, and increases linearly with f_{tc}/f_{cu}, the design rule corresponds to line OB on fig. 8.3, which is conservative in comparison with the computed increases in midspan moments in end spans. The rule is only 'unsafe' if cracking exceeding 10% of the span occurs when the 'uncracked' stress at an internal support is $0.1f_{cu}$ or less – an unlikely combination. The same argument gives line OD on fig. 8.3 for midspan moments in internal spans (as these are increased to allow for cracking at two adjacent supports), which agrees well with all the computed results except one.

CROSS-GIRDERS

Reduction of stiffness due to cracking does not influence the bending moments in an isolated composite cross-girder with cantilever ends unless it is supported on more than two main girders. The initial global analysis for the distribution properties of such a deck would be based on uncracked sections. If the results showed that high transverse tensile stresses occurred over the internal main girders, the analysis should be repeated with reduced transverse flexural stiffness in these regions. The narrow cracks likely to occur in service have little effect on the torsional stiffness of the deck slab, which is maintained by the effects of dowel action and aggregate interlock, so it is probably more accurate to use the 'uncracked' torsional stiffness in the second analysis.

8.2.2 Ultimate limit state

The distribution of longitudinal moments in a continuous beam at the ultimate limit state may differ from that given by elastic analysis of the uncracked structure for the following reasons:

(1) cracking of the concrete slab (unless prestressing is used)

(2) yielding of steel at sections designed as compact

(3) redistribution of stress from steel tension flanges of non-compact sections to the slab reinforcement.

CRACKING

Two methods of allowing for cracking are given in Part 5 of the Bridge Code: either the bending moments found by an 'uncracked' analysis should be redistributed from the supports to midspan regions by 10%, or a 'cracked' analysis can be done, with the stiffening effect of the concrete neglected at the internal supports, over 15% of each relevant span length. The origin of the '15%' rule is explained above.

The redistribution of moments actually caused by cracking cannot be predicted accurately because it depends on the sequence of casting and the effects of temperature and shrinkage, as well as on the proportions of the continuous member. If redistribution is less than the designer assumes, the steel web or compression flange in a negative moment region may buckle prematurely. This sudden type of failure must be avoided. If redistribution is more than is assumed in design, the error is self-correcting in the structure, as additional inelastic curvature in the midspan region transfers some bending moment back to the internal supports.

For safety, the maximum amount of redistribution to midspan of the negative moments calculated in an 'uncracked' analysis of a slender member must therefore be lower than the minimum redistribution likely to occur in practice. This redistribution will certainly be greater at the ultimate limit state than for serviceability loads. From consideration of fig. 8.3, the maximum value for use in design was fixed at 10%.

YIELDING

Compact sections normally occur only in midspan regions. There is then some inconsistency in using elastic global analysis, design moments of resistance for midspan sections that cannot be reached unless inelastic curvature occurs, and limiting stresses for non-compact sections at internal supports that lie in the elastic range. If regions of both hogging and sagging bending are governed by the same load case, there is likely to be, at the design ultimate load, inelastic behaviour in regions of hogging bending, which therefore resist higher moments than expected. At midspan, only partial plasticity occurs, and moments are lower than expected. No account is taken in Part 3 of this possible source of overstress, mainly because the two regions are normally governed by different load cases, such that the critical loading for hogging regions can be resisted without inelasticity at midspan. Elastic global analysis, with allowance for cracking, should then give an accurate enough prediction of the peak hogging moments.

REDISTRIBUTION

The yielding of the steel tension flange at internal supports that is permitted under certain conditions (section 8.4.3) causes, in theory, some redistribution of bending moments to midspan. No allowance is made for this in the global analysis. If it were allowed for, the compression flange in this region could be made less strong. This would be dangerous because if, as is likely, the yield strength of the tension flange should exceed the design value, the redistribution would not occur, and the load factor against buckling of the compression flange would be reduced.

8.3 Reinforced concrete deck slabs

A concrete deck slab forming the top flange of a composite beam has to be designed in accordance with both Part 4 and Part 5 of the Bridge Code. In this section, requirements of Part 4 and Part 5 relevant to the design of the slab for a beam-and-slab deck will be reviewed together. The terms 'longitudinal' and 'transverse' are here used in relation to the direction of span of the main beams.

  The slab in a right (i.e. not skewed) bridge has to resist six types of stress resultant, three due to the overall or 'global' behaviour of the superstructure and three due mainly to the local effects of wheel loads:

(*1*) *Longitudinal global bending* causes compressive or tensile stresses which in a large structure vary little through the depth of the slab and so are effectively membrane forces. Their maximum values occur above the webs of longitudinal girders. They are lower elsewhere due to shear lag (section 8.4.1).

(*2*) *Transverse global bending* of the slab is related to the differences in the deflections and torsional rotations of the main girders, and is influenced by the distribution properties of the deck. It is most severe in midspan regions remote from cross-girders. Near a composite cross-girder it is replaced by a transverse membrane force due to the interaction with the steel member, which is a maximum above its web. The vertical shear forces associated with global bending of the slab are negligible in comparison with those due to point loads (paragraph (6) below).

(*3*) *Longitudinal global shear* is a maximum in the vicinity of the shear connectors and diminishes with distance from the nearest steel member much more rapidly than do the longitudinal bending stresses (section 8.4.1). It can be considered at a late stage in the design of the slab, as reinforcement provided for other purposes can often also be considered as effective in resisting longitudinal shear, and is usually more than sufficient for this purpose. Complementary horizontal shear forces of equal magnitude exist in the transverse direction. Where composite cross-girders are used, there is a second set of in-plane shear forces. Interaction between the two sets is discussed in section 8.3.2.

(*4*) *Transverse local bending* occurs in the slab due to wheel loads. Positive (sagging) moments are most severe when a wheel load acts midway between the steel girders. Maximum negative moments occur above the steel girders due to loads near the edge of a cantilever slab, or to pairs of wheels straddling the line of the girder. Bending moments calculated for a typical cross section can usually be assumed to be constant along the length of the deck, but in fact are lower near diaphragms and composite cross-girders.

(*5*) *Longitudinal local bending.* For slabs that span transversely, longitudinal local bending moments due to wheel loads are lower than the transverse moments. For highway loading, CP 117: Part 2 allowed the longitudinal moment to be taken as 50% of the coexisting transverse moment at the point considered, in absence of more exact analysis. For slabs that span longitudinally between cross-girders, the longitudinal moments are as described in (4) above, and the '50% rule' would apply to the local moments in the direction transverse to the span of the slab. This rule is not included in Part 5 of the Bridge Code. General guidance on this subject is given in Part 4.

(6) *Punching shear.* As wheel loads can occur anywhere on the deck, it is impracticable to use shear reinforcement, so the thickness of the deck slab (and hence its weight, which is significant in the overall design of the bridge) may be governed by the vertical shear stresses in the vicinity of wheel loads. It is shown in section 8.3.1 that for HB loading, the stresses due to a group of wheels are usually higher than for a single wheel.

The effects of temperature and shrinkage (section 8.7) can cause stresses of types (1), (3) and (5), and also of types (2) and (4) if composite cross-girders are used.

These six load effects appear in a typical design sequence for the deck slab as follows.

(1) Choose the concrete strength and density, assume typical values for cover and percentages of flexural reinforcement, and find the minimum thickness for punching shear.

(2) Consider stresses due to longitudinal global and local bending and decide the thickness of the slab and its longitudinal reinforcement.

(3) Consider stresses due to transverse global and local bending, and design the transverse reinforcement.

(4) Consider longitudinal shear and increase the transverse reinforcement if necessary.

(5) Consider crack widths and modify amount and spacing of reinforcement if necessary.

8.3.1 Combined global and local effects in deck slabs

As would be expected, Part 5 refers to the treatment in Part 4 of slabs for concrete beam-and-slab bridges. Relevant extracts from Clause 4.8 of Part 4 are as follows.

'... it is also necessary to consider the loading combination that produces the most adverse effects due to global and local loading where these coexist in an element.'

'Analysis of the structure may be accomplished ... by separate analyses for global and local effects ... the forces and moments acting on the element from global and local effects should be combined as appropriate.'

'Ultimate limit state ... the resistance to combined global and local effects is deemed to be satisfactory if each of these effects is considered separately.'

When the effects are considered separately, as is usual, then it is required in Clause 4.1.1.3 that stresses under combined global and local, direct and bending effects are checked at the serviceability limit state. This two-stage procedure is necessary because no simple methods exist for checking the resistance of a reinforced concrete deck to the combined effects at the ultimate limit state. For example, in applying yield-line theory, it would be necessary to allow for the coexisting membrane stress in the slab when calculating moments of resistance, so that the bending strength at a yield line would vary with both its position and its direction.

In practice, elastic analysis is likely to be used for local effects. The slab is assumed to be uncracked, isotropic, and continuous over the steel beams, which are assumed not to deflect relative to one another. The torsional stiffness of steel I-beams is too low to affect local moments in the slab. If computer programs are not available, the methods of Westergaard[100] and the influence surfaces of Pucher[56] are useful. Bending moments in cantilever slabs with and without edge stiffening are given by Sawko and Mills.[45a]

The results of local elastic analyses can be used at both limit states, by appropriate scaling, as shown in section X13.

THE ONTARIO BRIDGE CODE

This design code for highway bridges[193] allows much lighter reinforcement in the deck slab than does the Bridge Code. The new rules lead typically to 0.3% of reinforcement each way near each face, and are based on extensive full-scale testing on the Conestogo River Bridge.[194] This showed that wheel loads on the 190-mm deck slab were resisted by arching action, rather than by local bending, to such an extent that bottom reinforcement for local loading could be reduced from 0.95% (given by conventional design) to 0.2%. These conclusions need to be reviewed in relation to the heavier wheel loads for which bridge decks in the UK are designed.

PUNCHING SHEAR

At the ultimate limit state, the design wheel loads given in Part 2 of the Bridge Code, including γ_{f3}, are 165 kN for HA loading and 161 kN for 45 units of HB loading. When 25 or more units of HB loading are specified, the HB wheel loads usually govern design for punching shear because they occur in groups of eight, whereas only one HA wheel load is considered.

The maximum punching shear stress will usually be obtained when as much of the eight-wheel HB bogie as can be fitted onto the slab panel considered is treated as a group of point loads.

Apart from special cases near a free edge of the slab, the most adverse layout is when the slab spans longitudinally between cross-girders at 3.0 m or wider spacing. It is now shown that for 45 units of HB loading, the minimum thickness of slab for punching shear is about 190 mm. If the bottom cover is 35 mm, and 16-mm bars are used in each direction, the effective depths d_x and d_y are 131 mm and 147 mm. Here, the mean value of 139 mm is used. The mean shear stress v is calculated on a perimeter at a distance 1.5 d from the loaded area.

The most adverse loading is a complete eight-wheel bogie. Each wheel load acts on an area 0.32 m square, increased to 0.37 m by dispersion at a spread:depth ratio of 1:2 through surfacing assumed to be 50 mm thick.

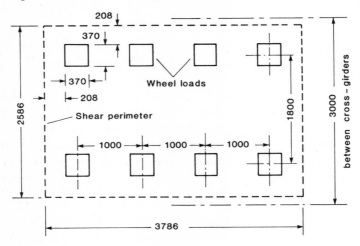

Fig. 8.4   Shear perimeter for bogie of HB vehicle

The dimensions of the shear perimeter are then as in fig. 8.4. Its length is 12.74 m, so that:

$$v = 8 \times 161/12.74 \times 139 = 0.73 \text{ N/mm}^2$$

For a 190-mm slab with Grade 40 normal-density concrete and only 0.5% of reinforcement in each direction, the design ultimate shear stress $\xi_s v_c$ given in Part 4 of the Bridge Code is $1.27 \times 0.59 = 0.75 \text{ N/mm}^2$.

Similarly it can be shown that for longitudinal girders at 2 m spacing, design for punching shear gives a minimum thickness of about 150 mm. This would probably be too thin for other purposes. In British practice, slabs are usually between 200 and 250 mm thick.

8.3.2 Analysis of cross sections of deck slabs

The overall thickness and reinforcement ratios for a deck slab are likely to be governed by load combination 1. In simply-supported bridges, global effects at the ultimate limit state usually govern, but in continuous bridges, limiting stresses at the serviceability limit state for the global-local combination may be critical. Serviceability limits on the width of cracks may determine the detailing of reinforcement provided for transverse local bending and for longitudinal global bending in negative moment regions.

SERVICEABILITY LIMIT STATE

Bending and membrane stresses in slabs are calculated by the usual elastic methods, neglecting concrete in tension and using the appropriate modular ratio for the type of loading considered. Longitudinal, vertical and punching shear in the deck slab are considered only at the ultimate limit state.

In Part 4 of the Bridge Code, compressive stress in concrete is limited to $0.5f_{cu}$ for near-triangular stress distributions, and to $0.38f_{cu}$ for near-rectangular distributions. Stress in reinforcement is limited to $0.75f_{ry}$. This is conservative in comparison with draft Eurocode 4[5] in which no limits on stress at the serviceability limit state are given, in the belief that other design criteria will ensure satisfactory behaviour in service.

The calculated longitudinal bending moment due to a wheel load at a given point on a cross section has to be combined with one of two types of global longitudinal stress.

(1) In midspan regions, there is global compression throughout the slab thickness. The local stresses can be calculated for the uncracked reinforced section. The stress limit in concrete can be taken as $0.5f_{cu}$, rather than the $0.38f_{cu}$ which may be appropriate for global bending alone.

If the span:breadth ratio of the main beams is low, the global longitudinal stresses in regions of highest local longitudinal moments are lower (due to shear lag) than they are above the steel webs. Allowance can be made for this using a formula given in Part 3 of the Bridge Code, which is explained in section 8.4.1. This applies to all cross sections at both limit states.

(2) Near internal supports, there is global tension throughout the slab thickness. If the design loading is to Part 2 of the Bridge Code, the HA wheel load is an alternative to the HA distributed and knife-edge loads. whereas the HB wheel load is an integral part of the HB loading. Two load cases then have to be considered.

The first consists of the loading that gives the maximum global tension in the slab. The bogies of the HB vehicle will then both be some distance from the relevant internal support, and the HA distributed load will be applied, so no local wheel load need be considered.

The second loading consists of the HB vehicle with one axle over the internal support, and the HA distributed loading on the adjacent lane or lanes. The local effects of an HB wheel then have to be considered in combination with a lower global tension than before. The relevant locations of the live loads are shown in fig. 8.5.

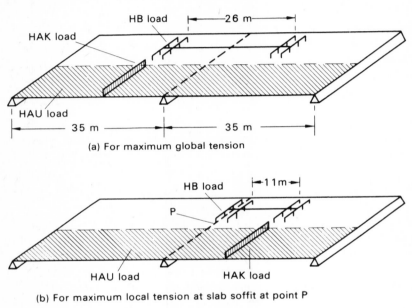

(a) For maximum global tension

(b) For maximum local tension at slab soffit at point P

Fig. 8.5 Loadings for maximum tension in slab at internal support

In both of these load cases, the intensity of HB loading specified in the Bridge Code is 25 units when checking crack widths, but is the full design value (up to 45 units) when checking stress in the reinforcement.

It will usually be found that the tensile strain in the longitudinal reinforcement due to the global tension is several times the cracking strain of concrete. The longitudinal local moments due to a wheel load should first be assumed to be resisted by the two layers of longitudinal reinforcement only, and the resulting stresses combined with the global tension. The total should not exceed $0.75 f_{ry}$. If it does, and the local compressive strain exceeds the global tensile strain at the same level, a calculation that allows for compression in concrete should be made.

For a wheel load, the local longitudinal sagging moment is larger than the local hogging moment, and the limiting crack width for the slab soffit is usually less than that for the top surface, which is protected by a waterproof membrane. A significant proportion (at least one-half in deep girders) of the total longitudinal reinforcement should be placed near the lower surface of the slab. Its global action is slightly less efficient, but it is better able to resist local bending and to control cracking.

The assumption that the local longitudinal moment is resisted by the reinforcement alone is conservative, particularly if the slab is assumed to be isotropic when the local moments are calculated. The cracked slab is in fact orthotropic, and a local wheel load is likely to be carried mainly in the transverse direction, unless the slab spans longitudinally. The calculated longitudinal moment is therefore too high. There is a need for a simple but more accurate design method for the local effects of wheel loads in hogging moment regions.

ULTIMATE LIMIT STATE

If global and local effects are considered separately, yield-line theory can be used for the local check. A worked example is given by Clark.[101] For longitudinal reinforcement, local effects will govern in midspan regions of the main beams, but not near internal supports. Transverse reinforcement in slabs that span transversely is likely to be governed by the combination of self weight and wheel loads if there are several cross-frames; but, if the slab is the only transverse member, except at supports, global bending also may influence the transverse reinforcement.

The following points are relevant to checks made at the ultimate limit state for combined global and local effects.

(1) If the global stress is compressive and below $0.1f_{cu}$, it can be neglected, according to Part 4 of the Bridge Code.

(2) The resistance of the slab for combined compression and bending can be assessed by the method given in Part 4 for cross sections of short columns.

(3) The preceding discussion of combined tension and bending is applicable also to the ultimate limit state, except that the limiting stress in reinforcement is $0.87f_{ry}$.

(4) If the relevant composite cross section is designed as slender, the design global tension in the slab at an internal support can be reduced by the use of unpropped construction and an appropriate sequence of casting.

BIAXIAL HORIZONTAL SHEAR

As noted above, horizontal shear stresses in the deck slab are a maximum in the vicinity of the shear connectors. The design procedure for transverse reinforcement (section 8.5.3) is intended to ensure that the slab has sufficient strength in in-plane shear to resist the mean shear stress, v_{hx} say, which is q_p/h_c on a plane of type 1-1 (fig. 8.24a) where q_p is the longitudinal shear per unit length and h_c the thickness of the slab.

Where two composite members intersect, calculations of longitudinal shear will in general give different values of mean shear stress v_{hx} and v_{hy} at a point in the slab such as C in fig. 8.6. Where the two members meet at right angles, it is easily shown (from equilibrium of a small element of slab) that v_{hx} and v_{hy} must be equal at any point in the slab, so that different values cannot both be correct. More accurate methods of global analysis show that the true in-plane shear at such a point is likely to be about $\frac{1}{2}(v_{hx} + v_{hy})$; thus, at a point such as B in fig. 8.6, where v_{hx} and v_{hy} are of opposite sign, it may be close to zero. No advantage should be taken of this in design because if the longitudinal shear is lower at one point along a beam, it will be higher somewhere else. A simple and conservative design rule for a region where v_{hx} and v_{hy} are different and neither is negligible, is to provide sufficient reinforcement transverse to both of the two intersecting beams to resist a shear corresponding to the greater of v_{hx} and v_{hy}.

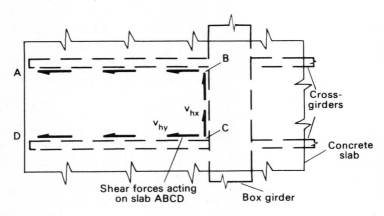

Fig. 8.6   Part plan of composite deck with cross-girders

8.3.3 *Control of cracking*

Local damage and decay of reinforced concrete occur when the reinforcement rusts, because it then increases in volume. The resulting pressure breaks up the surrounding concrete, so allowing the corrosion of the steel to spread. Rusting is prevented in practice by requiring all reinforcement to be surrounded by a minimum thickness of well-compacted concrete, with a cement content appropriate to the conditions of exposure, and by controlling the widths of cracks.

The first British code of practice to give a detailed procedure for the control of cracking was CP 110: 1972. This method is explained in Volume 1. The variation in the width of cracks in nominally identical situations makes it necessary to adopt a statistical approach in design. As in CP 110, the design crack widths given in both Parts 4 and 5 of the Bridge Code are limits which may be exceeded by not more than 20% of the cracks at the serviceability limit state. A safety margin is therefore included in the specified widths, to allow for these wider cracks.

Cracking in concrete components of composite members is checked using the method of Part 4, explained below, which was developed from that in CP 110. The formulae for crack widths include coefficients based on tests, such that they should also give values which have a 20% probability of being exceeded. Cracking in cased beams and filler beams is considered in section 12.6.

DESIGN LOADING AND CRACK WIDTHS

In a highway or railway bridge, cracks open and close continually with the passage of traffic. There is little data from research on the relationships between the onset of corrosion and the lengths of time over which cracks have various widths, but it is generally assumed that wider cracks are acceptable if they occur less frequently. There is evidence from fatigue tests of some increase in the widths of cracks under repeated loading of constant amplitude.

In Part 5 of the Bridge Code, crack widths were required to be checked for load combination 1 of Part 2, using the nominal values of dead plus highway or railway loading, with the partial safety factors γ_{fL} and γ_{f3} taken as 1.0. In the 1984 edition of Part 4, γ_{fL} for highway loading was increased to the value for the serviceability limit state (i.e. 1.2 for HA loading), and this change has been applied to composite structures also.[2] This and other changes, to be described, have increased the influence of crack-width control on the design process. They illustrate the growth in concern for the durability of concrete structures that occurred in the early 1980s.

The design loading for cracking does not include the effects of shrinkage of concrete because the design methods are based on tests on specimens in which shrinkage was already present. It is usual also to neglect the secondary effects of shrinkage, even though these were not present in the test specimens. The influence of shrinkage on crack widths in composite members is greater than in reinforced concrete[102] but the subject is complex, and most environments for bridges in the UK are humid enough for shrinkage to be small. Its influence is therefore neglected unless the free shrinkage strain is expected to exceed 0.006.[2]

As bridge decks are outdoor structures with a long design life, the specified crack widths are more severe than in CP 110. They range from 0.25 mm for internal or protected surfaces to 0.15 mm for surfaces exposed to the effects of de-icing salts. The lower limit is not normally applied to the soffit of a slab that carries wheel loads, for which the limiting crack width is 0.25 mm.

METHOD OF BS 5400 FOR CALCULATING CRACK WIDTH

It is convenient to consider first the effects of local loading on the deck slab, and then to take account of any coexisting global tension in the slab. As with limiting widths, the predicted widths are those expected to be exceeded by 20% of a random sample of measured widths. The crack width w at a point on the surface of a slab at a distance a_{cr} from the nearest longitudinal reinforcing bar is given in Part 4 as:

$$w = \frac{3a_{cr}\varepsilon_m}{1 + 2(a_{cr} - c_{nom})/(h - d_c)} \qquad (8.1)$$

where:

ε_m is the average longitudinal strain at the level considered, allowing for tension stiffening

c_{nom} is the required minimum cover, which may be less than the actual cover

h is the overall thickness of the slab

d_c is the depth of the elastic neutral axis from the compression face.

This is the well-known equation from Appendix B of Part 5, given originally in Appendix A of CP 110: 1972, except that c_{nom} has replaced the minimum cover, c_{min}. This relaxation is relevant to the inner layer of the mesh of bars normally placed adjacent to each surface of a deck slab.

For surface cracks due to local loading, the equation for tension stiffening given in Appendix A of CP 110 and in the 1978 edition of BS 5400: Part 4 becomes:

$$\varepsilon_m = \varepsilon_1 - 0.0012b_t h/A_r f_y$$

where:

ε_1 is the strain at the level considered, calculated neglecting tension stiffening

b_t is the breadth of the section at the level of the tension reinforcement

A_r is the area of reinforcement, of yield strength f_y.

A similar equation was given in Appendix B of Part 5. Both have been superseded by the equation given in Clause 5.8.8.2 of the 1984 edition of Part 4, which is as follows for cracks at the surface of the slab:

$$\varepsilon_m = \varepsilon_1 - 3.8 \times 10^{-9}(b_t h/\varepsilon_s A_r)[1 - (M_q/M_g)] \qquad (8.2)$$

but:

$$\varepsilon_m \not> \varepsilon_1$$

where:

ε_s is the strain in the tension reinforcement, calculated neglecting tension stiffening

M_q is the moment at the section considered due to live loads

M_g is the moment at the section considered due to permanent loads.

There is now evidence, summarised by Clark,[101] that the earlier equation overestimates tension stiffening. It was derived from short-term tests, and does not take account of repeated loading. This is allowed for by the new term $[1 - (M_q/M_g)]$, which has the effect of reducing tension stiffening to zero when $M_q \geqslant M_g$. Also, the earlier equation took no account of the influence of the strain in the tension reinforcement. Its value was probably assumed to be constant at $0.58f_y/E_r$. If f_y is replaced by $\varepsilon_s E_r/0.58$, with $E_r = 200 \text{ kN/mm}^2$, and the constant 0.0012 is increased to 0.0013, equation (8.2) is obtained.

The value of A_r must be the area of tension reinforcement, effective in the direction considered, in breadth b_t of the slab (taken as 1.0 m). Its calculation for skewed reinforcement is explained by Clark.[101]

The effect of tension stiffening, as given by equation (8.2), is greatest when $M_q = 0$. It is now illustrated, assuming the typical values: $\varepsilon_1 = 1.25\varepsilon_s$, and $A_r/h = 0.008$. Putting $\varepsilon_s = f_r/E_r$, where f_r is the calculated tensile stress in the reinforcement, and $E_r = 200 \text{ kN/mm}^2$, equation (8.2) becomes:

$$10^6\varepsilon_m = 6.25f_r - 95\,000/f_r \qquad (8.3)$$

This is plotted in fig. 8.7, together with a curve for lighter reinforcement $(A_r/h = 0.004)$ and the line for no tension stiffening, $\varepsilon_m = \varepsilon_1$. This shows that the lighter the reinforcement and the lower the stress in it, the greater is the effect of tension stiffening in reducing ε_m and hence the widths of cracks.

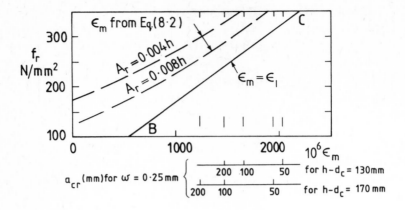

Fig. 8.7   Effect of tension stiffening on mean longitudinal strain

These results for ε_m can be related to spacing of bars as follows. We consider cracking at the soffit of a slab in Grade 40 concrete, for which $c_{nom} = 35$ mm and $w \not> 0.25$ mm. For a typical deck slab, the depth $(h - d_c)$ in tension lies between 130 and 170 mm. For these values, ε_m can be related to a_{cr} using equation (8.1). The results (fig. 8.7) show, for example, that if $A_r/h = 0.008$, $h - d_c = 170$ mm, and $a_{cr} = 200$ mm, the calculated crack width is less than 0.25 mm when $f_r < 255$ N/mm^2.

The maximum spacing of bars in tension is limited in Part 4 to 300 mm. For 20-mm bars with 35 mm cover, this gives $a_{cr} = 147$ mm. It is evident from fig. 8.7 that this spacing would be low enough to control crack widths due to local loads if M_q were zero. In practice however, M_q usually exceeds M_g, so that $\varepsilon_m = \varepsilon_1$ and line BC is relevant. The limiting value of f_r for $h - d_c = 170$ mm then falls to about 220 N/mm^2.

Global tension in the same direction as local tension increases crack widths. For global tension alone, causing stress f_r in the reinforcement adjacent to the surface considered, ε_m is still given by equation (8.2), but equation (8.1) is replaced by:

$$w = 3a_{cr}\varepsilon_m \qquad (8.4)$$

because at given ε_m, cracks are slightly wider for flanges in overall tension than for slabs in bending.

In practice, the design of reinforcement (for which f_y is usually 460 N/mm^2) is often governed by crack widths rather than by the ultimate limit state. The problem can be acute in some situations. For example, a composite beam-and-slab jetty was constructed in 1984 in the River Humber. If there were concrete surfaces exposed to salt spray, and the jetty had been designed to the 1984 revision of Part 4, then for Grade 40 concrete, the minimum cover would be 50 mm and the limiting crack width would be 0.15 mm. The closest practicable spacing of reinforcing bars is about 80 mm, and for 16-mm bars with 50 mm cover, a_{cr} is then 54.4 mm. If $M_q \geqslant M_g$ and the tension in the deck slab is mainly global, then equation (8.4) is relevant; whence

$$\varepsilon_m \ngtr 0.15/(3 \times 54.5) = 919 \times 10^{-6}$$

This corresponds to a stress of only 184 N/mm², at which the reinforcement is not being used efficiently. The requirements of Part 4 would be even more onerous for the equivalent reinforced concrete structure, as they would apply to the beams in midspan regions as well as to the slab at internal supports.

COMBINATION OF GLOBAL AND LOCAL EFFECTS

The design method of Part 5, as amended by the Department of Transport,[2] is to determine global and local crack widths separately, and check that their sum does not exceed the design crack width. This is a little less conservative than the method of the 1984 edition of Part 4 (which Part 5 now states *may* be used), and so is more economical. The method of Part 4 is that global and local tensile strains are added algebraically, and the crack width is then calculated from equation (8.4).

The widest cracks in the deck slab of a highway bridge may be those due to a combination of global and local effects near an internal support. When a check is made for 25 units of HB loading, longitudinal sagging bending due to a wheel load midway between two girders has to be combined with the global tension at the slab soffit. Advantage may be taken of the reduction in global stress in this region due to shear lag, given by equation (8.5).

BIAXIAL TENSION

Where the surface of a slab is in biaxial tension, separate checks should be done for cracking due to tensile strain in the two relevant layers of reinforcing bars. It is probable that the bars in the layer nearer to the surface act as crack initiators, and so modify the widths and spacings of the cracks caused by tension in the other layer of bars. Research on this subject is in progress at the University of Warwick. It has been shown that for a given arrangement of reinforcement, crack widths are mainly determined by the greater mean principal tensile strain at the surface of the slab; but no simple design method for taking account of biaxial interaction has yet been developed.

8.4 Design of cross sections for longitudinal moments and vertical shear

8.4.1 Effective breadth of concrete flange

The diverse treatments of this complex subject in codes of practice show it to be one where it is particularly difficult to find the right balance between accuracy (and hence economy of structure) on the one hand and simplicity (and hence economy in design time) on the other.

The rule for effective breadth given in CP 117: Part 2 is simple, but it can be much less accurate than is generally realised. As an example, we consider an uncracked slab at an internal support of a continuous beam-and-slab deck carrying distributed load. The outer (edge) girder has a cantilever flange with a projection beyond the girder web of one-tenth of the span; in the notation of the Bridge Code, $b/L = 0.1$. CP 117: Part 2 gives its effective breadth at the internal support as $0.82b$; Part 3 of the Bridge Code gives it as $0.40b$, or $0.60b$ if the slab is assumed to be cracked. For simply-supported spans, CP 117 is generally conservative in comparison with Part 3, which gives b_e/b ratios up to 30% higher (fig. 8.8). These examples are given not to criticise CP 117, but to show that it is impossible to find one simple rule that is even moderately accurate in all cases.

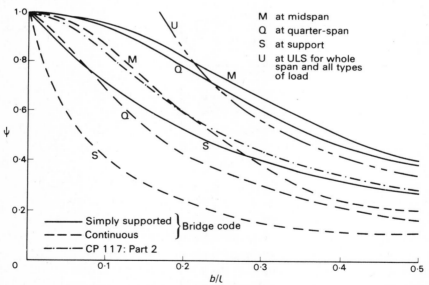

Fig. 8.8 Effective-breadth ratios for beams with distributed loading and no longitudinal stiffening

The rules for effective breadth in Part 3 of the Bridge Code are based on a parametric study of shear lag in steel box girder bridges[55] in which a very wide range of variables was considered. Linear-elastic theory was used throughout, and Poisson's ratio was taken as 0.3, so that the ratio E/G of the elastic moduli, which is relevant in shear lag studies, was 2.6. A more accurate value for an uncracked and unreinforced concrete flange would be 2.4, but variations due to reinforcement and cracking cause greater uncertainties than this.

All effective-breadth ratios given in Part 3 relate to calculations of longitudinal flexural stresses by elastic theory at cross sections of members, as explained in Volume 1. We have seen that shear lag has little effect on the results of a global analysis, but it has an important influence on stresses and deflections calculated by the elementary theory of bending, particularly for flanges with b/L ratios (fig. 8.8) exceeding 0.2, at internal supports, and directly under point loads (fig. 8.10). In this section, 'point load' refers to a column load or bearing reaction, not to a wheel or axle load.

In determining effective breadths, the overall breadth of the concrete flange is divided into areas associated with each steel web by lines above the webs and midway between them (fig. 8.9). The breadth:span ratio b/L for each such area is calculated, and the effective-breadth ratio ψ is found from a table. Shear lag is slightly greater in a slab strip with a free edge than for an internal strip, so ψ for edge strips is replaced by $k\psi$, where $k = 1 - 0.15b/L$. The cross-hatched area is thus the effective breadth of flange for the outer girder in fig. 8.9.

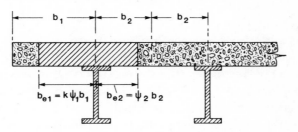

Fig. 8.9 Effective flange breadths

VALUES OF ψ GIVEN IN BS 5400: PART 3

An improved presentation of the material in the 1979 version of Part 5 is now given in Part 3. The tables are applicable to both stiffened steel flanges and concrete flanges. Account is taken of the increase in shear lag caused by longitudinal stiffeners by using a factor α. Concrete slabs are usually unstiffened, so the columns headed $\alpha = 0$ should be used. Values of ψ are given for midspan, quarter-span and supports of simply-supported, continuous, cantilever and propped cantilever beams, for ratios b/L ranging from 0 to 1.0. They are applicable for uniformly-distributed loads and standard highway and railway loading. End spans of continuous beams are treated as propped cantilevers.

Similar data for point loads are given in Appendix A of Part 3, together with a general method for finding ψ for loading and types of member not covered elsewhere.

Effective breadths at other cross sections can be found by linear interpolation, but the true variation along the span is more complex. Figure 8.10 shows this for a simply-supported box girder with $b/L = 0.2$, for various loadings. The values of ψ are lower than those in fig. 8.8 because these computations were for a plate with longitudinal stiffeners.

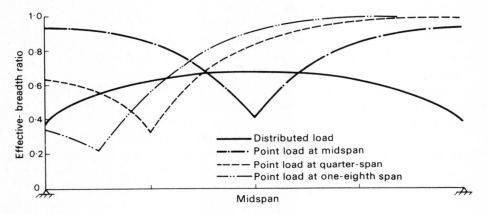

Fig. 8.10   Typical variation of effective breadth along span of simply-supported box girder with $b/L = 0.2$ and longitudinal stiffening

ACCURACY OF EFFECTIVE-BREADTH RATIOS FOR CONTINUOUS BEAMS

The values of ψ given in Part 3 for internal spans of continuous beams were calculated for fixed-ended members of uniform cross section. Effective breadths in real beams can be slightly different, for several reasons. Due to the flexibility of adjacent spans and the cracking of concrete in negative moment regions, internal supports provide less than full fixity, so negative moment regions are shorter than assumed; shear lag is greater, and the values of ψ are thus too high. The converse is true in midspan regions.

This overestimate of the degree of fixity (and consequent underestimate of the effects of shear lag) in a cracked negative moment region of a composite beam is more than offset by an effect of cracking of the concrete. Cracking tends to increase ψ, for it causes a larger reduction in the axial (extensional) stiffness of the deck slab than in its shear stiffness, due to the effects of aggregate interlock. The increase in ψ is given in Part

5 as $(1 - \psi)/3$, based on computer studies.[196]

SHEAR LAG AND THE ANALYSIS OF SECTIONS AT THE ULTIMATE LIMIT STATE

There is inconsistency between statements in various codes on this subject. In Clause 9.2 of Part 3 of the Bridge Code it is stated that the effects of shear lag '*may* be neglected'. In Part 5, Clause 6.3.1, which recommended that shear lag be considered, has been amended by the Department of Transport[2] to read: 'Longitudinal shear . . . *should* be determined . . . neglecting the effects of shear lag'.

This issue is significant at internal supports of beams with flanges of high b/L ratio. Tests on composite T-beams have shown[103-105] that at high hogging curvatures, the total tensile force in the slab reinforcement usually exceeds the predicted value neglecting shear lag. This is due partly to strain hardening, and does not occur until after the tensile force in the concrete has become negligible due to cracking. So in this respect, the amended Part 5 is correct.

It is also stated in Part 5 that in calculating longitudinal shear at the ultimate limit state, the slab should be assumed to be uncracked. This, coupled with the neglect of shear lag, can lead to the provision of excessive shear connection and transverse reinforcement in the slab, for shears that bear no relation to the calculated longitudinal stresses in this region.

In the draft Eurocode 4[5] it is recommended that the design longitudinal shear should be calculated neglecting concrete in tension and using the same effective breadths as for the bending resistances of the cross sections. In the original (1979) version of Part 5, shear was calculated for the uncracked section. For slender sections, the quarter-span value of ψ for distributed loading was used, and for compact sections a higher value ($\psi = L/6b$), which is shown as curve U in fig. 8.8. This was done partly to allow indirectly for the greater flexibility of shear connectors when in cracked concrete. It is still considered by the authors to be a good compromise between the extreme positions taken in the current (1985) version of the Bridge Code and in Eurocode 4.

COEXISTENT STRESSES IN DECK SLABS

The use of an effective breadth implies that the longitudinal bending stress in a thin flange has its maximum value throughout the effective breadth, and is zero elsewhere in the flange. Some more realistic assumption is necessary when considering the combination of global with local stresses (due, for example, to a wheel load).

For this purpose, Part 3 of the Bridge Code gives in Appendix A an empirical expression for the variation of global longitudinal stress σ_1

across the breadth b of a flange panel, also obtained by Moffatt and Dowling.[55] For regions between web centrelines the expression can be written

$$\sigma_1 = \sigma_{max}[\chi^4 + k(1 - \chi)^4] \tag{8.5}$$

where:

k $\quad$ $= 0.25\ (5\psi - 1)$
χ $\quad$ $= (b - x)/b$
x $\quad$ is the distance from the centreline of the nearer web
σ_{max} is the maximum stress given by elastic analysis using the effective breadth ψb.

The wording of the clause is appropriate for a very thin flange with negligible variation of σ_{max} through its thickness. It can be applied to thicker flanges by assuming that the level of the neutral axis of the composite member is constant across the full breadth of the flange, so that at every level in the slab, σ_1 varies across the breadth as given by equation (8.5), taking σ_{max} as the stress at that level given by the simple theory.

If σ_1 and σ_{max} are taken as mean stresses through the depth of a flange of thickness t, the total longitudinal force in the flange is obtained by integration of equation (8.5):

$$\int_0^b \sigma_1 t\ \mathrm{d}x = \sigma_{max} t b_e$$

This is the correct result, showing that no 'safety margin' has been included in equation (8.5) to allow for differences between the actual stress distribution and that assumed. The expression gives some negative values for σ_1 when $\psi < 0.2$; these should be replaced by zero.

The use of this clause can significantly reduce the influence of local loads on the design of concrete flanges, both for compressive stress and for control of cracking, when b/L exceeds about 0.1 in simple spans and 0.05 in continuous members.

8.4.2 Classification of cross sections of beams

Draft Eurocodes 3 and 4[4,5] have followed recent practice in many countries in defining four classes of slenderness for steel cross sections of beams. Their names and the associated design methods for bending resistance are given in table 8.2. Global analysis is by plastic hinge theory for Class 1 and by elastic theory for the other classes.

Plastic hinge theory is not at present accepted for analysis of bridge

Table 8.2 Classification of steel and composite cross sections

Eurocodes 3 and 4		Notes on calculation of bending resistance to Eurocodes 3 and 4	BS 5400: Part 3
Class No.	Name		Name
1	Plastic ⎱	⎰ Plastic analysis with check	—
2	Compact ⎰	⎱ on serviceability stresses	Compact
3	Semi-compact	Elastic analysis, limited by yield strength ⎱	⎰ Non-compact
4	Slender	Elastic analysis, limited by buckling stress ⎰	⎱ Longitudinally stiffened

decks, so the Bridge Code has no C ass 1. Methods of analysis for its 'compact' class are similar in character to those for Class 2 in the Eurocodes. Non-compact sections are subdivided in EC3 and EC4 according to their slenderness, class 4 sections being those so slender that buckling is essentially elastic, rather than inelastic. In the Bridge Code, the treatment of non-compact sections depends mainly on whether they have longitudinal stiffeners. The overall result is similar to that given by the Eurocodes because most sections that fall into Eurocode Class 3 have webs with vertical stiffeners only, and most class 4 sections occur in box or deep-plate girders, that have both vertical and longitudinal stiffeners.

Global analysis relates to systems of members, and design against lateral buckling to individual members, so classification is needed for these, as well as for cross sections. The general rule is that the class of a structure or member is that of its most slender cross section, but there are exceptions. It is not appropriate, for example, to require that procedures specified for 'longitudinally stiffened' members must be used throughout continuous composite beams such as those in the River Stour Bridge[23] where longitudinal stiffeners occur only in web panels adjacent to internal supports. It is understood that the present heading of Clause 9.10 of Part 3, 'Flanges in longitudinally stiffened beams', is to be revised for this reason. Longitudinal stiffening occurs mainly in long-span all-steel bridges, and so is not considered in this book, except in relation to composite plates (chapter 9).

DEFINITIONS OF COMPACT SECTIONS

Both the Eurocodes and the Bridge Code use limiting slendernesses for webs and flanges in compression related to the yield strength of the steel, but their datum yield strengths are different: 235 N/mm^2 in Eurocode 4 and 355 N/mm^2 in BS 5400. The relevant ratios are given in table 8.3 in the format of the Bridge Code. The Eurocode values are based mainly on tests on composite members.[154] Both b_o and d_c are measured excluding radiused roots or welds at web-flange intersections. The two ratios for webs are not directly comparable, because d_c is measured from the plastic neutral axis in Eurocode 4 and from the elastic neutral axis in BS 5400. For composite sections in sagging bending, the level of the latter axis depends on the modular ratio. Clause 6.7 of Part 3 states only that the 'appropriate' value should be used. In midspan regions, it is conservative to use the value for permanent load, which gives the greater depth of web in compression.

Table 8.3 Limiting slendernesses for compact cross sections

	Eurocode 4	BS 5400: Part 3
Outstand b_o of compression flange of mean thickness t_f; $b_o/t_f \not>$	$7.3(355/\sigma_y)^{\frac{1}{2}}$	$7.0(355/\sigma_y)^{\frac{1}{2}}$
Depth d_c in compression of web of thickness t; $d_c/t \not>$	$26.8(355/\sigma_y)^{\frac{1}{2}}$	$28.0(355/\sigma_y)^{\frac{1}{2}}$

For sequential construction, the neutral axis should be that of the cross section that is effective at the time considered.

A steel compression flange attached to a concrete slab is prevented from buckling by the shear connectors. For the flange to be compact, the spacing of these is limited in Part 3 to $12t_f(355/\sigma_y)^{\frac{1}{2}}$ in the direction of compression and twice this value in the transverse direction.

8.4.3 Resistance in bending and shear at the ultimate limit state

An outline is now given of the methods of the Bridge Code for the determination of the resistances to bending and vertical shear of cross sections of continuous composite beams. The resistances have to be shown to be not less than the highest of the design values given by the global analyses for the relevant load cases at the ultimate limit state. The critical sections are assumed to be at internal supports and near the centre of each span.

The main steps are further explained in sections 8.4.4 to 8.4.6. A good understanding of these methods and experience of their use are prerequisites for creative design work.

For simplicity, we assume that the nominal yield strengths of the structural steel in tension, σ_{yt}, and in compression, σ_{yc}, are equal. In worked examples, the value for Grade 50 steel, 355 N/mm², will normally be used. The characteristic yield strength of the slab reinforcement, f_y, will generally be taken as 460 N/mm². The notation will follow the practice of BS 5400, Parts 3 and 4.[1]

EFFECTIVE SECTION

One first defines the cross section to be analysed. Concrete in longitudinal tension is ignored. There may be a reduction in the effective area of the steel section due to holes or to the effects of local buckling. For example, the effective thickness of a longitudinally unstiffened web is taken in Part 3 of the Bridge Code as less than the actual thickness t_w if the depth of web in longitudinal compression exceeds $68t_w(355/\sigma_{yw})^{\frac{1}{2}}$, where σ_{yw} is the nominal yield strength of the web material.

The section is then classified as compact or non-compact, taking account of whether it is subjected to sagging or hogging bending. It is assumed here that the top flange is composite and the bottom flange is not.

RESISTANCE TO SAGGING BENDING

Except during construction (which is not now considered) the steel compression flange is restrained by the deck slab, so the section is 'stocky', a term used here to mean 'not susceptible to lateral buckling'.

For a compact section, plastic analysis is used. Account has to be taken of the different 'yield' strengths of the three materials in the cross section (concrete, reinforcement, and structural steel), and of the different values of γ_m applicable to them, which are explained in section 7.1.2. This is done in Part 3 by using a method of transformed sections analogous to that commonly used in elastic analysis. The transformed widths are calculated using ratios based on the strengths of the materials, rather than on their moduli of elasticity. In elastic analysis, the transformation can be into 'steel' or 'concrete'; here, it is into a material with a nominal yield strength equal to that of the steel compression flange, σ_{yc}.

The *plastic section modulus*, Z_{pe} in Part 3, is defined as the first moment of area of the transformed section about its plastic neutral axis. The design resistance is therefore:

$$M_D = Z_{pe}\sigma_{yc}/\gamma_m\gamma_{f3} \tag{8.6}$$

in which the partial safety factor γ_m, given in Part 3 as 1.05, is applied to all the materials. When calculating the transformed width of the concrete flange, its design compressive strength is therefore taken as $0.6f_{cu} \times 1.05/1.5$, which is $0.42f_{cu}$, (section 7.1.2), so that its partial safety factor γ_m and contribution to M_D are the same as they would be in a reinforced concrete T-beam designed to Part 4 of the Bridge Code.

The design strength of $0.72f_y$, given in Part 4 for reinforcement in compression, includes a γ_m factor of 1.15; so if the slab reinforcement is to be included, its transformed area A'_r (additional to the area of concrete displaced) for bars of total area A_r should strictly be:

$$A'_r = A_r[(0.72 \times 1.05f_y - 0.42f_{cu})/\sigma_{yc}]$$

For typical materials, the ratio A'_r/A_r ranges from about 0.85 to 1.2; A_r is usually less than 1% of the area of concrete in compression; and f_y/f_{cu} is about 11. The total compressive force in the slab will thus not be overestimated by more than about 2% if A'_r is taken as A_r, but as resistance to sagging bending rarely governs design, it is more usual to ignore the contribution from the reinforcement.

For a non-compact section in sagging bending, conventional elastic analysis is used, with a transformed section based on effective moduli of elasticity. The design resistance is given in Part 3 as the lesser of:

$$M_D = Z_{xc}\sigma_{\ell c}/\gamma_m\gamma_{f3} \tag{8.7}$$

and:

$$M_D = Z_{xt}\sigma_{yt}/\gamma_m\gamma_{f3} \tag{8.8}$$

where Z_{xc} and Z_{xt} are the relevant section moduli for the extreme fibres of the structural steel, $\gamma_m = 1.05$, and $\gamma_{f3} = 1.10$.

For a section not susceptible to lateral buckling, the limiting compressive stress $\sigma_{\ell c}$ is the lesser of $(D/2y_t)\sigma_{yc}$ and σ_{yc}. The factor $D/2y_t$ (which ranges from about 0.8 to 1.2) is defined and discussed in section 8.4.6. It is found that when $\sigma_{yc} = \sigma_{yt}$ and $D/2y_t < 1$, as is usual, equation (8.8) always governs.

The uncommon situation in which elastic analysis shows that M_D appears to be governed by compressive stress in the concrete slab is considered below, in the section on elastic-plastic analysis.

RESISTANCE TO HOGGING BENDING

The design methods of Part 3 are in principle as given above for sagging bending, with three exceptions.

(1) The factor γ_m in equations (8.6) and (8.7) is 1.20, not 1.05, because the scatter of test results for steel flanges in compression is greater when they are not composite, and therefore are liable to local and lateral buckling. In equation (8.8), γ_m remains at 1.05, because tension governs.

(2) Some hogging moment regions are susceptible to lateral buckling of the steel section. This is allowed for in Part 3 by using a reduced design compressive strength for the steel, $\sigma_{\ell i}$, but the reduction is applied differently to compact and to non-compact sections, as follows.

The lateral restraint to be provided to the bottom flange is first determined. This leads to a value of the lateral-torsional slenderness parameter λ_{LT}, as explained in section 8.4.4, and of the slenderness function β, defined by:

$$\beta = \lambda_{LT}(\sigma_{yc}/355)^{\frac{1}{2}} \qquad (8.9)$$

with σ_{yc} in N/mm^2. If $\beta \leqslant 45$, the compression flange and associated web are 'stocky', and $\sigma_{\ell i} = \sigma_{yc}$. Otherwise, $\sigma_{\ell i} < \sigma_{yc}$.

For compact sections, σ_{yc} in equation (8.6) is replaced by $\sigma_{\ell i}$, so that:

$$M_D = Z_{pe}\sigma_{\ell i}/\gamma_m\gamma_{f3} \qquad (8.6a)$$

In effect, the reduction is applied to the whole cross section, not just to steel in compression.

For non-compact sections, $\sigma_{\ell c}$ in equation (8.7) is taken as the lesser of σ_{yc} and $(D/2y_t)\sigma_{\ell i}$. The higher the value of β, the more likely it becomes that the design of the cross section at an internal support will be governed by equation (8.7) rather than equation (8.8).

(3) For the calculation of Z_{pe} for a compact section, the appropriate transformed area A'_r of reinforcement in tension with actual area A_r is:

$$A'_r = A_r(1.20 \times f_y/1.15)/\sigma_{\ell i}$$

because $\gamma_m = 1.20$ in equation (8.6a). For simplicity, $1.20/1.15$ is often taken as 1.0, as this underestimates A'_r by only 4%. The following method is then a convenient alternative to the use of equation (8.6a) for the calculation of M_D in hogging bending. Conventional rectangular-stress-block plastic theory is used, with yield strengths f_y for reinforcement and σ_{yc} for the steel section, and the resistance moment so found is multiplied by $\sigma_{\ell i}/\sigma_{yc}$, and divided by $\gamma_m\gamma_{f3}$.

ELASTIC-PLASTIC ANALYSIS

For non-compact composite sections, the strict use of elastic analysis can lead to over-conservative design for the ultimate limit state, so limited provision for the use of elastic-plastic analysis is made both in BS 5400 and in Eurocode 4. The 'elastic' equations (8.7) and (8.8) relate to the structural steel section only. For concrete and reinforcement the limiting stresses are as in Part 4, but the strains are not limited to the elastic values. In principle, the stress/strain curves given in Part 4 should be followed, but if hand calculation is used, it is simpler and accurate enough to use elastic-plastic theory based on bilinear elastic-plastic stress/strain relations. The following method is appropriate.

(1) Assume first that structural steel governs, and calculate M_D from equation (8.7) or (8.8).

(2) For concrete in compression, find the extreme-fibre stress at moment

M_D, using elastic theory. If this exceeds the limiting stress in Part 4, which is $0.67 f_{cu}/\gamma_m$, that is, $0.45 f_{cu}$, either reduce it to $0.45 f_{cu}$ by scaling down M_D, or recalculate M_D assuming that concrete is an elastic-plastic material with a yield stress of $0.45 f_{cu}$, up to a strain limit of 0.0035.

(3) Reinforcement in tension is most unlikely to reach its limiting stress, $0.87 f_y$, unless mild steel is used. If it does, follow a procedure similar to that in (2) above, with yield at $0.87 f_y$ and no limiting strain.

When unpropped construction is used, M_D at an internal support may be governed by equation (8.8), yielding of the tension flange, when the tensile stress in the slab reinforcement is still low. It is possible to allow, in calculations, for yielding of the tension flange and the resulting redistribution of stress to the reinforcement (section 9.2.2). The shear connection should be designed for the additional longitudinal shear.

VERTICAL SHEAR

The whole of the vertical shear is assumed to be resisted by the structural steel section, so the design method of Part 3 is given only in outline. For a web without holes, of thickness t_w and depth d_w, the design shear resistance is given by:

$$V_D = t_w d_w \tau_\ell / \gamma_M \tag{8.10}$$

where τ_ℓ is the limiting mean shear stress for the web panel considered and $\gamma_M = \gamma_m \gamma_{f3}$. If its slenderness ratio λ is less than 55, where:

$$\lambda = (d_w/t_w)(\sigma_{yw}/355)^{\frac{1}{2}} \tag{8.11}$$

and σ_{yw} N/mm² is the yield strength of the web, then design can be based on yielding, and:

$$\tau_\ell = \sigma_{yw}/\sqrt{3} \tag{8.12}$$

If $\lambda > 55$, τ_ℓ is less than the yield strength in shear, but greater than the elastic critical stress for web buckling, because of tension-field action. Its value depends on the length:depth ratio of the web panel, and on the ability of the steel flanges to anchor the tension field. This depends on the lower of the resistances of the two flanges to bending about an axis normal to the plane of the web. This resistance is assumed in Part 3 to be proportional to $b_{fe} t_f^2$ for the steel flange considered, where b_{fe} and t_f are its effective breadth and thickness. The top flange is usually the smaller, and so is found to govern the tension-field action available. The influence of the smaller flange is represented by a nondimensional ratio:

$$m_{fw} = (\sigma_{yf}/\sigma_{yw})(b_{fe} t_f^2 / 2 d_{we}^2 t_w)$$

where the subscripts f and w relate to the flange and the web respectively.

In design, the aspect ratio ϕ of the web panel considered is determined for an assumed spacing of vertical stiffeners. The ratio τ_ℓ/τ_y is then found, for given m_{fw}, ϕ and slenderness ratio λ, from graphs given in Part 3.

The method is based mainly on research on steel plate girders. For a composite member, the top flange is itself anchored into the concrete slab by shear connectors, so that $b_{fe}t_f^2$ is an over-conservative measure of its resistance. It was thought at one time that this additional anchorage would cause vertical tension in the shanks of stud connectors which would reduce their resistance to longitudinal shear. Tests on plate girders[184] have shown that the real behaviour is more complex. After yielding, the steel top flange develops additional local bending strength by shedding longitudinal tension to the slab reinforcement. Vertical restraint from studs is small because they are embedded in a slab that is cracked in longitudinal tension, if the beam is continuous.

A design method that takes account of this behaviour has been developed.[184] The resulting economy in practice (from the use of a thinner steel web) is sometimes insufficient to justify the use of the more complex design calculations, so the method is not given in the Bridge Code.

As explained below, a value is required also for V_R, the shear resistance of the section neglecting tension-field action. When $\lambda \leqslant 55$, as it generally is for rolled I- and channel-sections in Grade 43 or Grade 50 steel, $\tau_\ell = \sigma_{yw}/\sqrt{3}$ and $V_R = V_D$. When $\lambda > 55$, V_R is found from a value of τ_ℓ calculated assuming that $b_{fe}t_f^2$ for the flanges is zero.

RESISTANCE TO COMBINED BENDING AND SHEAR

In the Bridge Code, this resistance is given as a function of the resistances in pure bending and pure shear, as shown in fig. 8.11, not in terms of limiting stresses. The underlying idea is that the shear resistance of a web without flanges, V_R, can be combined with the bending resistance of two flanges without a web, M_R. This is point C on the diagram. Points B and D are conservative approximations to test data.[106]

To calculate M_R, each of the flanges is assumed to be uniformly stressed to the limiting stress used in the calculation of M_D, and the longitudinal forces in the flanges, F_c and F_t say, are found. It is stated in Clause 9.9.3.1 of Part 3 that

$$M_R = F_f d_f / \gamma_m \gamma_{f3} \tag{8.13}$$

where F_f is the lower of the two flange forces, and d_f is the distance between their lines of action.

For composite flanges of compact section, the forces and their lines of

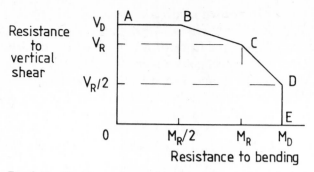

Fig. 8.11 Resistance to combined bending and vertical shear

action are easily found, but for a non-compact section, the effects of staged construction are relevant, as will be explained later. Usually, γ_m is 1.2 for a flange in compression and 1.05 for a flange in tension; so, in a beam where F_c exceeds F_t, but by less than about 15%, use of F_c in equation (8.13) would give a lower value of M_R. The writer would here follow the Code (i.e. use F_t, with $\gamma_m = 1.05$) for two reasons: the subject is moment–shear interaction, and γ_m for shear is 1.05; and the treatment of this subject in Part 3 is conservative for composite members.

BEAMS BUILT IN SEVERAL STAGES

The simplest example of sequential construction is when the whole of the concrete deck is cast at one time, so that its weight is carried by the steelwork alone. The mode of failure during construction that is most likely to govern design is lateral buckling of the steel compression flange at midspan. The relevant structure is not composite, so that design for this, to Part 3, is not considered here. It may involve provision of lateral bracing that is not required in the completed structure. If the bracing is not to be removed after the concrete has set, its influence on the global response of the composite structure should be considered, as it may provide unwanted local stiffness that leads to fatigue failures at points where the bracing is attached to the steel beams.

The methods for the analysis of the completed structure given in Part 3 depend on whether the section considered is compact or not. At a compact section, sufficient redistribution of stress can occur, before the ultimate limit state is reached, for the sequence of construction to be irrelevant. The entire load can be assumed to have been applied after the member became composite, and so the preceding methods are applicable.

For non-compact sections, bending moments due to load applied to steelwork alone (M_s, say) have to be kept distinct, in the global analyses, from moments due to load resisted by the composite structure (M_c, say).

These are combined, as shown below, to give a design bending moment, M_x say, which is compared with the resistance M_D given by equation (8.7) or (8.8). Let us assume, for example, that M_D is governed by the limiting stress $\sigma_{\ell c}$ for the steel compression flange. If Z_{xs} and Z_{xc} are the section moduli for this flange for the steel and composite sections, respectively, then the design condition is:

$$M_c/Z_{xc} + M_s/Z_{xs} \leqslant \sigma_{\ell c}/\gamma_M$$

where $\gamma_M = \gamma_m \gamma_{f3}$. From equation (8.7), $M_D = Z_{xc}\sigma_{\ell c}/\gamma_M$, whence:

$$M_x = M_c + M_s(Z_{xc}/Z_{xs}) \leqslant M_D$$

Creep of concrete can also be allowed for by this method. Suppose, for example, that M_c is the sum of a dead-load moment M_{cd} and an imposed-load moment M_{ci}. Let the relevant section moduli for the steel compression flange be Z_{xcd} and Z_{xci}, respectively, and assume that M_D has been calculated using Z_{xci}. Then the appropriate design bending moment is:

$$M_x = Z_{xci}[(M_{ci}/Z_{xci}) + (M_{cd}/Z_{xcd}) + (M_s/Z_{xs})] \qquad (8.14)$$

This expression is given in general form in Clause 9.9.5.3 of Part 3 of the Bridge Code. If M_D is governed by limiting stress in the steel tension flange, then M_x must be calculated using section moduli for that flange, and will usually have a different value.

CALCULATION OF M_R FOR NON-COMPACT CROSS SECTIONS OF BEAMS BUILT IN SEVERAL STAGES

In principle, M_R is the resistance of the section without its web, relevant to the type of design loading considered. It is calculated from a lever arm d_f, and the lesser value F_f of the flange forces F_c and F_t, equation (8.13). For a non-composite flange, calculation of the force and its line of action are straightforward.

Let us consider the top (composite) flange of a member subjected to hogging moments M_c and M_s, as defined in the preceding section. What is the value of F_t? The contribution from the steel flange is obviously $A_f \sigma_{yt}$. The stress in the reinforcement (of area A_r) due to M_s is zero, and it is conservative to assume that the stress in it due to M_c is the same as that caused by M_c in the steel flange. The total stress σ_{yt} in the steel is associated with a bending moment M_x, so that:

$$F_t \simeq A_f \sigma_{yt} + A_r \sigma_{yt}(M_c/M_x)$$

What is the lever arm? Let the centroid of the top flange and of the slab reinforcement be distances d_{fs} and d_{fr}, respectively, above the centre of the steel bottom flange. Then the accurate expression for d_f is:

$$d_f = \frac{d_{fs}A_f + d_{fr}A_r(M_c/M_x)}{A_f + A_r(M_c/M_x)}$$

but it can conservatively be taken as d_{fs}.

In a region of maximum sagging moment, vertical shear is unlikely to influence design of the section. If M_R has to be calculated, an estimate can quickly be found from the product of the force $A_f\sigma_{yt}$ in the steel bottom flange, and the distance from its line of action to the steel/concrete interface. The force F_c in the composite top flange is usually much higher than $A_f\sigma_{yt}$. Accurate expressions can be obtained by following the principles used in the method given for a hogging region.

8.4.4 Lateral torsional buckling to BS 5400

The design method of Part 3 is based on an extended analogy between classical lateral buckling, without distortion of the cross section of the member, and the Euler buckling of an axially loaded strut.[107] It is explained here in the context of the lateral buckling of a steel bottom flange in compression near an internal support of a continuous composite beam. It will be shown that the design method is rather conservative for this situation; essentially because this type of buckling can occur only if cross sections distort, and only in regions of steep moment gradient, where there is a variation in stress along the flange that does not occur in a strut. Also, the torsional stiffness of the flange and warping rigidity of the member are neglected. The adoption of a more realistic method would often enable bottom-flange bracing to be omitted. For unstiffened webs, such a method exists for the case now considered. It is explained in section 8.4.5, with an example calculated by both methods.

DEFINITION OF λ_{LT}

To define the slenderness parameter for lateral torsional buckling, λ_{LT}, we first use the classical Euler theory for the buckling load N_E for a strut of effective length ℓ_e, cross-sectional area A, least radius of gyration r, squash load $N_u = A\sigma_{yc}$, and Young's modulus E:

$$N_E/N_u = \pi^2 E A r^2/(\ell_e^2 A \sigma_{yc})$$

This gives the slenderness as:

$$\ell_e/r = (\pi^2 E N_u/\sigma_{yc} N_E)^{\frac{1}{2}} \tag{8.15}$$

The classical theory of lateral buckling relates to a beam of length ℓ_e

bent about the major $(x–x)$ axis by equal and opposite end moments M_E, and supported at both ends in a way that prevents lateral deflection and twist but allows warping and lateral and vertical bending. The critical compressive stress σ_{cr} is defined by:

$$\sigma_{cr} = M_E/Z_{xc} \qquad (8.16)$$

where M_E, the major-axis bending moment at which the compression flange buckles laterally, is analogous to N_E for the strut.

When lateral buckling governs the design of the cross section for major-axis bending, the limiting longitudinal stress is compressive, so the shape factor S is Z_{pe}/Z_{xc}, where Z_{pe} is the plastic modulus of the effective section, such that:

$$M_p = Z_{pe}\sigma_{yc}$$

The plastic moment of resistance for the beam, M_p, is assumed to be analogous to the squash load for the strut, N_u. Finally, the slenderness parameter λ_{LT} is defined as being analogous to ℓ_e/r.

Using these definitions and substitutions, equation (8.15) becomes:

$$\lambda_{LT} = (\pi^2 ES/\sigma_{cr})^{\frac{1}{2}} \qquad (8.17)$$

which is the equation given in Clause 9.7.5 of Part 3 for use when the critical stress σ_{cr} is known and the simplified methods (to be described) are not applicable.

In the medium range of slendernesses, buckling is influenced by plasticity and hence by the yield strength of the steel, in such a way that the effective slenderness can be assumed to be proportional to $\sigma_{yc}^{\frac{1}{2}}$. This is allowed for by replacing λ_{LT} by the slenderness function β, where, from equation (8.17):

$$\beta = \lambda_{LT}(\sigma_{yc}/355)^{\frac{1}{2}} = (\pi^2 E/355)^{\frac{1}{2}}(\sigma_{yc}S/\sigma_{cr})^{\frac{1}{2}} \qquad (8.18)$$

where σ_{yc} is in N/mm^2. Putting $E = 205$ kN/mm^2 gives:

$$\beta^2 = 5700(\sigma_{yc}S/\sigma_{cr}) \qquad (8.19)$$

DETERMINATION OF BUCKLING CURVE

Equation (8.19) defines the slenderness β in terms of the elastic critical buckling stress, σ_{cr}, which can be computed for many common situations. Assuming that σ_{cr} is known (section 8.4.5) or can be found from an empirical relationship (as in Clause 9.7.2 of Part 3), for the member, restraints and bending-moment distribution considered, the next step is to determine a representative ultimate hogging bending moment at the face of the internal support, M_u say, at which lateral buckling would be expected to occur if the member had the most adverse acceptable geometrical imperfections and residual stresses. The effect of these is represented by the parameter η, which has to be determined from analyses of test results.

It is convenient to relate M_u to the known M_p, defined above, and M_{cr}, the elastic critical value of the representative bending moment, given by $M_{cr} = Z_{xc}\sigma_{cr}$. Hence, using equation (8.19):

$$M_{cr}/M_p = Z_{xc}\sigma_{cr}/Z_{pe}\sigma_{yc} = \sigma_{cr}/\sigma_{yc}S = 5700/\beta^2 = k \quad \text{(say)} \quad (8.20)$$

By analogy with the usual method of allowing for imperfections in struts, these quantities are assumed to be related by a Perry-type equation:

$$(M_p - M_u)(M_{cr} - M_u) = \eta M_{cr} M_u \qquad (8.21)$$

For a perfect member ($\eta = 0$) this gives the expected result that the ultimate moment M_u is the lesser of the plastic moment and the critical moment. (It will be noted that the theory is being developed without consideration of local buckling, so it is applicable to compact sections and will need modification for non-compact sections.)

It is convenient to re-write equation (8.21) in nondimensional form:

$$(1 - \mu)(k - \mu) = \eta k \mu \qquad (8.22)$$

where $\mu = M_u/M_p$, and is the unknown variable, and k, defined in equation (8.20), can be calculated (if only approximately) for any given structure and materials.

The method is based on the results of tests on all-steel members, in which the usual mode of buckling involved lateral displacement and rotation of non-distorting cross sections. It will be shown later to be conservative for the distortional buckling that alone can occur in hogging regions of continuous composite beams, for which there are few test data.

All the reliable data from tests in which lateral buckling occurred have been plotted on a graph of μ against k. The value of η that gave the best-fitting curve, using equation (8.22), was found to be:

$$\eta = 0.005(\beta - 45) \qquad \text{with } \eta \not< 0 \qquad (8.23)$$

No distinction could be drawn between the results for members with compact and with non-compact cross sections.

Both η and k are functions of β, so equation (8.22) can be solved to give μ in terms of β. The results now have to be expressed in terms of the limiting compressive stress, $\sigma_{\ell i}$.

DESIGN OF COMPACT SECTIONS

From the definition of μ:

$$M_u = \mu M_p = \mu \sigma_{yc} Z_{pe}$$

The design bending moment M_D is $M_u/\gamma_m\gamma_{f3}$, and is given in Part 3 as:

$$M_D = \sigma_{\ell i} Z_{pe}/\gamma_m\gamma_{f3} \qquad (8.24)$$

Thus, μ can be re-defined as $\sigma_{\ell i}/\sigma_{yc}$. This is the form in which the solution of equation (8.22) is given in Clause G.7 of BS 5400: Part 3: 1982. The curve relating μ to β is given in Part 3 and in fig. 8.12, which also shows M_{cr}/M_p from equation (8.20).

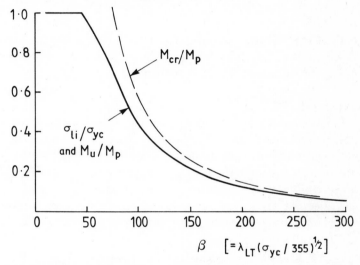

Fig. 8.12   Limiting compressive stress, critical moment, and slenderness function

DESIGN OF NON-COMPACT SECTIONS

For laterally braced members, M_D for cross sections governed by compressive stress is given by equation (8.6) if the section is compact, and by equation (8.7) if it is not. For a given section, the ratio of these values is Z_{pe}/Z_{xc}. This is the shape factor, S, and $S \simeq 1.15$ for a symmetrical steel I-section. In Part 3, this ratio was adopted also for sections subject to lateral buckling by using, for non-compact sections, a modified version of equation (8.24):

$$M_D = \sigma_{\ell i} Z_{pe}/1.15\gamma_m\gamma_{f3}$$

However, it was thought to be confusing to present a method for non-compact sections in terms of the plastic section modulus Z_{pe}, rather than the elastic section modulus for the compression flange, Z_{xc}. The method of presentation is essentially:

$$M_D = C\sigma_{fi}Z_{xc}/\gamma_m\gamma_{f3}$$

where C is a correction factor that should be $Z_{pe}/1.15Z_{xc}$, or $0.87S$, where S is the shape factor related to the compression flange. In Part 3, the simpler expression $D/2y_t$ is used, to avoid calculation of Z_{pe}, which is not otherwise needed for a non-compact section. Essentially, D is the overall depth of the section, and y_t is the depth in tension. More precise definitions are given in section 8.4.6, where the accuracy of the factor $D/2y_t$ and the effects of using it are examined.

The expression for M_D given above therefore appears in Part 3 as:

$$M_D = Z_{xc}\sigma_{fc}/\gamma_m\gamma_{f3} \tag{8.7) bis}$$

where σ_{fc} is the lesser of $(D/2y_t)\sigma_{fi}$ and σ_{yc}. The same assumption appears in Clause 9.7.3 of Part 3, where $0.87Z_{pe}/Z_{xc}$ is replaced by $D/2y_t$ when calculating λ_{LT} for non-compact box sections.

DETERMINATION OF λ_{LT}

Equation (8.17) shows that λ_{LT} depends on the elastic critical stress σ_{cr}. The situation considered in Part 3 is the non-distortional buckling of an isolated I-section with unequal flanges, for which σ_{cr} depends on many variables: the end conditions; the four elastic stiffnesses EI_x, EI_y, GJ and EH (warping rigidity); the bending-moment distribution along the member; its effective length ℓ_e; and the relative stiffness of the two flanges, represented by their second moments of area about a vertical $(y - y)$ axis, I_c and I_t.

When σ_{cr} is not known, the expression for λ_{LT} given in Clause 9.7.2 of Part 3 can be used:

$$\lambda_{LT} = (\ell_e/r_y)k_4\eta v \tag{8.25}$$

This is based on a close approximation to the elastic critical buckling moment for the situation considered in Part 3. Values are given for k_4 in terms of the type of steel section; for η (which is not the η of equation (8.21)) in terms of the bending moment distribution; and for v in terms of the geometry of the cross section and the effective length of the member.

For steel beams without web stiffeners, design is likely to be to Clause 9.6.6 of Part 3, which is for a beam continuously restrained by the deck. The expression for λ_{LT} can then be simplified. Both k_4 and η are taken as 1.0. The expression given for v (below table 9 of Part 3) includes variables λ_F, I_c and I_t. From Clause 9.7.2, $\lambda_F = 0$, and I_c and I_t are the second moments of area of the compression and tension flanges about a vertical axis. In a hogging moment region, $I_t \gg I_c$, because of the effect of the longitudinal reinforcement in the slab, and $I_c = t_f B_f^3/12$, where t_f and B_f relate to the steel bottom flange (fig. 8.16).

The equation for v then simplifies to:

$$v = (2i)^{-0.5}$$

where $i = I_c/(I_c + I_t)$. The minor-axis second moment of area of the whole cracked composite section, Ar_y^2, is equal to $I_c + I_t$, because the contribution of the web is negligible. Substitution of these results into equation (8.25) gives:

$$\lambda_{LT} = (6A\ell_e^2/t_f B_f^3)^{0.5} \tag{8.26}$$

This result depends mainly on ℓ_e and the dimensions of the bottom flange, as it should. An increase in the breadth of the composite top flange increases the total area of the section, A, and so makes it more slender. This reflects the reduction in the stiffness of the torsional restraint provided by the slab when the spacing of the main beams is increased.

DETERMINATION OF THE EFFECTIVE LENGTH, ℓ_e

In the absence of lateral bracing, the steel bottom flange is restrained laterally by the web. To assume that the web is fixed at its upper edge is to neglect the flexibility of the slab, and therefore is unsafe, so it is considered as part of an inverted U-frame (ACDE in fig. 8.13).

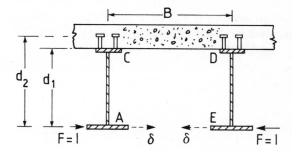

Fig. 8.13   Inverted U-frame

If the webs are unstiffened, the frame provides continuous but fairly flexible restraint. Its flexibility is determined by calculating δ, the elastic deflection due to a pair of unit horizontal forces per unit length of the beam, as shown.

If the webs have vertical stiffeners, the sections where they occur can be considered as discrete and fairly stiff U-frames, which provide intermittent restraint. The deflections per unit horizontal force are calculated per frame, and multiplied by the frame spacing to give a value corresponding to δ in fig. 8.13.

The equation for δ given in Clause 9.6.6 of Part 3 (equation 8.34), below) contains two terms, which represent the flexibilities of the webs and of the slab. The flexibility of the shear connection in the transverse plane is thus assumed to be negligible in comparison with the other flexibilities. Typical calculations show that this assumption is realistic when the webs are unstiffened, but consideration should be given to this extra flexibility when stiffened webs and discrete U-frames are used, particularly in regions of longitudinal hogging bending. Research on this subject is in progress at the University of Warwick.

For continuous restraint and a compression flange free to rotate about a vertical axis at the supports, the effective length is given in Part 3 as:

$$\ell_e = 2.5(EI_c\delta)^{0.25} \tag{8.27}$$

where I_c is the second moment of area of the compression flange about a vertical axis.

This result is obtained by finding an elastic critical load, as follows. A length L of the flange is represented by an initially straight pin-ended elastic strut (fig. 8.14) of stiffness EI_c subjected to uniform axial load N, which is increased until buckling occurs. The strut is restrained vertically by the web and laterally by a force y/δ per unit length, where y is the infinitesimal displacement that occurs when the critical load N_{cr} is reached. This is a conservative model, because in the real flange, N decreases rapidly with distance from the internal support.

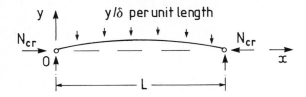

Fig. 8.14   Laterally restrained pin-ended strut

An approximate solution can be obtained by assuming a sinusoidal buckling mode. The analysis then gives:

$$N_{cr} = (L^2/\pi^2\delta) + (\pi^2 EI_c/L^2) \tag{8.28}$$

For minimum N_{cr}, $dN_{cr}/dL = 0$, which gives $L = \pi(EI_c\delta)^{0.25}$, whence, from equation (8.28):

$$N_{cr} = 2\pi^2 EI_c/L^2 = 2(EI_c/\delta)^{\frac{1}{2}} \tag{8.29}$$

The effective length is defined by:

$$N_{cr} = \pi^2 EI_c/\ell_e{}^2 \tag{8.30}$$

whence, from equation (8.29):

$$\ell_e = (\pi/\sqrt{2})(EI_c\delta)^{0.25} \qquad (8.31)$$

which is slightly below result (8.27) because of the approximation that the displacement is sinusoidal.

SUMMARY OF CONTINUOUS U-FRAME METHOD WHEN σ_{cr} IS NOT KNOWN
Find δ from the properties of the cross section, ℓ_e from equation (8.27), λ_{LT} from equation (8.25) or (8.26), and β from equation (8.18). The buckling stress $\sigma_{\ell i}$ is then found from fig. 8.12, and used as explained in section 8.4.3. There is a worked example at the end of section 8.4.5.

8.4.5 Elastic critical stresses for lateral buckling

When the elastic critical stress σ_{cr} is known, equation (8.17) provides a method that is simpler than the use of equation (8.25). Computed values of σ_{cr} are available for continuous composite beams where the steel member is a rolled or built-up I-section symmetrical about both axes. For the UB range of sections available in the UK, such members provide a maximum span of almost 30 m, and are convenient for use in viaducts.

The beams analysed[108] were fixed-ended for major-axis bending, but free at the supports to warp and to rotate about a vertical axis. The deck slab was assumed to be restrained against lateral movement and rotation about a longitudinal axis. The web was unstiffened, except at supports, so that the critical buckling mode was found to be as shown in section in fig. 8.15(a), with variation of the lateral displacement u over a half span as shown in fig. 8.15(b).

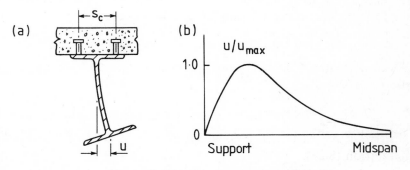

Fig. 8.15   Distortional lateral buckling

The initial state of stress was found by elastic analysis, in accordance with Part 3, of the major-axis bending of the beam due to unit uniformly-distributed load, allowing for cracking of concrete. This gave a conservative (high) value for the length of bottom flange in compression in a typical continuous member. This stress distribution was scaled up until buckling occurred, giving the critical compressive stress σ_{cr} in the bottom flange at the supports, that is, at the location where design would be governed in practice by the limiting stress $\sigma_{\ell i}$.

A parametric study of 20 beams was done, using I-sections with web slendernesses (d/t_w in fig. 8.16) ranging from 39 to 100, flange B_f/t_f ratios from 9.6 to 15, length/breadth ratios of bottom flanges (L/B_f in table 8.4) from 48 to 90, and three sizes of concrete flange. The only one of these variables that had a significant influence on the critical stress was the web slenderness, which was based on the clear depth between flanges because fillets were neglected in the theory.

A lower bound to the results is given by:

$$\sigma_{cr}/\sigma_{yc} = 600(d/t_w)^{-1.4} \tag{8.32}$$

where $\sigma_{yc} = 355 \text{ N/mm}^2$. Substitution in equation (8.17) gives:

$$\lambda_{LT} = 3.08S^{\frac{1}{2}}(d/t_w)^{0.7}$$

The shape factor S ranged from 1.23 to 1.43, and is now conservatively taken as 1.45. This gives a possible design equation:

$$\lambda_{LT} = 3.7(d/t_w)^{0.7} \tag{8.33}$$

Results from this method are given in table 8.4 for some of the members analysed, and compared with results obtained using the methods explained in section 8.4.4. The slabs for beams B2 to B11 were 3.0 m wide and 0.22 m thick, with 1.5% of longitudinal reinforcement. The slab for B12 was smaller, with 1% of steel, and that for B13 was larger, with 2% of steel. Three values of $\sigma_{\ell i}/\sigma_{yc}$ are given, all obtained from the slenderness curve of Part 3 (fig. 8.12). The headings show how λ_{LT} was determined, as follows.

(1) New, equation (8.17). The computed values of σ_{cr} and S were used in equation (8.17).

(2) New, equation (8.33). This is a simplified method, which is a safe approximation to the previous method, for use when σ_{cr} is not known.

(3) Part 3. λ_{LT} was found from equations (8.25) and (8.26).

For all three methods, the limiting compressive stress ($\sigma_{\ell c}$ in Part 3) depends on whether the section is compact or not. For compact sections (beams B2 to B7), $\sigma_{\ell c} = \sigma_{\ell i}$, but the plastic section modulus is used. For non-compact sections (B9 to B13), $\sigma_{\ell c}$ is the lesser of $(D/2y_t)\sigma_{\ell i}$ and σ_{yc}.

The results show that the new design compressive stress is so close to σ_{yc} for all of these beams that no lateral bracing is needed, whereas with the existing method of Part 3, limiting stresses for all the beams are below $0.6\sigma_{yc}$, unless bracing is used.

Table 8.4 Limiting stresses $\sigma_{\ell i}$ for beams with unstiffened UB sections

Beam No.	Steel section (mm, mm, kg/m)	B_f/t_f	d/t_w	L/B_f	$D/2y_t$	S	$\sigma_{\ell i}/\sigma_{yc}$ New equation (8.17)	New equation (8.33)	Part 3
B2	920.5 × 420.5	11.5	39.0	54.7	1.03	1.29	1.00	0.97	0.59
B3	UB 388	11.5	39.0	65.0	1.03	1.29	1.00	0.97	0.59
B7	926.6 × 307.8 UB 289	9.6	43.7	90.0	1.08	1.35	0.97	0.93	0.42
B9	918.5 × 305.5 UB 253	10.9	49.5	90.0	1.11	1.36	0.93	0.89	0.37
B11	903.0 × 303.4	15.0	56.4	90.0	1.17	1.40	0.89	0.83	0.32
B12		15.0	56.4	90.0	1.00	1.31	0.92	0.83	0.37
B13	UB 201	15.0	56.4	90.0	1.39	1.43	0.88	0.83	0.26

In the computations on which these higher values of $\sigma_{\ell i}/\sigma_{yc}$ are based, no account was taken of possible local buckling of the steel compression flange. This question has also been studied. It has been found[188] that for most laterally unbraced plate-girder sections with unstiffened webs and for a few rolled UB sections, local buckling precedes lateral buckling in fixed-ended beams under distributed loading, even when the compression flange satisfies the shape requirements of Part 3. It should thus be considered in design. The proposed method consists of calculating an equivalent slenderness β_L for local buckling, comparing it with the distortional-buckling slenderness β_d, taken as $3.1(S\sigma_{yc}/355)^{\frac{1}{2}}(d/t_w)^{0.7}$ from equations (8.17) and (8.32), and then finding $\sigma_{\ell i}/\sigma_{yc}$ for the higher of these two slendernesses.

WORKED EXAMPLE

Calculations are given for the first beam listed in table 8.4, as an example of the method currently used and of the new method.

Details of the cross section of the beam and some of the notation are shown in fig. 8.16. The slab has 1.5% of longitudinal reinforcement, which for this purpose is smoothed into two layers each 1.65 mm thick. The beam is continuously restrained laterally by the deck, but the compression flange is assumed not to be restrained against rotation in plan at the supports.

The effective length is first calculated in accordance with Clause 9.6.6.2 of Part 3, so equation (8.27) is applicable, with δ given by:

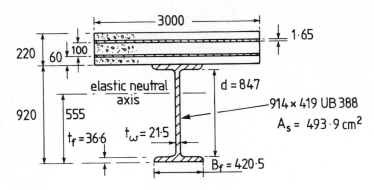

Fig. 8.16   Cross section of beam B2

$$\delta = (d_1^3/3EI_1) + (uBd_2^2/EI_2) \tag{8.34}$$

where:

d_1 and d_2	are as shown in fig. 8.13
I_1	$= t_w^3/12$
u	$= 0.33$, as the beam is assumed not to be an edge beam
B	is the beam spacing, assumed to be 3.0 m
I_2	is the second moment of area of unit length of deck slab, here assumed to be $h_c^3/12\alpha_e$, with the modular ratio α_e taken as 13.6.

Substitution in equation (8.34) gives: $\delta = 1.52$ mm per N/mm. For the compression flange, $I_c = B_f^3 t_f/12$. Equation (8.27) then gives:

$$\ell_e = 7.25 \text{ m} \tag{8.35}$$

The area of the cracked section is: $A = 493.9 + 3.3 \times 30 = 593$ cm². From equation (8.26):

$$\lambda_{LT} = (6 \times 59\,300 \times 7250^2/(36.6 \times 420.5^3))^{\frac{1}{2}} = 82.9$$

Assuming that $\sigma_{yc} = 355$ N/mm², fig. 10 in Part 3 (or fig. 8.12 here) gives:

$$\sigma_{\ell i}/\sigma_{yc} = 0.59 \tag{8.36}$$

For the new method, equation (8.33) is used. The web depth between flanges is 847 mm, so $d/t_w = 39$. From equation (8.33):

$$\lambda_{LT} = 3.7(39)^{0.7} = 48.1$$

From fig. 8.12:

$$\sigma_{\ell i}/\sigma_{yc} = 0.97 \tag{8.37}$$

This result is 64% above the previous result. It is based on a more realistic mathematical model, and is easier to calculate. It is in accordance with Clause 9.7.5 of Part 3 provided that this type of buckling is deemed to be a 'case not covered' by Clauses 9.7.2 to 9.7.4.

8.4.6 The factor $D/2y_t$

In this section, a fuller explanation is given of the use and effects of this factor, which were referred to in sections 8.4.3 and 8.4.4. The notation is as used there and in BS 5400: Part 3. For a composite section in hogging bending, D and y_t are defined in Part 3 as follows:

D is the overall depth from the top of the tension reinforcement to the bottom of the steel section

y_t is the depth from the top of the tension reinforcement to the elastic neutral axis of the cracked composite section.

The 'top of the tension reinforcement' is usually taken to mean the centre of the top layer of reinforcing bars.

The factor is applicable only to non-compact sections. It is used to modify the limiting compressive stress for lateral torsional buckling, $\sigma_{/i}$, to enable calculations for bending resistance to be done using the elastic section modulus Z_{xc} rather than the plastic section modulus Z_{pe}. The factor, for which the symbol s will now be used, is an approximation to $0.87S$, where S is the shape factor relevant to the extreme compression fibre of the steel section.

WORKED EXAMPLE

We now consider an asymmetric steel I-section, fig. 8.17, and study the way in which use of the factor s influences its calculated resistances in sagging and hogging bending, over the practical range of lateral-torsional slenderness. This slenderness is represented by the ratio $\sigma_{/i}/\sigma_{yc}$, denoted by μ. Its range (fig. 8.12) is from 1.0, when $\beta \leqslant 45$, down to about 0.4, when $\beta = 105$.

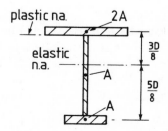

Fig. 8.17   Idealised monosymmetric steel cross section

The slenderness of the steel web is assumed to be such that this section lies on the borderline between compact and non-compact. For such sections, the relationship between the elastic and plastic properties is of interest, because a small change of web thickness can cause Z_{xc} to be used in design in place of Z_{pe}.

The area of the cross section is $4A$. The thicknesses of the flanges are assumed to be negligible in comparison with the depth D between their centres, so that the plastic neutral axis can be assumed to be a distance $3D/8$ above the elastic neutral axis. The section moduli are:

$$Z_t = 2.05AD, \qquad Z_b = 1.23AD, \qquad Z_p = 1.5AD \qquad (8.38)$$

where t, b and p mean top, bottom and plastic, respectively.

Nondimensional bending resistances are defined by:

$$m = M/AD\sigma_{yc} \qquad (8.39)$$

where M is the resistance given by the Bridge Code, omitting the factors γ_m and γ_{f3}. The variation is studied of m_s, m_h and m_p with μ, where s, h and p mean sagging, hogging and plastic, respectively, and μ is as given by fig. 8.12.

The values of s are:

$$s_s = 0.80, \qquad s_h = 1.33 \qquad (8.40)$$

The shape factors for the compression flange are:

$$S_s = 1.5/2.05 = 0.73, \qquad S_h = 1.5/1.23 = 1.22$$

so that in this example, the values of $0.87S$ are 20% below $D/2y_t$, for both sagging and hogging bending. It will be shown that this error, which is on the 'unsafe' side, is similar to the largest error that occurs when using $D/2y_t$ in place of $0.87S$ for composite sections in hogging bending.

If the section is compact, the resistances in both sagging and hogging bending are:

$$m_p = 1.5\mu$$

If the section is non-compact, then if s were not used, the sagging resistance would be governed by compression along line AB (fig. 8.18) and by tension along BC. The hogging resistance, governed always by compression, would be as shown by line DC. When s is used, in accordance with Part 3, the resistances m_s and m_h both have the same values, line EFGC, but different definitions:

m_s is the lesser of $2.05s_s\mu$ (but $s_s\mu \not> 1$) and 1.23
m_h is the lesser of $1.23s_h\mu$ (but $s_h\mu \not> 1$) and 2.05.

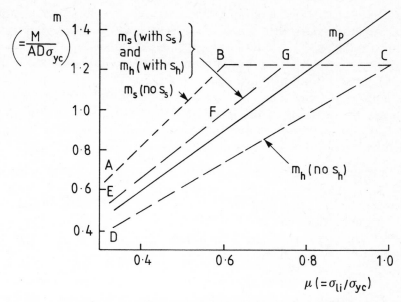

Fig. 8.18 Variation of bending resistance with limiting compressive stress

The results on fig. 8.18 illustrate the following points.

(1) For stocky members ($\mu = 1$), the resistances of a non-compact section are governed by the stress in the smaller flange, and $m_s = m_h = 1.23$. The shape factor for the compact section, 1.22, is within the normal range (1.2 to 1.4) for hogging moment regions of composite beams.

(2) The line labelled m_p gives the lateral buckling resistance, as predicted from analyses of test data, for both compact *and non-compact* sections, as explained in section 8.4.4. Design of compact sections to Part 3, using $\sigma_{\ell i}$ and Z_{pe} but not factor s, correctly gives this line.

(3) For non-compact sections:

(i)   If the correction factor s were omitted, elastic design to limiting tensile stress σ_{yt} (taken as equal to σ_{yc}) and compressive stress $\sigma_{\ell i}$ would give resistances that are significantly too high for sagging bending, and too low for hogging bending.

(ii)   When the larger flange is in compression (i.e. for sagging bending), $Z_{xc} > Z_{xt}$, and $s < 1$. Use of s_s reduces the non-conservative error along line AB from 37% to 9% (line EFG). Use of the intended factor, $0.87S$, would give a line 13% below that for m_p, for the reason explained in section 8.4.4.

(iii) When the smaller flange is in compression (hogging bending),

$Z_{xc} < Z_{xt}$, $s > 1$, and compressive stress always governs. If s were omitted, the resistance would be governed by buckling (line CD), and would be 18% too low over most of the range of μ. The effect of using s_h ($= 1.33$) is that the resistance is governed by yield along line GC ($\mu s_h > 1$), and by buckling along line EFG ($\mu < 0.75$). As before, the use of s rather than $0.87S$ gives results (along EFG) that are 9% above m_p rather than the intended 13% below. This is more than outweighed by conservative approximations elsewhere, particularly in the calculation of λ_{LT}, and hence of μ.

(4) At a given value of μ, the resistances m_s and m_h are equal when the methods of Part 3 are used, whether the section is compact or not. One would expect a member of the section shown in fig. 8.17 to be more prone to lateral buckling when the smaller flange is in compression. This is so; the effect appears not in the calculation of m at given μ, but in the calculation of λ_{LT} and hence of μ. This is evident in equation (8.26), in which reduction in the size of the compression flange increases λ_{LT}.

NON-COMPACT COMPOSITE CROSS SECTIONS

Unlike the steel section in fig. 8.17, composite sections in hogging bending have effective tension flanges at two or three levels, because of the reinforcement in the slab. The effect of this is now considered. The preceding example showed that the greatest errors occur at moderate and high values of λ_{LT} ($\mu < 0.75$ in fig. 8.18), when the ratio of the design resistance m_h to the appropriate value, $0.87m_p$, is simply the ratio of $D/2y_t$ to $0.87S_h$, where S_h is the shape factor for the steel compression flange. The error is zero for a rolled steel I-section, because $D/2y_t = 1.0$ and $S_h \simeq 1.15$, so $0.87S_h = D/2y_t$.

No systematic study has been done of the range of errors to be expected for composite sections, but examples show that it can be quite wide. One complication, absent from steel sections, is that the depth of the elastic neutral axis, and hence both S_h and $D/2y_t$, is influenced by the sequence of construction, creep of concrete, and shear lag in the slab. None of these factors could be considered when the curve that relates σ_{ri} to β was deduced from the test data.

For hogging bending of composite sections, the ratio $s_h/0.87S_h$ is usually between 1.0 and 1.1. The extreme values found in a study by the authors of non-compact composite I-sections are as follows.

(1) The ratio is 0.945 (5% conservative) for a section consisting of a 3600×300 mm slab with 1.3% of reinforcement ($f_y = 400$ N/mm^2) composite with a plate girder with 800×60 mm flanges and a 1754×30 mm web ($\sigma_{yc} = 338$ N/mm^2).

(2) The ratio is 1.19 (19% non-conservative) for a section with a 3000 × 220 mm slab with 1.5% of reinforcement composite with a plate girder with 394 × 15.8 mm flanges and a 867 × 8 mm web, and strengths of materials as above.

BOX GIRDERS

Lateral-torsional buckling is not discussed in chapter 9 because box girders usually have proportions such that $\lambda_{LT}(\sigma_{yc}/355)^{\frac{1}{2}}$, or β, is far below 45. The stress $\sigma_{\ell i}$ is then taken as equal to σ_{yc}.

The requirement in Part 3 that the limiting compressive stress $\sigma_{\ell c}$ is taken as the lesser of σ_{yc} and $(D/2y_t)\sigma_{\ell i}$ applies also to box girders. A wide box girder will have $\lambda_{LT} = 0$, and may have a cross section in hogging moment regions with $D/2y_t < 1$. The Code then gives the incorrect result that $\sigma_{\ell c} < \sigma_{yc}$. If the factor $D/2y_t$ were to be omitted whenever $\beta < 45$, there would be a sudden change in the value of $\sigma_{\ell c}/\sigma_{yc}$ at $\beta = 45$, which may be why Part 3 is as it is. This aspect of box-girder design would be more rational and less conservative if there were a smooth transition. This could be achieved by a new rule that when $\beta < 45$, $D/2y_t$ is replaced by $(D/2y_t)[(185 - \beta)/140]$, but not exceeding 1.0.

CONCLUSIONS ON DESIGN TO PART 3 FOR LATERAL BUCKLING IN HOGGING REGIONS

(1) The design method, taken as a whole, is over-conservative when applied to composite members, because it is based on test data for non-distortional buckling of steel I-sections. No cases have been found where it appears to be unsafe.

(2) Significant errors can occur in the value of the limiting compressive stress in steel, when the factor $D/2y_t$ is used as an approximation to 0.87S. The error is usually non-conservative, and partly offsets the conservative errors in the determination of λ_{LT} and $\sigma_{\ell i}$, explained in section 8.4.4.

(3) There are few relevant test data for composite members. Numerical analyses are being used to develop better methods, as explained, for example, in section 8.4.5.

(4) The factor $D/2y_t$ is incorrectly applied in Part 3 to box-girder members with low values of the slenderness parameter β.

8.4.7 Restraints to compression flanges

It was shown in section 8.4.4 that when an inverted U-frame is used to restrain a bottom flange against lateral buckling, and the design method of Clause 9.7.2 of Part 3 is used, the slenderness λ_{LT} is proportional to the effective length ℓ_e (equation (8.26)), which in turn depends on the stiffness of the U-frame (equation (8.27)).

It is necessary to ensure that the U-frame is strong enough to retain this stiffness at the ultimate limit state. For a continuous U-frame, the relevant requirement is in Clause 9.12.3.2 of Part 3. The frame has to be designed for forces F per unit length (fig. 8.13) given by:

$$F = 1.5(\ell_e/1000\delta)\left[\frac{1}{(\sigma_{cr}/\sigma_{fc}) - 1}\right] \qquad (8.41)$$

where:

δ is the elastic deflection (fig. 8.13) when $F = 1$

σ_{fc} is the maximum compressive stress in the flange

σ_{cr} is the elastic critical stress, given by equation (8.17).

Equation (8.41) can be derived as follows. The bottom flange is represented by a laterally-restrained pin-ended strut, as shown in fig. 8.14, of length ℓ_e. It has an initial lateral deflection y_0 (fig. 8.19) and is subjected to axial compression $B_f t_f \sigma_{fc}$, so that the maximum axial compressive stress is σ_{fc}.

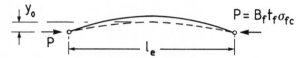

Fig. 8.19   Pin-ended strut with initial deflection

The maximum initial bow of a flange is required in Part 6 of the Bridge Code not to exceed $L/1000$. For design a safety margin of 50% is added, to give:

$$y_0 = 1.5(\ell_e/1000)$$

The force P is assumed to increase the bow in the flange by the factor k_f, where:

$$k_f = 1/[(\sigma_{cr}/\sigma_{fc}) - 1]$$

Unit lateral force causes deflection δ, so the lateral restraining force per unit length when the deflection is $1.5(\ell_e/1000)k_f$ is that given by equation (8.41).

This result is very conservative for the applications considered here, for two reasons.

(1) The U-frame has to be designed for force F per unit length

throughout its length, whereas the maximum deflection, for which F is calculated, occurs only at one point.

(2) The factor k_f is for constant axial force, whereas in a flange, fig. 8.15 shows that the maximum deflection occurs at about $0.12L$ from each end, where the force in the bottom flange is much less than at an internal support.

For continuous restraint, equation (8.41) can be simplified by using equation (8.27) to eliminate δ and putting $I_c = B_f{}^3 t_f/12$ and $E = 205\ \text{kN/mm}^2$. This gives:

$$F = 1001 t_f (B_f/\ell_e)^3/[(\sigma_{cr}/\sigma_{fc}) - 1] \qquad (8.42)$$

WORKED EXAMPLE TO CLAUSES 9.6.6 AND 9.7.2 OF PART 3

From the previous calculations (section 8.4.5) for beam B2 of Table 8.4, $\ell_e = 7.25\ \text{m}$, and for the bottom flange, $B_f = 420.5\ \text{mm}$ and $t_f = 36.6\ \text{mm}$. The shape factor $S = 1.29$, so from equations (8.17) to (8.19):

$$\sigma_{cr}/\sigma_{yc} = 5700 \times 1.29/\lambda_{LT}{}^2 = 1.07 \qquad (8.43)$$

because $\lambda_{LT} = 82.9$.

If the member is fully stressed, the maximum stress at the face of the support is $S\sigma_{fi}$, because the section is compact, but for this purpose, it is taken as σ_{fi}, because of comment (2) above relating to moment gradient. Equation (8.36) gives $\sigma_{fi}/\sigma_{yc} = 0.59$, so from equation (8.42):

$$F = 1001 \times 36.6(420.5/7250)^3/[(1.07/0.59) - 1]$$

$$= 8.79\ \text{kN/m} \qquad (8.44)$$

The depth d_1 (fig. 8.13) is $920.5 - 18.3 = 902.2\ \text{mm}$, so the transverse bending moment at the shear connection is:

$$M_t = 8.79 \times 0.902$$

$$= 7.93\ \text{kNm per metre run of beam} \qquad (8.45)$$

The web is 21.5 mm thick, so the vertical bending stress at the top of the web is a little less than:

$$\sigma_t = 6 \times 7930/21.5^2 = 103\ \text{N/mm}^2 \qquad (8.46)$$

It is unrealistic to assume that this occurs at the face of an internal support, where stiffeners prevent transverse bending of the web. It would be appropriate to check the web for this stress at a distance of $\ell_e/3$ from the support. In this example, the span was 23 m, so $\ell_e/3$ is close to the point of maximum lateral deflection (fig. 8.15).

If the shear stress at the top of the web is τ, and the longitudinal stress is σ_f, the appropriate check is that given in Clause 9.11.3 of Part 3:

$$\sigma_f^2 + \sigma_t^2 - \sigma_f\sigma_t + 3\tau^2 \leqslant (\sigma_{yw}/\gamma_m\gamma_{f3})^2 \qquad (8.47)$$

If, as is likely, σ_f is negligible, then for $\sigma_t = 103$ N/mm^2, $\sigma_{yw} = 355$ N/mm^2, $\gamma_m = 1.05$, and $\gamma_{f3} = 1.1$, we find for this example:

$$\tau \leqslant 167 \text{ N/mm}^2$$

To illustrate the effect of M_t on the shear connection, we now assume that two rows of 19-mm studs are provided at 200 mm pitch, with a lateral spacing (s_c in fig. 8.15) of 350 mm. Assuming a lever arm equal to s_c, the uplift force per stud is:

$$T_u = 7.93 \times 0.2/0.35$$

$$= 4.5 \text{ kN per stud}$$

The design shear resistance at the ultimate limit state (table 7.5, with $\gamma_m = 1.4$) is 77.9 kN per stud. It is obvious that the reduction in the shear resistance due to tension (equation (7.12)) is negligible.

WORKED EXAMPLE USING COMPUTED CRITICAL STRESSES

From the calculations in section 8.4.5, $\lambda_{LT} = 48.1$ and $\sigma_{\ell i}/\sigma_{yc} = 0.97$. From equation (8.43):

$$\sigma_{cr}/\sigma_{yc} = 5700 \times 1.29/48.1^2 = 3.18 \qquad (8.48)$$

Assuming, conservatively, that $\sigma_{fc} = \sigma_{\ell i}$, as before:

$$\sigma_{cr}/\sigma_{fc} = 3.18/0.97 = 3.28$$

The effective length can be estimated by the method used for the restrained strut in fig. 8.14, taking P_{cr} as the value corresponding to the critical stress found above. From equation (8.30):

$$\ell_e = \pi(EI_c/N_{cr})^{\frac{1}{2}}$$

Putting $N_{cr} = 355B_f t_f(\sigma_{cr}/\sigma_{yc})$, $I_c = B_f^3 t_f/12$, and $B_f = 420.5$ mm gives:

$$\ell_e = 21.8B_f(\sigma_{yc}/\sigma_{cr})^{\frac{1}{2}} = 5130 \text{ mm}$$

As before:

$$\delta = 1.52 \text{ mm per N/mm}$$

From equation (8.41):

$$F = 1.5 \times 5130/(1520 \times 2.28) = 2.22 \text{ kN/m} \qquad (8.49)$$

This is lower than result (8.44) mainly because the critical stress is much higher than that given by the other method, so that the axial force factor k_f is reduced from 1.23 to 0.44. The checks on stresses in the web and forces on the shear connectors are as before.

8.4.8 Serviceability limit state

The situations in which checks should be made at the serviceability limit state (SLS) are summarised in section 8.1.1. For the deck slab, local and global effects have to be combined. The relevant checks are given in section 8.3.2.

For structural steelwork, the checks relating to bending and shear are considered in this section. The relevant partial safety factors are given in section 7.1.3 for γ_{fL} and in section 7.2.6 for γ_m. The factor γ_{f3} is always taken as 1.0.

COMPACT CROSS SECTIONS

The effects to be considered are as for the ultimate limit state (ULS) and in addition:

(1) shrinkage of concrete as modified by creep (in analyses both of the structure and of cross sections)
(2) shear lag (in cross sections)
(3) settlement of supports
(4) for load combination 3, the primary and secondary effects of differential temperature.

The methods of analysis and the limiting stresses and resistances in Part 3 of the Bridge Code are as specified for non-compact sections at the ultimate limit state, but using the partial safety factors for the SLS. It follows that in unpropped construction, the bending moments in the steel section and in the composite section have to be kept separate, because the sums of the bending stresses in steel, given by elastic theory, are limited by Clause 9.9.5.2 of Part 3 to σ_{fc}/γ_M in compression and σ_{yt}/γ_M in tension. The notation is explained in section 8.4.3.

The question of whether the check at SLS is likely to govern design is now examined, using bending in a simply-supported beam as an example. If it does govern, then:

$$\gamma_{fLs}M_s = \sigma_{yt}Z_{xc} \tag{8.50}$$

and, at ULS:

$$\gamma_{fLu}M_u < \sigma_{yt}SZ_{xc}/\gamma_m\gamma_{f3} \tag{8.51}$$

where subscripts s and u indicate SLS and ULS, respectively; M is the bending moment due to the nominal loads; and the other symbols are as defined earlier.

The ratio $\gamma_{fLu}/\gamma_{fLs}$ is likely to lie between 1.1 and 1.2 (table 7.1), and is now taken as 1.15. The shape factor S is about 1.3; $\gamma_{f3} = 1.1$ and $\gamma_m = 1.05$ if tension governs. The condition for SLS to govern is then:

$$M_s/M_u > 1.15 \times 1.05 \times 1.1/1.3 = 1.02 \qquad (8.52)$$

The additional effects listed above will almost certainly increase M_u by more than 2%, so in this example, SLS is likely to govern. This was so in a set of model calculations published in 1984[49] for a simply-supported span of 25 m. The design midspan bending moment at ULS was 7726 kNm, and a beam with $M_D = 7900$ kNm was provided, giving 2.2% more resistance than required. At SLS, the calculated stress in the steel bottom flange was 354 N/mm², only 0.3% below the yield strength.

At an internal support, γ_m is likely to be 1.2, not 1.05, which increases M_s/M_u from 1.02 to 1.17, but this is partly offset by the effect of the 10% redistribution of moments at internal supports that is permitted at ULS but not at SLS. No general conclusion can be reached on which limit state will govern.

It is recommended that the initial design be done for the ultimate limit state. The sections should be so chosen that a margin of extra resistance is provided. In choosing this margin, account should be taken of whether any of the extra effects are unusually large in the span considered.

NON-COMPACT CROSS SECTIONS

Of the additional effects listed above for compact sections, all except shear lag are required by Part 3 to be considered at the ultimate limit state for non-compact sections. Also, the shape factor S has to be omitted from equation (8.51), so the preceding estimate of the minimum value of M_s/M_u for SLS to govern becomes:

$$M_s/M_u > 1.15 \times 1.05 \times 1.1 = 1.33 \qquad (8.53)$$

It is obvious that the serviceability limit state is most unlikely to govern, either in midspan regions or at internal supports.

VERTICAL SHEAR

The serviceability limit state never governs for vertical shear alone, because there is no equivalent to the shape factor, and no influence of sequence of construction. For compact sections, vertical shear need be considered only when sections are checked at SLS for combined bending and shear.

DEFORMATIONS IN SERVICE

The limitation of the stress in steel to the yield stress does not imply that no yielding will occur when a member is first subjected to a high proportion of its design load for the serviceability limit state. Some local yielding is likely, because of the residual stresses, which are neglected in design. The resulting irreversible deformation of the member ('permanent set') is often assumed in the design of steel structures to be 10% of the calculated deflection due to the design load.

Similarly, slight buckling may be evident in slender webs designed for tension-field action. It is clear from the definition of the serviceability limit state in Clause 4.2.2 of Part 3 that such deformations are acceptable.

8.5 Design of shear connection for static loads

8.5.1 The governing limit state

Design methods for shear connection are first discussed in relation to three criteria: strength, stiffness and simplicity.

STRENGTH

The possible modes of failure of the shear connection of a composite bridge girder and the ways in which they are prevented in design are as follows:

(1) Local failure of all the connectors, or the concrete surrounding them, between the cross section considered and a free end of the beam, so reducing the flexural strength at that section to that of the steel member alone. This requires a check on the static strength of the connection at the ultimate limit state. It is explained below why this never governs the design of the shear connection in slender cross sections.

(2) Local failure of a single connector due to an excessive longitudinal force, causing transfer of load to other connectors and hence progressive failure as in (1). This is avoided by the use of connectors with sufficient flexibility to redistribute load before failure.

(3) Fatigue failure of a single connector due to excessive number or amplitude of load cycles, with consequences as in (2). The design procedure for repeated loading (section 10.4) is intended to prevent this mode of failure. It rarely governs near supports, but may do so in midspan regions.

(4) Fatigue failure of a single connector due to the expected number and range of load cycles, but coupled with an excessive level of mean shear force, with consequences as in (2). This is prevented by limiting the static load per connector at the serviceability limit state.

STIFFNESS

In current practice, slip at the steel/concrete interface is neglected in all calculations for deflections, fatigue, and resistance to bending. The shear connection must therefore be stiff enough for the errors caused by slip to be negligible in the elastic range of behaviour. Much greater slip is acceptable at the ultimate limit state in compact sections, where plastic analysis is used, and advantage is taken of this in the design methods used for beams in buildings. But, excessive slip in non-compact sections could lead to premature buckling, so a more conservative method is needed for bridges.

SIMPLICITY

Ideally, design is done for the governing limit state only, so that checks on the others can be omitted. The following examples show the influence of different types of loading.

For dead load applied to a simply-supported composite member of uniform section, design of the shear connection by elastic theory using the $VA\bar{y}/I$ formula gives the correct number and distribution of connectors to develop the forces in the slab due to that loading, if the materials remain elastic. As explained in section 7.2.6, the design loads per connector at the serviceability and ultimate limit states are given in the Bridge Code as $0.54P_u$ and $0.714P_u$, respectively. The mean factor $\gamma_{fL}\gamma_{f3}$ for the weight of the structure is 1.0 at SLS and about 1.26 at ULS. If elastic analysis of cross sections is used at both limit states, and N_s and N_u are the numbers of connectors needed in a half span at the two limit states, the ratio N_s/N_u for dead loading is:

$$N_s/N_u = 0.714P_u/(1.26 \times 0.54P_u) = 1.05 \qquad (8.54)$$

This ratio is only relevant to propped construction. The corresponding ratio for superimposed dead load, 0.82, is relevant for all methods of construction, but this loading is only a small fraction of the total load.

The majority of the loading on shear connectors comes from imposed load, for which the comparison is complicated by the effects of travelling loads. It is illustrated (fig. 8.20) by a single point load W crossing a simply-supported span of length L. The design envelopes of bending moment and vertical shear (which is proportional to longitudinal shear) are given in fig. 8.20(a). At the serviceability limit state, and assuming

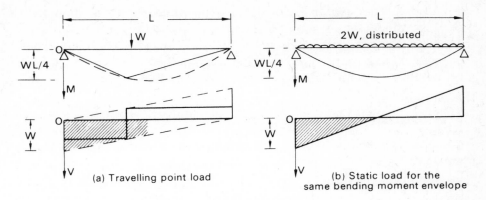

Fig. 8.20   Vertical shear due to travelling point load

that no slip or inelastic action occurs, the maximum longitudinal shear at each section is correctly given by the $VA\bar{y}/I$ formula applied to the shear force envelope, so that the number of connectors to be provided in a half span is proportional to the cross-hatched area in fig. 8.20(a). This is the method of design given for the serviceability limit state in Part 5 of the Bridge Code.

At the ultimate limit state, one can rely on slip to redistribute load between shear connectors, to such an extent that it is assumed in design of beams for buildings that all the connectors between the section considered and the nearer support develop their design resistance before any one of them fails. Mode of failure (1), above, is then relevant, and has to be checked at every cross section. It can be prevented, in the example of fig. 8.20, by providing shear connection by the $VA\bar{y}/I$ method for the shear force diagram for the static loading that gives a bending moment *diagram* identical with the bending moment *envelope* for the travelling load. This is shown in fig. 8.20(b). The total number of connectors, proportional to the hatched area, is only 67% of that for the previous method. Using the values of γ_{fL} for HA loading, the ratio N_s/N_u for a single travelling load is:

$$\frac{N_s}{N_u} = \frac{1.2 \times 0.714 P_u}{0.67 \times 1.1 \times 1.5 \times 0.54 P_u} = 1.44 \qquad (8.55)$$

Similar results are obtained for continuous spans and other types of travelling load.

The ability of connectors to redistribute load by slip diminishes at high concrete strengths ($f_{cu} > 40$ N/mm^2), and is less for 'stiff' connectors of the welded-bar type than it is for studs. Also, the slip needed to achieve a given redistribution increases with span, which is why partial shear connection is not used in buildings for spans exceeding about 20 m. Even so, the value 1.44 in equation (8.55) so far exceeds 1.0 that it was assumed in drafting BS 5400: Part 3: 1979 that the serviceability limit state would nearly always govern the design of shear connection. A check at the ultimate limit state was required only in three special situations:

(1) When the whole of the load on a beam built unpropped is assumed to be resisted by the composite section.

(2) When account is taken of redistribution of stress from a steel tension flange to the reinforcement in the adjacent slab.

(3) When the connectors are designed for combined shear and vertical tension.

It is inconvenient in practice to design initially for the serviceability limit state, because most other aspects of design are considered first, or only, at the ultimate limit state. The problem can be avoided by doing the initial design for the shear connection at the ultimate limit state, but using the method appropriate to the serviceability limit state, that is, working from the factored shear force envelope, as in fig. 8.20(a). For that example, this gives 50% more connectors, so that a typical ratio N_s/N_u for imposed load drops from 1.44 (equation (8.55)) to $1.44/1.5 = 0.96$. The method is a safe approximation to design for the serviceability limit state, because $N_u > N_s$.

For HB loading, the ratio N_s/N_u is 1.02, but the serviceability limit state is unlikely to govern when HB loading is combined with HA loading and superimposed dead loading ($N_s/N_u = 0.82$). It is only for propped construction that a check at the serviceability limit state appears to be necessary.

The authors therefore recommend that the presentation of this subject in Part 5 of the Bridge Code should be changed. Design of shear connection should be based on shear force envelopes for the ultimate limit state, with a requirement for a check at the serviceability limit state when propped construction is used (unless experience shows this to be unnecessary).

Finally, a check for fatigue should be made (section 10.4), as this may govern near midspan.

8.5.2 *Longitudinal shear at the steel/concrete interface*

It is often forgotten that the equation given by elastic theory for longitudinal shear flow, $q = VA\bar{y}/I$, is valid only for members of uniform cross section. The general expression is:

$$q = \frac{\mathrm{d}}{\mathrm{d}x}(MA\bar{y}/I) \tag{8.56}$$

For members of varying cross section, both the bending moment M and the section property $A\bar{y}/I$ are functions of x, so that equation (8.56) becomes:

$$q = \frac{VA\bar{y}}{I} + M\frac{\mathrm{d}}{\mathrm{d}x}\left(\frac{A\bar{y}}{I}\right) \tag{8.57}$$

If the effective flange breadths for bending stresses (section 8.4.1) were used also for calculations of longitudinal shear, the second term in equation (8.57) would be non-zero, even in members of uniform cross section. It would also be inaccurate, because the variation of effective breadth (and hence $A\bar{y}/I$) with x is itself simplified for design purposes.

To enable the second term of equation (8.57) to be omitted from all calculations for longitudinal shear in members where the steel cross section and overall breadth of concrete flange are constant within each span, two approximations are made:

(1) The effective breadth of the concrete flange throughout any span is assumed to be the value given in Part 3 of the Bridge Code for uniformly-distributed load at quarter span.

(2) The deck slab is assumed to be uncracked and unreinforced throughout its length.

It would be consistent with these assumptions also to neglect small changes in the neutral axis level and the second moment of area of the steel section due, for example, to a change in the thickness of a steel web or flange.

Assumption (2) enables the design of the shear connection to be simplified in another way. As explained in Volume 1, it has been found that connectors are less stiff and slightly less strong when set in cracked concrete. No allowance need be made for this when shear flows are calculated for uncracked concrete, because cracking, if it occurs, is likely to reduce the shear flow at the steel/concrete interface more than it reduces the strength of the connectors.

With these simplifications, the design shear flow for dead and imposed load for a member of uniform section is proportional to the vertical shear due to loading on the composite section, and so is easily calculated from the envelope of vertical shear force. The nominal strengths of shear connectors given in the Bridge Code are applicable for loads in both longitudinal directions, so account has to be taken of the sign of the vertical shear force only when combining the effects of different loadings.

Shear flows due to the effects of temperature and shrinkage (section 8.7) should be considered before the spacing of the shear connectors is calculated.

8.5.3 Detailing of the shear connection

The choice of the type and size of shear connector to be used is influenced by the rules on detailing discussed below. For a given type of connector it is a simple matter to calculate the required spacing along the member, p, from the local value of the design shear flow, q per unit length, and the design resistance of the connector, P_u/γ_m. If q has been calculated in accordance with Part 3 of the Bridge Code, it will not include γ_{f3}; the required spacing is then given by:

$$p \not> (P_u/\gamma_m\gamma_{f3})/q$$

When connectors are placed in rows across the breadth of the steel flange, rows at spacing np with n connectors per row of course satisfy this requirement, provided that np does not exceed the maximum permitted spacing.

The required spacing will vary along the span, and so is inconvenient for detailing. CP 117: Part 2 therefore allowed the spacing to be kept constant over lengths where the shear flow was within $\pm 5\%$ of the mean value.

During the preparation of Part 5 of the Bridge Code, Dr D. M. Osborne-Moss studied the consequences of increasing this allowance to $\pm 10\%$ or $\pm 15\%$. He analysed the loads on connectors and stresses in the steel section due to HA and temperature loadings on a simply-supported bridge girder of 60 ft (18.3 m) span, allowing for slip at the interface and shear lag in the concrete flange. It was found that the end connectors become overloaded as the allowance is increased, due to undercutting of the end of the shear envelope (e.g. line AB in fig. 8.21). This led to the rule in Part 5 of the Bridge Code that the spacing at the end of each span should be sufficient for the maximum shear, and should be maintained for at least 10% of the span (line CD). When this was done, Osborne-Moss found that with $\pm 15\%$ variation elsewhere, no connectors were overloaded, and the effect on maximum stresses in the steel section was negligible. In view of the limited scope of this study, this was reduced to $\pm 10\%$.

In fact, the negative tolerance is irrelevant, because it does not matter if the connector spacing is kept constant over any length where the actual shear flow is less than the design shear flow. For example, the shear flow at E in fig. 8.21 is more than 10% below the mean shear flow over length

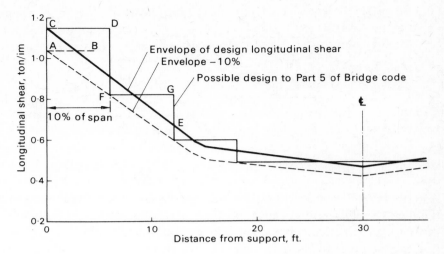

Fig. 8.21   Use of shear connectors at uniform spacing

FG of the beam, and the consequence is a slight over-provision of shear connectors. The -10% limit has therefore been omitted from Part 5, which enables designers to use longer blocks of uniformly-spaced connectors.

A possible design of shear connection to the new rule is shown in fig. 8.21. With four different spacings in the half span, the total number of connectors required is the same as when the $\pm 5\%$ rule is used with seven different spacings. When repeated loading is included, the variation in spacing along a span is usually less than in this example (e.g. plates 2 and 6). The use of constant spacing over the end 10% of each span is probably conservative, but it can be used to justify approximations made in the calculation of loads on connectors due to temperature and shrinkage (section 8.7.1).

The other rules for detailing of the shear connection are mainly based on experience, for the number of relevant variables is so great that the cost of systematic research on the subject could exceed the resulting benefits. The rules for stud connectors in slabs at least 125 mm thick are summarised in fig. 8.22, and the reasons for the rules are given below. The design of the local reinforcement is treated in section 8.6.

DETAILING OF STUD SHEAR CONNECTORS

The placing of connectors in widely-spaced groups could cause excessive stress concentrations in the slab, or allow uplift to occur between groups. It would invalidate the assumption made in design that the transfer of shear across the interface is continuous. Detailing of connectors and of

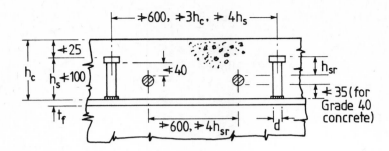

Fig. 8.22 Detailing rules for shear connectors and transverse reinforcement

transverse reinforcement should ideally be linked, because the position of the bars should relate to the stress concentrations caused by the studs[80] but no simple rule has been found that would ensure this.

The spacing of the bars is related to the projection of the studs above them and a minimum projection is specified, to ensure that the bars intersect any surface of longitudinal shear failure that passes above the studs. When there is a crossfall in the slab, it is usually found that the studs have to be longer than 100 mm.

In draft Eurocode 4[5], only minor changes were made to these rules. The maximum spacing of studs was given as 600 mm or $4h_c$. The limit of $4h_s$ was omitted; but studs are required to have an overall height of not less than $3d$ and heads at least $0.4d$ thick and $1.5d$ in diameter. The minimum projection of the underside of the head above the reinforcement is reduced to 30 mm for unhaunched slabs but kept at 40 mm for haunches.

Eurocode 4 and the 1979 edition of Part 5 both make an exception to the rules for maximum spacing of studs, to facilitate the use of precast deck slabs. Connectors can be placed in groups at a wider spacing, provided that 'consideration is given in design to the non-uniform flow of longitudinal shear and the greater possibility of slip and vertical separation between the slab and the steel member'. In practice, this means that the local reinforcement in the slab should be related to the positions of the groups. The force transferred by a group may be as large as that at an anchorage of a large prestressing tendon, and existing data on stresses in such a region[109] may be useful for the design of the reinforcement. Particular care is needed where groups of connectors are used in haunched slabs. It may be necessary to check local stresses in the steelwork. This exception was not accepted by the Department of Transport.[2]

A second exception given in the 1979 edition of Part 5 relates to shear connection for a composite cross beam near its intersection with a main beam (as shown, for example, in plate 6). These connectors are subjected to both longitudinal and transverse loading. There is evidence[110] that biaxial stressing may reduce fatigue life, so it was stated in Part 5 that no connectors on a cross-girder flange of breadth b_f should be placed within a distance $2b_f$ of the nearest connectors on the main girder. It has been found that this rule is impracticable for composite crossheads at piers, so it has been deleted. Biaxial fatigue of studs is further discussed in section 10.5.2.

No limits are given in Part 5 for the minimum spacing of connectors. At close spacings, it is difficult to ensure that the concrete surrounding the studs is properly compacted, and to satisfy the requirements for transverse reinforcement. Draft Eurocode 4 gives the minimum spacing of stud connectors as $5d$ in the direction of loading and $4d$ in the transverse direction, where d is the diameter of the shank.

WELDED JOINTS BETWEEN CONNECTORS AND THE STEEL FLANGE

For studs, the welding process ensures that the area of weld is at least as great as the cross-sectional area of the stud. For other connectors, minimum lengths and sizes of welds are given in Part 5. Minimum thicknesses of flange plates to which studs can be welded are given in section 9.4.2.

Connectors may not be placed within 25 mm of the edge of a steel plate, to limit the stress concentration at the flange edge and to improve the protection of the connector from corrosion. The minimum concrete cover to the side of a connector is also specified, as 50 mm, because the breadth of a concrete haunch may be different from that of the adjacent steel flange. This limit is not related to corrosion, but to the danger of local splitting or spalling of the concrete. This subject is discussed in section 8.6.

FORCES TENDING TO CAUSE UPLIFT

The ways in which significant uplift forces can occur, and their effect on the shear strength of connectors, are discussed in section 7.2.3. The detailing rules are intended to ensure that the deck slab is 'positively tied to the girder' in such a way that significant separation cannot occur under longitudinal shear forces until the design shear strength of the connection is reached. Even where there is a net vertical compressive force between slab and girder, the local deformation of a stud connector as failure is approached causes tension both in the shank and in the concrete on the 'unloaded' side of the connector. This is shown in fig. 8.23, which is based on photographs of sawn cross sections of push-out specimens after failure.[79]

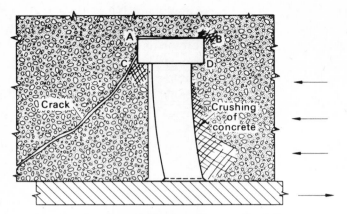

Fig. 8.23 Section through shear connector after failure

Reference is made to slabs 'connected to the girder by two separate elements, one to resist longitudinal shear and the other to resist forces tending to separate the slab from the girder', to cover the use of horseshoe-type connectors (as in 'Preflex' beams) which have no surface that resists uplift.

The criterion given in Part 5 for such a two-part connector is that the separation measured in push tests should not exceed half the longitudinal slip, at up to 80% of the nominal shear strength of the connector. In draft Eurocode 4, the requirement is more general: 'All other types of shear connector (i.e. other than headed studs) should incorporate or be supplemented by anchoring devices designed for a tensile force, perpendicular to the plane of the steel flange, of at least 0.1 times the design shear resistance of the connectors'.

8.6 Transverse reinforcement

Local stresses in the concrete slab in the vicinity of the shear connectors are too complex to calculate, so that, as with bolted connections in steelwork, the detailed design of this region has to be based on tests to failure.

Design for longitudinal shear consists essentially of identifying possible surfaces of shear failure, and making sure that the mean ultimate shear resistance of each surface exceeds the shear to be transferred across it at the ultimate limit state. A corresponding surface in a reinforced concrete T-beam is a horizontal plane through the web between the neutral axis and the tension reinforcement. Its strength is augmented by intersecting it with reinforcing bars (the stirrups) that are fully anchored

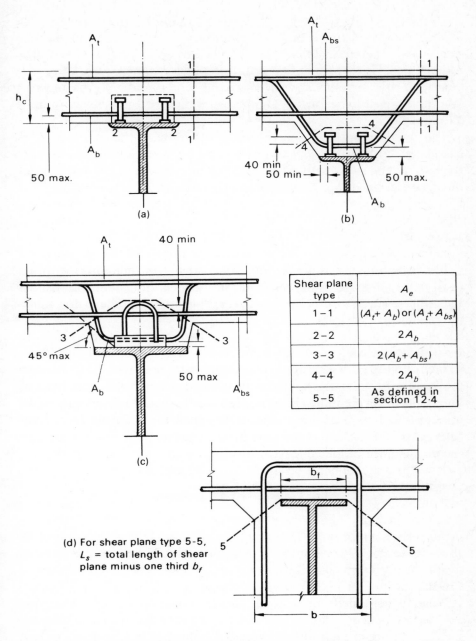

Shear plane type	A_e
1-1	$(A_t + A_b)$ or $(A_t + A_{bs})$
2-2	$2A_b$
3-3	$2(A_b + A_{bs})$
4-4	$2A_b$
5-5	As defined in section 12·4

(d) For shear plane type 5-5, L_s = total length of shear plane minus one third b_f

Fig. 8.24   Shear planes and transverse reinforcement

on both sides of the plane considered. Similarly, transverse reinforcement is used in composite beams as may be necessary to supplement the shear strength of the concrete on each plane considered. The design rules given in Part 5 of the Bridge Code refer only to bars of the types shown in fig. 8.24, but bars inclined to the longitudinal axis (cf bent-up bars in reinforced concrete beams) could be used in regions where the longitudinal shear is mainly in one direction. Such inclined bars, known as 'anchors', are sometimes welded to connectors, and may be inclined laterally as well as vertically. Guidance on their design is given in draft Eurocode 4.

The longitudinal shear force per unit length at the interface, q, is calculated in the same way as for the shear connection at the ultimate limit state. The shear per unit length for which a potential failure surface must be designed, q_p, is q multiplied by the ratio of the area of the effective concrete flange 'excluded' by the surface considered to the total effective area. For example, if surface 1-1 in fig. 8.24(b) is midway between the centreline of the flange and its effective edge, then q_p can be taken as $q/4$ (which is conservative because it ignores the haunch).

It will be found in practice that when the bottom reinforcement in a deck slab that spans transversely is continued without reduction across the supporting steelwork, it is usually more than sufficient for longitudinal shear.

DESIGN OF TRANSVERSE REINFORCEMENT

The development of a design method for transverse reinforcement from the results of tests is explained in detail in Volume 1 and consists essentially of one equation and some rules for its use. The equation is used also in Part 5 of the Bridge Code, but with minor changes in the rules associated with it, which are now explained.

The method is applicable to beams having any of the various types of cross section shown in fig. 8.24. Checks are made to ensure that q_p for any possible surface of shear failure does not exceed the shear resistance of that surface. It is assumed (for simplicity) that the transverse cross section of each potential failure surface is constant along the beam. These surfaces can then be represented by lines, such as 1-1, 2-2, etc., in fig. 8.24. They are referred to in Part 5 as 'shear planes' although only surfaces of type 1-1 are in fact plane.

For shear planes where the effects of transverse bending moment in the slab can be neglected, the design equation appears in Part 5 in the form:

$$q_p \not> k_1 s L_s + 0.7 A_e f_y \tag{8.58}$$

The notation is as follows:

k_1 is a constant; 0.9 for normal-density concrete and 0.7 for lightweight-aggregate concrete (assumed to be of density not less than 1400 kg/m^3)

s is a unit stress, included to make the equation nondimensional

L_s is the length (in the plane transverse to the span of the member) of the shear surface considered

A_e is the cross-sectional area per unit length of girder of the transverse reinforcement that crosses the shear plane considered, is fully anchored on both sides of it, and is assumed to be effective in resisting longitudinal shear

f_y is the nominal yield strength of the transverse reinforcement

f_{cu} is the characteristic cube strength of the concrete.

Limits are placed on the maximum values that can be used for f_y and f_{cu}, because of the lack of tests on specimens made with stronger materials.

The criteria for determining A_e are discussed in Volume 1, and typical values for shear planes of types 1 to 4 are given in fig. 8.24, together with other rules for the detailing of the shear connection. In order to provide design methods for the various types of shear plane, haunches, and transverse bending that can occur in practice, it was necessary to define four classes of fully-anchored transverse reinforcement:

A_t is top steel, including any provided for flexure

A_b is bottom steel (including any provided for flexure) that is close enough to the steel/concrete interface to contribute to the strength of the slab in the highly stressed region near the base of each shear connector, and far enough below the top of each connector to resist uplift forces

A_{bs} is steel at the bottom of the slab that is more than 50 mm above the steel/concrete interface (due for example to the use of a haunch)

A_{bv} is bottom steel (as for A_b) but excluding any provided for flexure.

The various definitions of A_e given in fig. 8.24 are applications of the following general principles:

(1) For a shear plane crossing the full thickness of the slab, or entirely within a haunch, all reinforcement crossing the plane is effective.

(2) Other shear surfaces (e.g. 2-2 and 3-3 in fig. 8.24) pass just over the tops of the shear connectors. Bars that intersect such a surface are assumed to be effective if they are placed at a clear distance of not less than 40 mm below the surface of the shear connector that resists uplift forces, and are not required to resist tension due to transverse sagging bending of the slab.

In certain situations, particularly in L-beams, it may be difficult to provide full anchorage for a transverse bar on both sides of a possible shear plane. One can consider as effective the appropriate proportion of the area of a partly anchored bar, but it is better to use U-bars.[111]

In practice, equation (8.58) is likely to be used in the form:

$$A_e \not< (\gamma_{f3}q_{ps} - k_1sL_s)/(0.7f_y) \tag{8.59}$$

where q_{ps} is the shear flow due to the loading calculated in accordance with Part 3, and so not including γ_{f3}. It will often be found that the reinforcement provided for other purposes exceeds A_e. For consistency, we continue to use here the symbol q_p to mean the shear flow calculated in accordance with Part 4, and including γ_{f3}.

Other requirements of Part 5 are that the area per unit length of beam of transverse reinforcement in the slab must be not less than $0.8sh_c/f_y$, where h_c is the thickness of the slab. There is also a limit on the maximum shear that can be resisted:

$$q_p \not> k_2L_sf_{cu} \tag{8.60}$$

where k_2 is 0.15 for normal density concrete and 0.12 for lightweight-aggregate concrete. These rules are based on the evidence from tests (Volume 1) and are similar to those in CP 117: Part 2.

Detailed modifications to this method are given in Part 5 to enable it to be used with concrete of cube strength less than 20 N/mm² (relevant during construction), and in haunched members with various arrangements of reinforcement. The reasons for these will be self-evident.

INTERACTION BETWEEN LONGITUDINAL SHEAR AND TRANSVERSE BENDING

Transverse sagging bending of the slab can occur above a steel web due to non-uniform loading. As this causes transverse tension at the slab soffit, it is assumed to reduce the shear strength of planes of type 2-2 in fig. 8.24, but not that of planes of type 4-4 (because the lateral tension is assumed not to affect a deep haunch) or 1-1 (because the effect of tension is offset by an equal transverse compression in the upper half of the slab). For design, equation (8.58) is replaced by:

$$q_p \not> k_1sL_s + 1.4A_{bv}f_y \tag{8.61}$$

which has the effect of excluding from A_e the area of bottom reinforcement that is required to resist (at its design yield stress) the tension due to transverse bending at the ultimate limit state. For a plane of type 3-3 (fig. 8.24), the area A_{bv} would consist of A_b plus that part of A_{bs} (if any) not required for transverse bending.

Transverse hogging bending of the slab has a beneficial effect on the strength of the shear connection in unhaunched members. Part 5 gives an optional method allowing for this, which is explained in Volume 1. It allows some reduction in the area of bottom transverse reinforcement, and is of value only when the design of the deck slab for bending does not govern.

SHALLOW HAUNCHES

If the depth of a haunch tapers out to zero in the longitudinal direction, the question arises: when does it cease being a 'haunch' for the purposes of this clause? This can be answered by considering the rule for planes of type 3-3, mentioned above. A_b is bottom reinforcement that passes within 50 mm of the steel section, and may or may not be kinked (e.g. figs 8.24(a) and (c)), and A_{bs} is bottom slab reinforcement that is not kinked. When the haunch becomes shallow enough for unkinked bottom reinforcement to pass within 50 mm of the steel section (i.e. when its depth is between 10 and 20 mm, depending on the concrete cover) then the member should be treated as 'unhaunched', so that a plane of type 3-3 is treated as type 2-2.

The many different details that can be used for shear connectors, reinforcement, and haunches made these clauses particularly difficult to draft. There is room for some application of 'engineering judgement' in their interpretation, provided that their objectives are clearly understood. These objectives are stated in Part 5 (Clause 6.3.2), with the rider that 'special consideration should be given to details not in accordance with these clauses'. The nature of this consideration will depend on the problem and could at one extreme include full-scale testing under both static and repeated loading. An example where 'judgement' should suffice is now given. Part 4 of the Bridge Code specifies the minimum cover for deck soffits in Grade 30 concrete as 45 mm. The clauses now discussed were based on the assumption that the bottom longitudinal bars in the slab would lie above the transverse bars. If for some reason it is essential to place them below, it is not possible for bottom transverse reinforcement to be placed 'at a clear distance not greater than 50 mm from the nearest surface of the steel beam'. In view of the limited test data on this subject and the tendency for general clauses to be conservative in particular cases, the designer might decide that 60 mm was 'near enough'.

OTHER DESIGN METHODS FOR TRANSVERSE REINFORCEMENT

Since Part 5 of the Bridge Code was written, two other codes have given rules for resistance to longitudinal shear that are similar in presentation to those in Part 5, but give different results. The differences for normal-density concrete are shown in fig. 8.25 for one representative set of properties of the materials.

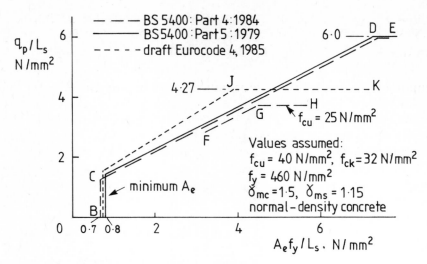

Fig. 8.25   Design rules for transverse reinforcement

As explained in section 11.4.1, Part 4 gives equations for resistance to longitudinal shear of the interface between precast and *in situ* concrete, in terms of the area of reinforcement that crosses unit area of the interface (A_e/L_s in fig. 8.25). These apply also to 'monolithic construction' and are almost identical to those in Part 5 (lines BCDE) when $f_{cu} = 40$ N/mm². At lower cube strengths, such as 25 N/mm², the cut-off level for q_p (i.e. the maximum amount of reinforcement that can be assumed to be effective) is much lower than in Part 5, as is shown by lines GH and DE.

The treatment of transverse reinforcement in draft Eurocode 4 of course relates to composite members, but it is based on the equations in draft Eurocode 2 (for concrete structures), which are the European equivalent of those in Part 4 discussed above, because it was thought that EC2 and EC4 should be consistent. In the comparisons, the cylinder strength f_{ck} is assumed to be $0.8f_{cu}$. The 'reinforcement' term in the equivalent of equation (8.58) is taken as $0.87A_e f_y$, to be consistent with design of links in beams by the truss analogy. This is why line CJ is steeper than line CD, for which the term is $0.7A_e f_y$. A more significant difference is that the cut-off line JK is much lower than DE. As in Part 4, its level is proportional to concrete strength, so this difference occurs for all grades of concrete.

All of these design rules are based on interpretations of test data and on assumed analogies between different situations. The equations in Part 5 are the only ones to have been deduced mainly from tests on composite T-beams; the other rules are based on research on reinforced concrete. It is hoped that the recent evaluation of Eurocode 2 may lead to resolution of the discrepancy between it and Part 4 of the Bridge Code.

DETAILING

The limiting size of haunch to which the rules in Part 5 for transverse reinforcement are appropriate is indicated by the dimensions '45° max' and '50 min' in figs 8.24(b) and (c). They are as discussed in Chapter 2 for beams in buildings, except that the 40-mm side cover has been increased to 50 mm for bridges, as recommended by Teraszkiewicz.[91]

Transverse reinforcement is particularly necessary at the ends of composite cantilevers. Figure 8.26 shows, in plan, a cantilever AB in which the deck slab also has a free edge along DC. The horizontal forces

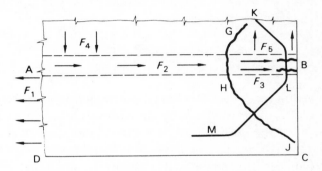

Fig. 8.26   Reinforcement at the end of a cantilever

acting on the concrete slab ABCD are F_1, longitudinal tension in the reinforcement; F_2 from the shear connectors; and an additional end force F_3 due to shrinkage, further increased if the slab is colder than the steel. These forces cause a clockwise couple which must be resisted by transverse forces F_4 and F_5. The combined effects of F_3 and F_5 tend to cause vertical splitting of the slab at B, and this is often found to occur at quite an early stage in laboratory tests. Forces F_3 can also cause cracking along lines such as GHJ. Reinforcement equivalent to the bar KLM should be designed for these forces.

8.7 Design for temperature and shrinkage

The temperature changes and differences given in Part 2 of the Bridge Code are summarized in section 7.3.4, and the values of the free shrinkage of concrete and the associated creep coefficients are given in section 7.3.5. In this section, they are referred to as TSC effects. The variation in shrinkage strain with ambient humidity is small in comparison with the initial drying shrinkage, so its effects are treated as permanent and considered in all five load combinations. Peak temperatures however occur very rarely, so temperature loading is in combination 3 only. Methods of allowing for these effects in design are now considered.

Both the primary and secondary effects (section 7.3.4) of temperature and shrinkage are diminished by inelastic behaviour of the structure. They are unlikely to influence the ultimate strength of members of compact section, and cannot be included in plastic analysis of a cross section. For non-compact members, where elastic analysis is used at both limit states, TSC effects are considered at the ultimate limit state. This is obviously correct for members that are so slender that the buckling stress is well below the yield stress, because adverse TSC effects would erode the margin of safety against buckling.

For members that are only just slender, the buckling stress may exceed the yield stress, and here the inclusion of factored TSC effects is conservative, because the reduction in these effects caused by inelastic behaviour is ignored.

Both types of non-compact member are treated alike, both in Part 5 and in draft Eurocode 4, because no simple method has been found for taking account of the greater safety of just-slender members.

Particularly in continuous members, it is impracticable to analyse the interaction between the cracking of concrete due to global or local loading and the strains and internal forces due to TSC effects. These are normally calculated assuming the slab to be uncracked and unreinforced. Research has shown[112] that cracking significantly reduces the primary effects of shrinkage, so this assumption is conservative in hogging moment regions, particularly at the ultimate limit state.

Shrinkage of the concrete slab is equivalent to a fall in its temperature relative to that of the steel member, so that the same design formulae can be used for both temperature and shrinkage effects. The only difference is that the short-term modular ratio is used for temperature effects, since peak values are likely to be of short duration. For shrinkage, appropriate allowance for creep is made (section 7.3.5).

In the following more detailed account, it is necessary to distinguish between our understanding of the real behaviour and the simplifications of it made in the Bridge Code and, less conservatively, in the draft Eurocode 4.

8.7.1 Primary effects in simply-supported and continuous members

The following simplified model of the sequential casting of a simply-supported composite beam illustrates the primary effects of shrinkage. A midspan length AB is cast first (fig. 8.27(a)), and its shrinkage causes two units, say, of shear force to be transferred across the interface at each of its ends. An end section BC is then cast, and at point B its shrinkage is restrained half by reinforcement projecting from AB and half by the steel girder. The internal forces are then as shown in fig. 8.27(b). The process continues, and eventually the interface forces consist of two units at each end of the steel member, and equal and opposite pairs of single units acting over short lengths at each construction joint, fig. 8.27(c).

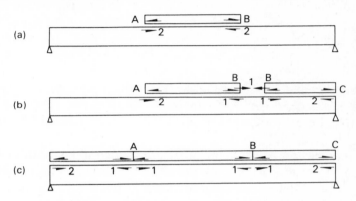

Fig. 8.27   Shrinkage forces in deck slab cast in three sections

Similar pairs of internal forces occur in continuous members, due both to the sequence of casting and to cracking of the concrete slab at internal supports. The continuity of the reinforcement and the flexibility of the shear connection ensure that only a small proportion of the shrinkage tension is transferred to the steel member when the concrete cracks. These self-equilibrated pairs of forces are further reduced by inelastic action and slip as ultimate load is approached. It follows that these internal effects can be neglected.

The primary interface shears due to both temperature and shrinkage can be considered to be concentrated forces near the end supports only, in both simple and continuous members, and irrespective of the sequence of casting. Primary bending stresses due to TSC effects at cross sections remote from the ends of members can therefore be calculated by full-interaction theory. The analysis is given in Appendix A. The results for shrinkage are now summarised and discussed.

PRIMARY STRESSES DUE TO SHRINKAGE OF THE CONCRETE SLAB

We consider the response of a member, with the cross section shown in fig. 8.28(a), to a free shrinkage strain ε_{sc} in the concrete slab, which is assumed to be uncracked. Tensile stresses, strains and forces and sagging moments are assumed to be positive, as in Appendix A, so that numerical values of ε_{sc} are negative. The elastic properties A and I of the composite section are calculated in 'steel' units.

The effects of shear lag are explained in Appendix A. Here, they are neglected, so the effective-breadth ratio ψ_p can be omitted, and the cross-sectional area of concrete is given by $A_c = bh_c$. The modular ratio α_e ($= E_s/E'_c$) takes account of creep of concrete.

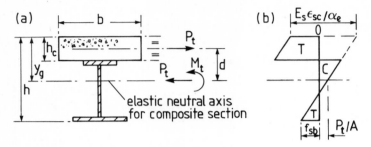

Fig. 8.28   Primary shrinkage stresses in composite section

The slab has longitudinal reinforcement, but its contribution to the values of A and I is usually neglected. It can then also be ignored when calculating the longitudinal force $\bar{F}$ (or P_t in the notation of BS 5400: Part 3) that would be needed to prevent shortening of the (free) slab when it shrinks. From Appendix A, this force is:

$$\bar{F} = E_s\varepsilon_{sc}A_c/\alpha_e \qquad (A.18) \text{ bis}$$

To restore longitudinal equilibrium, an equal and opposite force acts at the centroid of the composite section, and the bending moment $\bar{M}$ (or M_t in BS 5400) is:

$$\bar{M} = -\bar{F}(y_g - h_c/2) \qquad (A.19)$$

The uniform sagging curvature of the member is:

$$\phi = \bar{M}/E_sI \qquad (A.16)$$

and the longitudinal stresses at any depth y below the top of the slab are, in concrete:

$$f_c = [-E_s\varepsilon_{sc} + \bar{F}/A + \bar{M}(y - y_g)/I]/\alpha_e \qquad (A.15)$$

and in steel:

$$f_s = \bar{F}/A + \bar{M}(y - y_g)/I \qquad (A.14)$$

so that the bottom-fibre tensile stress (fig. 8.28(b)) is:

$$f_{sb} = \bar{F}/A + \bar{M}(h - y_g)/I \qquad (8.62)$$

It follows from equation (A.15) that the total longitudinal tensile force in the slab, except near its ends, is:

$$Q = (A_c/\alpha_e)[-E_s\varepsilon_{sc} + \bar{F}/A + \bar{M}(h_c/2 - y_g)/I] \qquad (A.20)$$

This force is significantly less than the force $\bar{F}$ (or P_t), as is evident from equations (A.18), (A.19) and (A.20).

Full-interaction theory gives no information on how the force Q is transferred across the interface at the ends of the member. Partial-interaction theory shows that the interface shear per unit length, q, increases towards a simple support as shown by curve OB in fig. 8.29,

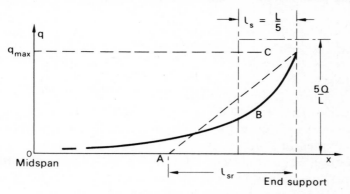

Fig. 8.29 Interface shear due to shrinkage

which is based on equation (A.22). The area under this curve is the total longitudinal force due to temperature or shrinkage in the slab in regions remote from a free end of the member. It is a little less than Q, due to the effects of slip, but this difference is neglected in design.

In Part 5 of the Bridge Code, this curve is replaced for design purposes by the line AC, where the transfer length ℓ_{sr} is given by the conditions that the end shear per unit length, q_{max}, is that given by partial-interaction theory and the area under AC (i.e. the total force transferred) is equal to Q. The partial-interaction analysis (Appendix A) gives the result:

$$\ell_{sr} = 2(Qp/k\varepsilon_{sc})^{\frac{1}{2}} \qquad (8.63)$$

where k is the ratio of load to slip for the shear connectors, which are assumed to be at uniform spacing p.

As explained below, the precise value of ℓ_s is not important, so the same length ℓ_s can be used for temperature effects; alternatively, ε_{sc} can be replaced by the difference between the free thermal strains at the centroids of the slab and the steel section.

The connector modulus k is a very variable quantity.[113] For example, the load–slip curves shown in fig. 7.3 give values (based on slip at half the maximum load) ranging from 60 kN/mm for 16-mm studs to 700 kN/mm for the 25-mm square bar connectors. The spacing p will probably not be uniform along the span, and may not be known by the designer when ℓ_s is calculated.

The ratio p/k in equation (8.63) is essentially the flexibility of the shear connection and any value used for it in design will obviously be approximate. Provided that it is not so large that $\alpha L/2$ (Appendix A) falls below about two, the longitudinal force Q is independent of p/k. The design method correctly provides for the force Q even if p/k is wrong, because the peak shear flow q_{max} is given by:

$$q_{max} = 2Q/\ell_s \tag{8.64}$$

and will be too large if ℓ_s is too small. For these reasons, approximate values of $K\ (=p/k)$ are given in Part 5: 0.003 mm^2/N for stud connectors and 0.0015 mm^2/N for other connectors. These are for normal-density concrete; for lightweight-aggregate concrete, they are doubled. Comparisons with data from tests on connectors and actual designs show that these values are high, even after allowing for some creep, so that ℓ_s may be overestimated in practice.

For stud connectors only, advantage is taken of their proven ability to redistribute longitudinal force by giving, in Part 5, a simpler alternative method, also shown in fig. 8.29. A uniform rate of shear transfer $5Q/L$ (where L is the effective span) is assumed over a length $L/5$, so that the connectors resist a total force Q, as before.

The span is in fact an irrelevant parameter; it does not appear in equation (8.63). In Eurocode 4, a different simplification is made. A triangular distribution of interface shear per unit length is allowed for all types of connector, and a rectangular one for studs. In both cases the total area is Q, and the base length is the effective breadth of the slab. In fact, any reasonable assumption about this distribution is good enough because the shear forces due to TSC effects are small in comparison with the total values.

At simple supports, the shear force q per unit length due to shrinkage acts in the opposite direction to the shears due to other loads, and so can be neglected. At the ends of cantilevers, it acts in the same direction, and should be considered.

Temperature effects cause interface shear at end supports in the same direction as that due to live load, so account should be taken in load combination 3 of the interface shear due to temperature difference, except at the ends of cantilevers.

8.7.2 Secondary effects in continuous members

In practice, the deck slab of a continuous member is usually cast in stages, but it is convenient to give, first, two methods of analysis for secondary effects in members where the whole of the slab is cast at one time, and then to discuss their applicability in practice.

When the continuous member (assumed to be uncracked) is of uniform section within each span, moment distribution can be used. Let the sagging curvature along the whole length of the rth span due to the primary effects of temperature and shrinkage be ϕ_r. The continuous member can be straightened by applying hogging moments $EI_r\phi_r$ at both ends of the rth span and similarly for other spans. These make the beam compatible with its supports. The applied end moments are then treated as 'fixed-end moments' and are redistributed by the method of moment distribution in the usual way.

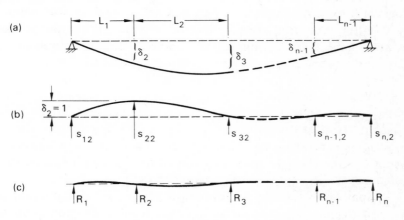

Fig. 8.30 Secondary effects in a continuous member

Where the cross section of the member varies along its length, the Mohr moment-area method (Appendix X) is a better hand method than moment distribution. Or, computer analysis based on the following theory can be used.

The continuous girder is first made statically determinate by removing all its supports except two. The deflections due to the primary TSC effects at the points that are no longer supported are then calculated from the known curvatures. For example, there are $(n - 2)$ such deflections δ at the positions of the internal supports for the beam of $(n - 1)$ spans shown in fig. 8.30(a). A set of stiffness coefficients s_{pq} is then obtained for each internal support, by calculating the forces at the supports for unit deflection at that support only, with all other supports at the correct level, as shown for support 2 in fig. 8.30(b). In these calculations it would be appropriate to take account of variations of flexural rigidity along the member, if this is done in the analyses for other types of loading.

The reactions R_1 to R_n that reduce all the deflections δ_2 to δ_{n-1} to zero (fig. 8.30(c)) are given by the matrix equation

$$
\begin{bmatrix} R_1 \\ R_2 \\ \vdots \\ R_n \end{bmatrix} = \begin{bmatrix} s_{12} & s_{13} & \cdots & s_{1,n-1} \\ s_{22} & s_{23} & & \cdot \\ \vdots & \vdots & & \vdots \\ s_{n2} & s_{n3} & \cdots & s_{n,n-1} \end{bmatrix} \begin{bmatrix} \delta_2 \\ \delta_3 \\ \vdots \\ \delta_{n-1} \end{bmatrix} \tag{8.65}
$$

where s_{pq} is the upwards reaction at support p when load at the position of support q causes unit upwards deflection at q, all other supports not deflecting.

The 'secondary effects' are these support reactions R_1 to R_n and the associated moments, shears, stresses, and deflections. The term 'secondary' does not imply that the stresses are lower than the primary stresses; the ratio of the two depends entirely on the geometry of the structure, and the secondary effects may be the more important.

INFLUENCE OF SEQUENCE OF CONSTRUCTION OF THE DECK SLAB

The preceding methods of analysis have the disadvantage that no account is taken of the sequence of construction of the deck slab. As is shown in section X.7, they may give large negative moments at each internal support, which are assumed to be resisted by the composite section and so contribute to the widths of cracks.

In practice, the deck slab above internal supports is usually cast some time after the slab in midspan regions, so that the secondary effects of most of the shrinkage of the midspan regions can be assumed to be resisted at the internal supports by the steel section alone. The secondary support reactions R_1 to R_n then have to be calculated separately for the shrinkage due to each stage (or group of stages, depending on the construction timetable) of concreting, taking account of the variation of stiffness along the member at the time considered. The second method of the preceding section can be used for each calculation, and the resulting bending moments, etc., in the steel and composite sections are then summed separately. An example is given and discussed in section X.10.

It is obvious from the shape of the curve in fig. 8.30(a) that secondary moments due to shrinkage are negative over most of the length of a continuous member, and so increase the tension in the slab at internal supports. Calculations for continuous bridge girders, neglecting cracking, have found[114] that the total of the primary and secondary shrinkage stresses in the slab at internal supports is typically about 1 N/mm^2 per 100×10^{-6} free shrinkage strain. In practice this stress is sometimes reduced by jacking down the internal supports after construction of the deck (section 6.2.3 and plate 9).

8.7.3 Design for temperature and shrinkage to BS 5400

The following summary relates to conventional simply-supported and continuous beams, including box girders, with the concrete slab above the steel member, in which any difference between the coefficients of thermal expansion for steel and concrete can be neglected. For the structural steel, there is the usual association between the sign of the longitudinal stresses and the type of curvature stated.

The effects of shrinkage of concrete, analysed in section 8.7.2, can be summarised as:

(1) *Primary*: sagging curvature, but tension in the concrete.
(2) *Secondary*: hogging curvature.

The temperature differences type 1 and type 2 specified in Part 2 of the Bridge Code are such that type 1 governs, and has these effects:

(1) *Primary*: hogging curvature, but compression in concrete.
(2) *Secondary*: sagging curvature.

The requirements of Part 3 for checks on the temperature and shrinkage (TSC) effects are essentially as follows:

(1) At the ultimate limit state, sections are checked unless they are compact and have $\beta \leqslant 45$, where β is the slenderness function defined in section 8.4.4.

(2) At the serviceability limit state, sections are checked unless they are non-compact.

The TSC effects and the requirements of Part 3 are summarised in table 8.5. The entry 'adverse' means that the effects have to be considered in design to Part 3. No guidance is given in Part 3 on what to do when an effect is 'favourable'. It can of course be ignored, but if it is a consequence of an action that also produces an adverse effect, it may be more rational to include it. For example, the favourable secondary effects of shrinkage of concrete in a steel member at midspan can be used to offset the unfavourable primary effects.

On the other hand, it may not be appropriate at the ultimate limit state to use the favourable primary effects of shrinkage of concrete at an internal support to offset the unfavourable secondary effects. This is because local cracking of the slab may have almost eliminated the primary effects[112] (Appendix A), but the secondary effects, which are proportional to the integral of the primary curvature along the member, may be little altered.

Table 8.5 Design of composite beams for the effects of shrinkage and temperature

Section:		Compact		Non-compact	
At:		Internal supports	Midspan	Internal supports	Midspan
Shrinkage					
ULS	Prim.	Negligible due to cracking and yielding	Adverse in steel (1)	Negligible due to cracking (2)	Adverse in steel bottom flange
	Sec.	Adverse (1)	Favourable	Adverse	Favourable
SLS	Prim.	Favourable in steel	Adverse in steel	No check required	
	Sec.	Adverse	Favourable		
Temperature difference, type 1					
ULS	Prim.	Negligible due to cracking and yielding (3)	Adverse in concrete but neglect (4)	Negligible due to cracking (2)	Adverse in concrete but neglect (4)
	Sec.	Favourable	Adverse (1)	Favourable	Adverse
SLS	Prim.	Adverse in steel Unlikely to govern	Adverse in concrete but neglect (4)	No check required	
	Sec.	Favourable	Adverse Unlikely to govern		

Notes
(1) But may be neglected at ULS (Part 3) if $\beta \leqslant 45$.
(2) But required by Part 3 to be considered.
(3) But required by Part 3 to be considered if $\beta > 45$.
(4) Part 3 does not require the stress in concrete to be considered.

The methods of Part 3 are similar for the two limit states (with appropriate γ values). The primary effects, calculated including γ_{fL}, are split into three groups:

(1) The effects of the restraining forces applied to various parts of the section, as may be necessary to prevent longitudinal strain (e.g. tensile force P_t applied to the concrete slab in fig. 8.28).

(2) The equilibrating force P_t applied at the centroid of the composite section.

(3) The bending moment M_t applied to the composite section.

The 1982 amendment[2] to the treatment of this subject in Part 3 requires items (2) and (3) to be 'combined with other load effects as appropriate' and, for the serviceability limit state only, item (1) to be allowed for by modifying the design strengths of the steel member, σ_{yt} and $\sigma_{\ell i}$. This method is unsatisfactory for these reasons. Firstly, inclusion of item (2) means that the member is subjected to combined bending and axial load, and presumably has to be designed following the rather complex Clause 9.9.4 of Part 3. This is not necessary, for the total axial force in the member, including item (1), is zero. Secondly, items (1) to (3) are components of a single state of stress at a cross section that satisfies internal equilibrium. To take account of items (2) and (3), but not (1), is to design for a set of stresses that are not in equilibrium.

It is therefore correct (as in Part 3) to include item (1) in the checks at the serviceability limit state, but it should also be included whenever checks are made at the ultimate limit state. For shrinkage, the relevant stress is a tension P_t/A_c in the concrete only, and it is not clear from Part 3 whether the design strength of the concrete should be modified. For concrete in compression, it is on the safe side not to do so; for concrete in tension, cracking will relieve any overstress.

Temperature effects cause type (1) stresses in the steel member, as well as in the concrete slab. If, for example, the design (factored) temperature rise causes a type (1) compressive stress in a steel flange of σ_{tem}, the modified design strengths are:

for a tension flange, $\sigma'_{yt} = \sigma_{yt} + \gamma_m \gamma_{f3} \sigma_{tem}$
for a compression flange, $\sigma'_{\ell c} = \sigma_{\ell c} - \gamma_m \gamma_{f3} \sigma_{tem}$

The γ factors are included here because the calculated resistances are later divided by them.

CONCLUSION

It is considered that the appropriate treatment of the effects of temperature and shrinkage, in the context of BS 5400, is as follows. (Effects of shear lag and cracking are further discussed in Appendix A.)

(a) Members with compact cross sections only
Check at the serviceability limit state only.

In midspan regions, include items (1) to (3), and check against excessive stress in structural steel only, neglecting secondary effects.

Near internal supports, ignore items (1) to (3), but include secondary effects with the design moments and shears.

(b) Members with one or more non-compact cross sections
Check non-compact sections at the ultimate limit state only.

In midspan regions, as in (a) above, except that the secondary effects of temperature should be included.

Near internal supports, as in (a) above.

8.7.4 Design for temperature and shrinkage to draft Eurocode 4

The methods of Part 3 may be considered to be unnecessarily complex for effects that are small in comparison with those of the dead and imposed loads. They take no account of the progressive disappearance of primary effects that is associated with inelastic behaviour. The effects of shrinkage of concrete are likely to be much reduced by the phased casting of the deck slab, in a way that is too complex to follow in design calculations, and are further relieved by cracking and shear lag in the concrete (Appendix A).

Effects of temperature contributed to at least one of the box-girder collapses of the early 1970s[115] but no examples are known to the authors of failure in a member, not slender enough to need longitudinal stiffeners, that was influenced by primary TSC effects.

For these reasons, a simpler treatment of the subject is given in draft Eurocode 4. Checks are made only for secondary effects, only at the ultimate limit state, and only for non-compact members.

8.8 Deflection and vibration of composite bridge decks

The need to predict or to control the deflection or vibration of a composite superstructure varies from job to job. In very light structures, such as footbridges, design to limit vibration due to imposed loading (section 8.8.2) will usually ensure sufficient control of deflection, because pedestrians are more sensitive to vibration than to deflection. Highway and railway bridges usually have a high enough ratio of dead to imposed load for vibration to be no problem, because passengers in vehicles are less sensitive to it than are pedestrians.

8.8.1 Deflection

In certain bridges, excessive deflection may reduce clearance under the bridge to below some specified limit, or may prevent the drainage of surface water. Deflections due to dead and imposed loading must then be calculated. The elastic theory, neglecting concrete in tension, and nominal loads are used. For the reasons discussed below, accurate prediction of deflection is difficult, so a margin of clearance or drainage gradient should be provided, based on an estimate of the likely error in the prediction.

In large structures, prediction of deflection to the highest practicable accuracy may be needed during construction, particularly if prestressing or cantilever construction is used. The comparison of predicted and measured deflections during erection is sometimes an important method of checking that the structure is behaving as predicted. Here two sets of calculations may be done, based on 'optimistic' and 'pessimistic' assumptions, as this gives an indication of the accuracy of the result.

No general limits on deflections are given in Part 3 of the Bridge Code, other than one related to appearance rather than function. This is that the sagging deflection of a nominally straight soffit should not exceed span/800. This may make it necessary to camber rolled sections used for simply-supported spans. Plate girders are normally fabricated with camber.

SIMPLY-SUPPORTED MEMBERS

Where unpropped construction is used, and the span is short enough for the deck to be cast at one time, most of the dead load is carried by the steelwork. The influence of creep on deflections is small, so quite accurate prediction is possible. If deflection is critical for some reason, careful assessment of the effects of temperature and shrinkage should be made, as discussed in sections 7.3.4, 7.3.5 and 8.7. In uncased members, it would be prudent to increase the calculated deflection due to load carried by the composite member by up to 10% to allow for slip at the steel/concrete interface. There is some evidence[116] that relief of residual stresses in steelwork can increase deflection, particularly if the bridge has been subjected to a high proportion of its design load, or has been overloaded, and some designers make an allowance for this (section 8.4.8).

EFFECTIVE BREADTH

The use of an effective flange breadth for a T- or box-section (section 8.4.1) is intended to ensure that the extreme-fibre bending stress in the plane of each web is correctly calculated. The longitudinal curvature should also be correct (if slip is neglected) so in principle the integration of these curvatures along the span should give the deflection. But no simple expressions are available for the variation of effective breadth along a span, and the greatest complexity tolerable in practice, even when global analysis is by computer, is to use a constant effective breadth along any equivalent simply-supported span, and another constant value in each negative-moment region. Part 5 of the Bridge Code recommends that the effective breadth for distributed load at quarter span (and at the quarter-point near the support for cantilevers) be used throughout each span, as for the calculation of longitudinal shear.

CONTINUOUS MEMBERS

Deflection is usually less critical, but accurate prediction of it is more difficult. The comments above for simply-supported members are relevant, and, in addition, allowance must be made for the secondary effects of temperature and shrinkage, for phased casting of the deck, and for cracking of concrete.

Cracking modifies both the distribution of longitudinal bending moments (section 8.2) and the deflections due to a given set of moments. Rigorous calculations neglecting all concrete in tension would involve trial and error and be very tedious, and would give an overestimate of deflection, due both to the tensile strength of concrete and to the effect of tension stiffening between cracks, even though research[102] has shown the latter to be less in tension flanges of composite T-beams than it is in rectangular concrete beams. The following simpler method is suggested.

The negative moments at internal supports of continuous beams should be assumed to be as calculated in the analysis for the serviceability limit state, scaled down for those loads (e.g. HA) for which factors γ_{fL} greater than 1.0 have been used. These moments (M_1 and M_2 in fig. 8.31) will be based on the assumption that certain negative moment regions are 'cracked'. If this has been allowed for in design by increasing the midspan moments (the 'simple' method, section 8.2.1) without reducing the end moments, then a conservative estimate of deflection should be obtained if the end moments *are* reduced, for this purpose only, by amounts statically equivalent to the increases made in the midspan moments. If the analysis has been repeated, allowing for cracking over 15% of the span on each side of a support where the slab is 'cracked', then the moments so found need no further adjustment.

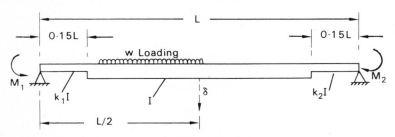

Fig. 8.31 Midspan deflection of non-uniform beam

In calculating deflections, it is suggested that the changes of flexural rigidity due to cracking and to change of effective breadth should both be assumed to occur at 15% of the span from an internal support. The analysis of every continuous span is thus reduced to the problem shown in fig. 8.31, for which a computer program can easily be written. End spans can be included by putting $M_1 = 0$ and $k_1 = 1$.

For a uniformly-distributed load w per unit length over the whole span, the deflection at midspan is given by:

$$\delta = \frac{5wL^4}{384E_sI}\left[1 + 0.0192\left(\frac{1}{k_1} + \frac{1}{k_2} - 2\right)\right]$$
$$- \frac{L^2}{16E_sI}\left[0.91(M_1 + M_2) + 0.009\left\{M_1\left(\frac{9}{k_1} + \frac{1}{k_2}\right) + M_2\left(\frac{1}{k_1} + \frac{9}{k_2}\right)\right\}\right]$$

$$(8.66)$$

where:

I	is the second moment of area of the uncracked composite section at midspan in 'steel' units
k_1, k_2	are factors that take account both of cracking and of the different effective breadths in negative moment regions
M_1, M_2	are the end moments, taken as positive when hogging.

The use of equation (8.66) for a typical internal span with $k_1 = k_2 = 0.60$ showed that cracking has a much greater effect on midspan deflection (increase of up to 20%) through the reduction that it causes in end moments than through the reduction in flexural stiffness near internal supports.

8.8.2 Vibration

Excessive vibration of a bridge superstructure due to imposed loading is a serviceability limit state. It is not considered in Parts 3 or 5 of the Bridge Code, because recommendations for bridges of all materials are given in an appendix to Part 2. Composite decks are usually too stiff to be susceptible to wind-excited oscillations, which are not considered further.

The dynamic effects of highway and railway loading are assumed to be adequately covered by the impact factors that are included in the nominal loads specified in Part 2. This is consistent with the conclusion reached from a study of steel bridges in the United States[118] which states that: 'there is no evidence of bridge motions producing discomfort of occupants of moving vehicles, so there appears to be no need for limits on deflections or accelerations of bridges which do not carry pedestrian traffic under normal conditions'.

Composite bridges intended primarily for use by pedestrians should be designed against the risk of excitation in the fundamental modes of vertical and horizontal vibration by groups of people moving in unison.

The fundamental undamped natural frequency of vibration, f, of a uniform simply-supported beam with relevant stiffness $E_s I$ and span L is given by elastic theory as:

$$f = (\pi/2)(gE_s I/WL^3)^{\frac{1}{2}} \qquad (8.67)$$

where I is calculated for the uncracked section using the short-term modular ratio and full breadth of the deck slab. The vibrating mass (W/g) should be based on the nominal dead load, and include finishes. In footbridges, the contribution of the parapets to the flexural stiffness may not be negligible. Equation (8.67) will give a low result ('safe' in this context) if the member is continuous.

If the natural frequency exceeds 5 Hz, no vibration check is necessary, as this is above the frequency range within which resonance can occur with people running or jumping, or with the vertical oscillations of lorries on their springs and tyres (fig. 8.32).

Human response to vibration has been studied by Allen (Volume 1) for floors in buildings and by Wright and Walker[118] for bridge decks. They conclude that maximum acceleration is the most consistent criterion of annoyance or discomfort.

For simple harmonic motion at frequency f and amplitude A, the maximum acceleration a_{max} is given by:

$$a_{max} = \pm 4\pi^2 f^2 A \qquad (8.68)$$

so this criterion corresponds to constant Af^2. For a given level of annoyance the limiting value a_{max} is much higher for a transient vibration than for steady motion. The static component of deflection due to a load crossing a bridge can be neglected, as it is a half cycle only, and the dynamic response can be considered as transient. This led Wright and Walker to propose the limit $a_{max} = 2.5$ m/s^2, which is shown as line AB on fig. 8.32. This corresponds roughly to the boundary between the responses 'perceptible' and 'unpleasant to some', and the authors comment that 'this acceleration will require that pedestrians be educated to expect to feel bridge motions'. It is ten times the acceleration given by line CD (from chapter 3) for composite floors in buildings with a damping ratio (logarithmic decrement) of 0.04.

The limit that was given in CP 117: Part 2 is $aA \not> 32.3$ cm^2/s^2, shown by line EF in fig. 8.32. The maximum velocity v is given by:

$$v = 2\pi f A$$

so from equation (8.68):

$$aA = v^2 \qquad (8.69)$$

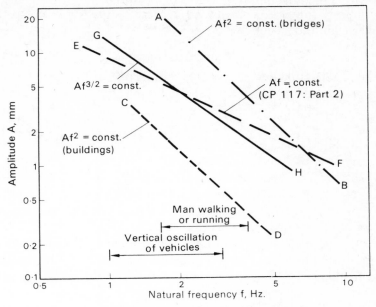

Fig. 8.32   Limitations on vibration

Thus line EF is a limit on velocity, rather than acceleration, and can be expressed as Af = constant. The limit given in Part 2 of the Bridge Code is:

$$a_{max} \not> \pm 0.5(f)^{\frac{1}{2}} \quad \text{m/s}^2 \tag{8.70}$$

This is of the form $Af^{\frac{3}{2}}$ = constant (line GH), so its slope lies between those in reference 118 and in CP 117: Part 2. It is more conservative than Wright and Walker's proposal for bridges, and gives similar values to CP 117 for the limiting amplitude. This limit is appropriate for normal use. It may be considered that a higher acceleration is acceptable when the bridge is deliberately excited by a group of people.

A method is given in Part 2 for estimating a_{max} in simple and in continuous spans of uniform section, due to the types of disturbance that can be applied by pedestrians. It can be used with equations (8.67) and (8.70) for checking whether vibration is likely to be excessive. Methods have been developed[119] for reducing the liveliness of footbridges.

The draft Eurocode 4 gives the following guidance on composite footbridges. Calculation of the maximum vertical acceleration is not required, but the logarithmic decrement may have to be estimated or measured:

'If $f \leqslant 5$ Hz, vibration sufficient to cause discomfort to some users is likely if two or more of the following conditions apply.

(1) The frequency f lies within the range 1.7 Hz to 2.2 Hz. (This is unlikely in spans less than 35 m.)
(2) The vertical stiffness of the deck at midspan is less than 8 kN/mm.
(3) The logarithmic decrement is less than 0.03.

'The relevant value of logarithmic decrement is that for the maximum displacement amplitude excited by a pedestrian walking across the span with paces at frequency f. A typical value is 0.04, but it can be as low as 0.02.'

Chapter Nine

Box Girders and Composite Plates

9.1 Introduction

Typical composite box-girder bridge decks and the appropriate methods of global analysis for them have been discussed and illustrated in chapter 6. There are two main types of box: 'open-top', in which a steel trough section is closed at the top by the concrete deck slab; and 'closed top', in which the top plate of the steel box acts compositely with the concrete deck in resisting imposed loads.

Little need be said here about the design of open-top boxes, because in most respects the appropriate procedures for steel or concrete structures are applicable, and the design of the shear connection between each web and the deck slab is essentially as for a beam-and-slab bridge (chapter 8). The two design problems discussed in this chapter are: the influence of torsional shear on the cracking of the slab, and of this cracking on torsional stiffness; and the need to consider the effect of uplift forces and torsional shear on the strength of the connectors in longitudinal shear.

Closed-top boxes occur in composite bridges in a wide range of sizes. The terminology used here is that a 'small' box is one small enough not to need longitudinal stiffeners (that is, less than about 1.4 m wide, as in plate 2 and fig. 9.2), and a 'large' box is longitudinally stiffened, as in the Raith Bridge, plate 4, and the Tinsley Viaduct, plate 6.

Large boxes can be up to 6 m wide and at least 8 m deep, and the all-steel boxes are the subject of an extensive literature. This chapter is concerned with aspects of their behaviour relevant to composite construction, and with small closed-top boxes.

9.2 Design for longitudinal moments, torsion and vertical shear

9.2.1 Analysis of the structure

For preliminary design, it is possible to neglect distortion of the cross

181

sections of the boxes, which reduces torsional stiffness, and the effects of warping restraint, which increase it. As box-girder decks have good distribution properties, static distribution of loading to each longitudinal steel web will certainly be conservative. At this stage, it is possible to consider each box and the full breadth of its associated concrete flange as an isolated girder, and to calculate longitudinal moments and vertical shears in continuous members by the methods given in the Bridge Code for beam-and-slab decks, which have been discussed in chapter 8.

The torsional moments in each box due to eccentric loading on single-box decks, or to curvature of the deck in plan, can be estimated by simple approximate methods.[46]

The preliminary design for the superstructure is then analysed, for the various loadings expected to be critical, by one of the methods of global analysis discussed in section 6.4, and is then modified and re-analysed until all of the various design criteria are satisfied. More detailed analysis of parts of the structure are then carried out, to check that the additional stresses due to any secondary effects neglected in previous analyses are not excessive.

INFLUENCE OF CRACKING OF CONCRETE AND OF SLIP

The extent to which these effects need be considered in global analyses of box structures is now discussed. Interactions between flexural and torsional moments depend mainly on ratios of these stiffnesses in each region, and are not sensitive to their precise values. As the following example shows, the contribution of a deck slab to the torsional stiffness of a closed box is small, but for an open-top box it is significant.

Box members are usually non-compact, as defined in the Bridge Code, so their steelwork is designed for the ultimate limit state on an elastic basis. Yield of steel occurs only in regions too limited to have much effect on flexural or torsional stiffness. Flexural cracking of concrete is almost certain to occur over internal supports, so that appropriate reductions in flexural stiffness should be made. It is likely to be caused mainly by symmetrical loadings. Maximum torsional moments in these regions occur under different arrangements of loading, which cause lower flexural moments. The pre-existing flexural cracks then do not cause much loss of torsional stiffness, because of aggregate interlock; there will be loss of stiffness only if the torsional shears are themselves large enough to cause further cracking.

For closed-top boxes in which, as is usual, the thickness of the steel top plate is at least as great as that of the webs, it is accurate enough to neglect cracking and slip, and include the uncracked unreinforced slab in calculations of torsional stiffness.

For open-top boxes, use of the method for closed-top boxes should be accurate enough at the serviceability limit state, but is likely to overestimate torsional stiffness at the ultimate limit state. Calculated torsional moments are then too high, and bending moments too low.

There is little relevant research. It was concluded[120] from research on the open-top box-section members used in the Mersey tunnel approach viaducts that cracking and slip should be considered in assessing torsional stiffness. Warren[121] tested three open-top and three closed-top model box girders in combined hogging bending, shear and torsion, and did numerical analyses. He found that for one open-top box, the reduction in torsional stiffness, after cracking occurred, increased from 22% in pure torsion to 37% when the maximum bending moment was about double the torsional moment. The increase was attributed to the influence of the ratio of torque to bending moment on the directions of the cracks. It was found that warping rigidity was sensitive to the stiffness of the shear connection, and that in the model tests, torsional stiffness fell more rapidly than flexural stiffness as ultimate load was approached.

It seems unlikely that the in-plane shear stiffness of a slab could drop at the ultimate limit state to below about one-third of the uncracked value. The form of equation (9.1) shows that for an open-top box of square section, this change would reduce the torsional stiffness of the box, C, by about 30%. It is concluded that the design of such a structure should be robust enough to accommodate the effects of an uncertainty of this order.

9.2.2 Analysis of cross sections and of plate panels

Most, but not all, composite box girders have longitudinal stiffeners that span between crossframes. Their treatment in Part 3 is based on these principles:

(1) The crossframes must be strong and stiff enough to prevent overall buckling of the relevant flange or web and to resist end reactions from the longitudinal stiffeners.

(2) The longitudinal stiffeners must be strong and stiff enough not to buckle, so that the edges of the plate panels between them are either simply supported against out-of-plane deflection, for open-section stiffeners, or supported and restrained against rotation, for closed-section stiffeners.

(3) Account must be taken of local buckling of slender plate panels.

The design methods given in Part 3 for crossframes, stiffeners and plate panels are as relevant to steel girders as to composite girders, and are mostly outside the scope of this book. Reference is made only to subjects where composite behaviour is relevant.

It is recognised in Part 3 that stable yielding in tension, in both flanges and webs, and stable local buckling in stiffened webs, are acceptable at the ultimate limit state. Provision is therefore made in the design of longitudinally stiffened beams for redistribution of both longitudinal force and bending stresses from webs to flanges, and of tensile force from steel flanges both to webs and, in composite members, to reinforcement in the slab. Redistribution from webs is also allowed at the serviceability limit state.

REDISTRIBUTION OF WEB STRESSES

In a composite member, redistribution is likely to be done, at either limit state, because the combination of longitudinal and shear stress in a web at a web-flange junction, as calculated by elastic theory, is found to be excessive.

Clause 9.5.4 of Part 3 allows longitudinal stress to be reduced by up to 60% by shedding 'moment and/or axial force'. The percentage reductions are required to be 'uniform within any one panel'. This appears to mean within the length of a panel, not within its depth, because the latter would require axial force and bending moment to be reduced in the same proportion.

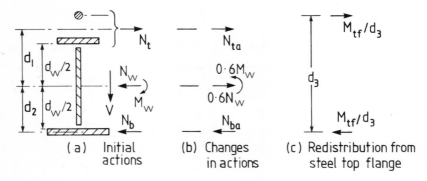

Fig. 9.1   Redistribution of longitudinal stresses

As an example, we consider a composite section subjected to hogging bending. The relevant steel cross section is separated into three areas. The actions on each area before redistribution are shown in fig. 9.1(a), with their lines of action. The maximum changes that can be made in accordance with Part 3 are shown in fig. 9.1(b). The changes must satisfy equilibrium, so that:

$$N_{ba} - N_{ta} = 0.6N_w \qquad (9.1a)$$

and:

$$N_{ta}d_1 + N_{ba}d_2 = 0.6M_w \qquad (9.1b)$$

These equations enable N_{ba} and N_{ta} to be found. The flanges must be capable of resisting the increased forces without exceeding the relevant limiting stresses. There is no requirement that the stresses must be consistent with a linear distribution of longitudinal strain.

The background to this method has been explained by Horne.[122] The limit of 60% results from the limitation of the shear strain in the web to twice the yield strain, to avoid invalidating the assumptions on which the design methods for stiffened flanges are based.

REDISTRIBUTION OF TENSION-FLANGE STRESSES

This type of redistribution is applicable at the ultimate limit state only. It is useful in hogging moment regions when the slab is cast at a late stage of the construction sequence, so that most of the dead-load effects are resisted by the steel section alone. It is required that the stresses after redistribution should be consistent with a linear strain distribution and the relevant stress–strain curves, and should satisfy the other limitations on stress at the ultimate limit state; also, the strain in the tension flange should not exceed twice the yield strain divided by $\gamma_m\gamma_{f3}$.

If, as is likely in a composite member, the web at the section considered is already highly stressed, it is preferable not to redistribute longitudinal stress to it. As an example, we consider the cross section shown in fig. 9.1(a). Suppose that when account is taken of sequence of construction, the steel top flange is found to reach its limiting stress, $\sigma_{yt}/\gamma_m\gamma_{f3}$, when the live-load bending moment falls short of the design value by M_{tf}, and that stresses elsewhere are then below their limiting values. The extra bending moment can be assumed to be resisted by equal and opposite forces, M_{tf}/d_3, in the slab reinforcement and steel bottom flange (fig. 9.1(c)). The associated extra tensile strain in the steel top flange results from yielding at constant stress. The additional strains in the web are consistent with a linear distribution of longitudinal stress, which is assumed to have been shed to the flanges in the manner explained above.

It is required in the Bridge Code that the shear connection be designed for the additional longitudinal shear. One could, for example, calculate the length over which redistribution is needed, and divide the maximum additional force in the slab reinforcement by this length to find the extra longitudinal shear per unit length. In fact, no adjustment is necessary, for two reasons.

(1) The maximum vertical shear, for which the shear connection is designed, invariably exceeds the shear associated with the maximum hogging bending moment, so a margin already exists in relation to the load case for which redistribution is done.

(2) When the strain in the top flange exceeds the yield strain, the tensile force in the slab will be far less than that calculated for the uncracked section, for which the shear connection is designed.

EFFECTIVE BREADTH

The rules for effective flange breadth given in Part 3 of the Bridge Code (discussed in section 8.4.1) are based on elastic analyses of box sections.[55] They are given for general use because checks showed that they could also be applied with little error to concrete slabs acting compositely with steel I-beams. Where the top flange is a composite plate, shear lag effects are least when the longitudinal shear is transferred to the concrete slab above the steel web (as was assumed in reference 55). The effects of spacing the shear connectors across the breadth of the plate are discussed in section 9.3.1.

EFFECTIVE CROSS SECTION FOR BENDING STRESS ANALYSIS

The bending resistance of a web without longitudinal stiffeners is assumed to be reduced by local buckling if the slenderness ratio for the region in flexural compression, y_c/t_w, exceeds $68(355/\sigma_{yw})^{\frac{1}{2}}$. Account is taken of this in design to Part 3 by using an effective web thickness given by:

$$t_{we}/t_w = 1.425 - 0.00625(y_c/t_w)(\sigma_{yw}/355)^{\frac{1}{2}} \qquad (9.2)$$

The resistance of a slender steel compression flange is also reduced by local buckling if the lines of restraint to out-of-plane deformation (webs or stiffeners) are too widely spaced. Account is taken of this in design to Part 3 by using either a reduced design compressive stress, or an effective area less than the actual area of the flange between lines of restraint. This effective area, applicable at the ultimate limit state, should not be confused with the effective breadth that allows for shear lag (which may be neglected at the ultimate limit state).

As an example, we consider a midspan region of a typical box girder for the Saltings Viaduct,[36] and assume that the design yield strength of the steel was 355 N/mm^2. The structure has an *in situ* concrete deck composite with four boxes of the cross section shown in fig. 9.2, which have centre-to-centre spacings of 5.0, 6.2 and 5.0 m. The maximum span is 31 m. When carrying the wet concrete slab, the unstiffened steel

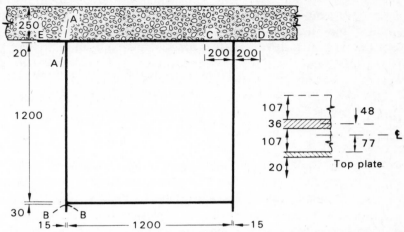

Fig. 9.2   Typical cross section of one box girder for Saltings Viaduct

top plate between the webs has a relevant slenderness, $(b/t_f)(\sigma_{yf}/355)^{\frac{1}{2}}$, of 1200/20, which is 60. The spacing of cross beams was greater than 2.4 m, so the lowest buckling mode for the plate consists of nearly-square dimples of side about 1.2 m. The buckling coefficient K_c for this slenderness is then given by fig. 5 of Part 3 as 0.49. It can be assumed that lateral buckling cannot occur ($\lambda_{LT} < 45$). For design, the plate is assumed either to have an effective area of $0.49bt_f$ and a limiting compressive stress of σ_{yc}, or its full area and a limiting stress of $0.49\sigma_{yc}$.

This section is non-compact during construction. For the composite section, the elastic neutral-axis is so high that the webs become compact, and the top flange can easily be made so by appropriate spacing of the lines of shear connectors. The relevant limits to b/t_f are 7 for the outstands and 24 for the region between the webs. This is why Part 3 limits the lateral spacing of rows of connectors to $24t_f(355/\sigma_{yf})^{\frac{1}{2}}$, or 480 mm in this structure. The longitudinal spacing must not exceed half the value, or three-quarters of it if adjacent lines are staggered. When these requirements are satisfied, Part 3 gives K_c as 1.0, so that no account need be taken of local buckling of the plate in design of the composite section.

TORSION

The classical St. Venant and Bredt theories are applicable. Except where the boxes are small, the torsional rigidity of the slab between boxes and of the cantilever edge strip is likely to be negligible in comparison with that of the steel box and the strip of slab between its webs. The calculation of the torsional rigidity and the torsional forces for which the shear connection must be designed is best illustrated by an example, again using the Saltings Viaduct. The influence of the deck slab on the torsional stiffness of a typical box is calculated, assuming that the top and bottom flange plates have the thicknesses shown in fig. 9.2 (they vary along the span) and that the modular ratio is 7. It is convenient to work in 'steel' units, using the method of transformed sections.

The torsional rigidity of a closed thin-walled elastic box is:

$$C = \frac{4GA^2}{\int (ds/t)} \tag{9.3}$$

where A is the area to the centreline of the walls and t is the wall thickness which is a function of s, the co-ordinate round the box perimeter. The small steel flanges cut off by the lines A-A and B-B in fig. 9.2 can obviously be neglected, being 'open' sections, and the slab between boxes is also neglected initially. The equivalent thickness of the top plate is $20 + (250/7) = 56$ mm. Assuming the transformed concrete to be concentrated at the mid-depth of the slab, calculation of first moments of area gives the effective centreline of the top plate as 77 mm above the slab soffit (inset on fig. 9.2). The depth of the box is thus $77 + 20 + 1200 + 15 = 1312$ mm. Its breadth is 1215 mm, so its area is 1.59 m^2, and:

$$\int \frac{ds}{t} = \frac{2 \times 1312}{15} + \frac{1215}{30} + \frac{1215}{56} = 237$$

For steel, $G = 78.9$ kN/mm^2, so from equation (9.3),

$$C = \frac{4 \times 78.9 \times (1312 \times 1215)^2}{237 \times 10^6} = 3.38 \times 10^6 \text{ kNm}^2$$

This is 28% higher than the value for the steel box alone. The torsional stiffness of the strip of deck slab 3.8 m wide between adjacent boxes is:

$$C_d = \frac{1}{3} Gbt^3 = \frac{78.9 \times 3.8 \times 250^3}{3 \times 7 \times 10^3} = 0.22 \times 10^6 \text{ kN m}^2$$

It should be noted that the full concrete thickness has been used, with the shear modulus for concrete, here taken (approximately) as one-seventh of that for steel. The figures show that for these relatively small boxes, the contribution of the deck to the total torsional stiffness is small but not negligible.

The stresses due to a torque of 1000 kNm are now calculated, considering only the box. The mean shear stress τ in a wall of thickness t is given by $T = 2A\tau t$.

The highest stress in the steel occurs in the web, and is:

$$\tau_{web} = \frac{1000}{2 \times 1.59 \times 15} = 21.0 \text{ N/mm}^2$$

For the steel top flange:

$$\tau_{flange} = \frac{1000}{2 \times 1.59 \times 56} = 5.6 \text{ N/mm}^2$$

The mean shear stress in the slab is found by dividing by the modular ratio:

$$\tau_{slab} = 5.6/7 = 0.8 \text{ N/mm}^2$$

The slab is 0.25 m thick, so the torsional shear flow to be resisted by the shear connection is $0.8 \times 250 = 200$ kN per metre run. In this calculation it is implied that the shear connection is concentrated above each web. If the shear connectors for each web were in fact spread over a width such as CD (fig. 9.2), the error would be negligible, but if they were uniformly spaced across the whole flange ED, those near the webs would be overstressed (as explained below) and the torsional stiffness would be less than calculated.

If the slab in this example were cracked sufficiently to halve its stiffness in shear, the torsional rigidity C would not be much altered, because the main contribution comes from the steel box. The effective thickness of the top plate would be reduced from 56 to 38 mm, but it remains greater than those of other walls of the box.

In open-top boxes resisting primary (i.e. statically determinate) torsional moments, the shear flow in the concrete slab is not reduced at all by cracking, so that torsional effects must be considered carefully when designing the slab reinforcement.

9.3 Longitudinal shear

In calculations for longitudinal shear due to the overall (global) loading, it is assumed that the shear is transferred to the slab only along lines above the webs of the steel girder. In practice, connectors must be distributed across the breadth of the steel flange and in closed-top boxes, allowance is made for the resulting additional shear lag, as explained in section 9.3.1.

If a simple method of global analysis has been used, design values of vertical shear and torsional moment for the loading considered will be available for each cross section of the box girder. The longitudinal shears due to bending are calculated from the vertical shears in the same way as for beam-and-slab decks (section 8.5.2) and their distribution between webs is easily found by considering the effective breadth of flange associated with each web. The shears due to torsion are found by the usual Bredt theory for thin-walled boxes, as in the preceding example. The total design shear per unit length of web is the algebraic sum of these, plus the shears due to restrained warping and to local loading, if these are significant.

A more elaborate global analysis, that takes account of shear lag and the effects of warping, can give the required longitudinal shear directly, and also the coexisting transverse shear and uplift (if any) per unit length of web.

If the transverse shears are not negligible, it is convenient to use studs or high-strength bolts as shear connectors, as these are the only standard types that are equally stiff and strong in all directions in the plane of shearing. The longitudinal and transverse shears per unit length are added vectorially to give the design shear per unit length. Any uplift force is allowed for as explained in section 7.2.3.

In open-top box girders, where the steel flange is relatively narrow, the shear connection and transverse reinforcement can then be designed in the usual way, exactly as for composite I-beams.

9.3.1 Design of shear connection in closed-top box girders

It is usual to provide shear connection over the whole breadth of the top plate of a closed-top box girder (e.g. as shown in plate 6), so that the plate can act compositely with the deck slab in resisting local loads, and to enable the slab to prevent the steel plate from buckling in midspan regions. These subjects are discussed in section 9.4. It is convenient to consider first the design of the shear connection for the longitudinal shear due to global bending and torsion.

The way in which a given number of shear connectors are arranged across the breadth of the steel plate has an important influence both on the way in which the longitudinal shear is shared between the connectors, and on the shear lag in the steel plate and concrete slab.

Both slip and shear lag cause loss of interaction in composite closed-top box girders, and their effects are interrelated, as shown by the following example.

The compression flange of a closed-top box consists of the steel plate AB and a thin concrete slab CD shown (separated) in fig. 9.3. We assume first that the shear connectors are all placed over the webs, and then that they are all placed at point E. The loss of interaction between

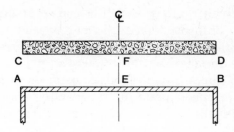

Fig. 9.3   Shear lag and slip in a composite plate

steel and concrete depends essentially on the longitudinal displacement, relative to the web, of the slab above the web. When the shear connectors are placed at A and B, this displacement is simply the slip. When they are placed at E there are in addition the effects of the shear lag in breadth AE of the plate and in breadth FC of the slab. Also, the shear lag in the plate is greater than before, because the shear stresses include the whole of the shear transmitted to the slab.

If half of the connectors are placed at A and B, and half at E, analysis shows that the connectors at A and B transmit nearly all the shear, and so carry almost twice their design load.

The design method of Part 5 of the Bridge Code is based on the results of a detailed study[123] of this overloading. Finite-element partial-interaction elastic analysis was used to calculate the longitudinal and lateral forces on shear connectors in simply-supported box and T-section girders in which the top flange was a composite plate. Different fractions (n_w/n) of the total number of shear connectors (n) were assumed to be concentrated above the webs, and the remainder were distributed uniformly across the whole breadth of flange associated with the web considered.

The distribution of longitudinal force on the connectors across the breadth of the flange was found to vary little with the breadth:span ratio of the girder, the position along the span, or the type of loading. It is given approximately by

$$Q_x = \frac{q}{n}\left[K\left(1 - \frac{x}{b}\right)^2 + 0.15\right] \qquad (9.4)$$

where Q_x is the load on a connector at a distance x from the web (fig. 9.4), q is the longitudinal shear per unit length at the steel-concrete interface and, for the web considered, K is the function of n_w/n, given in fig. 9.5, where n is the total number of connectors per unit length associated with the web considered. It is assumed that all the connectors are of the same type and size.

For the symmetrical girders analysed, the transverse forces on the shear connectors over the webs were always zero, and reached significant values only near the edges of the plates furthest from the web. The longitudinal forces are least on the connectors placed in these regions, so that transverse forces on the shear connectors over the webs were always zero, and reached significant values only near the edges of the plates furthest from the web. The longitudinal forces are least on the connectors placed in these regions, so that transverse shear forces due to global loading are unlikely to influence the design of box-girder decks without composite cross-girders.

It was also concluded[123] that in order to limit the loss of interaction, due to slip, to a level consistent with that found to be acceptable in T-beams, all the connectors provided to resist longitudinal shear should be placed within the effective breadth of each flange. The design method given in the Bridge Code appears not to conform with this last recommendation. This anomaly will be discussed after the method itself has been explained.

In girders of the type shown in fig. 9.4, the effective breadths b_{e1} and b_e are calculated separately, but for the present purpose it is necessary to consider as one the whole breadth of slab (ABC) associated with the web BD, because the connectors for region AB have to be placed within BC. This causes some additional shear lag, not considered by Moffatt. In this respect the detail at the box corner shown in fig. 9.2 is better than that in fig. 9.4.

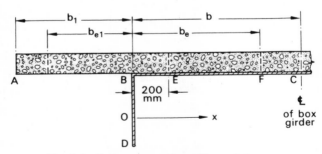

Fig. 9.4 Part cross section of box girder

The concentration of connectors 'over the web' is convenient in theory but not in practice, so in Part 5, n_w was defined as the number of connectors per unit length placed within 200 mm of the relevant web. This arbitrary distance was chosen such that all connectors on stub flanges (fig. 9.2) are likely to qualify for inclusion in n_w.

The design of the shear connection is likely to be based on the assumptions that the most heavily loaded connectors will be those placed over the web, and that these resist their design longitudinal shear force, P_d. Putting $x = 0$ and $Q_x = P_d$ in equation (9.4) gives:

$$n = q(K + 0.15)/P_d \qquad (9.5)$$

As shown in fig. 9.5, K ranges from 0.85 to 3.85, depending on n_w/n, so for

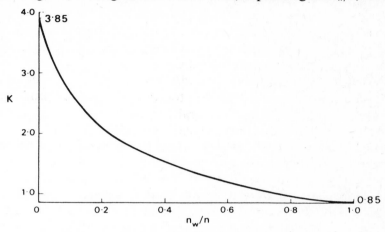

Fig. 9.5   Values of K for use in equation (9.4)

economy the highest possible proportion of the connectors should be placed over the web. This conflicts with the need to have some connectors spread uniformly across the plate to resist local loads. The following procedure for a typical cross section is suggested for use in practice.

DESIGN PROCEDURE

It is assumed that if there are significant forces tending to cause uplift or transverse shear forces along the centreline of the web, allowance will be made for these initially, by increasing the calculated design longitudinal shear q, as explained in section 7.2.3. The mean effective breadth of the composite plate (b_e in fig. 9.4) is also known initially.

(1) Calculate the number of shear connectors per unit area of composite plate, n_p say, required for local effects (section 9.4) and to prevent the steel plate from buckling, and provide these in regions outside the effective breadth (FC in fig. 9.4). The number per unit length is therefore:

$$n_0 = n_p(b - b_e) \qquad (9.6)$$

(2) Estimate n_w/n, find K from fig. 9.5, and calculate n from equation (9.5).

(3) Arrange $(n - n_0)$ connectors within the breadth b_e such that n_w/n is not less than the value assumed and the connector density (number per unit area) is nowhere less than n_p. If this is impracticable, return to (2) and choose a new ratio n_w/n. Normally, not more than two different densities are likely to be used, with the change of density at E or F (fig. 9.4), but in wide boxes it may be worth using different densities in BE, EF and FC.

(4) For approximate values of x, calculate the longitudinal shear per connector from equation (9.4) and combine it with the coexisting shear force due to local loading, to check whether connectors not above the web are overloaded. If they are, provide more connectors as appropriate.

TOTAL SHEAR PER UNIT LENGTH TRANSFERRED BY THE CONNECTORS

Moffatt's recommendation[123] that all the connectors provided to resist longitudinal shear should be placed within the effective breadth is obviously satisfied by this design method when the effective-breadth ratio ψ is 1.0. For ratios below 1.0 the method is still presented in terms of the connectors on the whole breadth of flange, because it is based on analyses which included shear flexibility of flanges and so had no need to use effective breadths. To prevent an unsafe error occurring due to the placing, outside the effective breadth, of a higher proportion of the shear connectors than was considered by Moffatt, it is required in Part 5 of the Bridge Code that the connector density provided in any area outside the effective breadth should not exceed the least density provided within the effective breadth. It will now be shown that when this condition is satisfied, the total shear transmitted by the connectors within the effective breadth is never less than about $0.9q$, where q is the full-interaction design shear, as defined above.

The n_w connectors placed within 200 mm of the web are each assumed to transfer a shear force P_d. The forces in the other connectors, assumed to be spread uniformly across the whole breadth b of the flange, can be obtained from equation (9.4). By integration it can be shown that q_p, the total shear transmitted by the connectors within the effective breadth, is given by:

$$q_p/q = v(K + 0.15) + (1 - v)[0.15\psi + 0.33K\{1 - (1 - \psi)^3\}] \quad (9.7)$$

where v is the ratio n_w/n, and K is as given in fig. 9.5. The ratio q_p/q is plotted in fig. 9.6 for values of ψ that cover the range used in practice. For $\psi = 1.0$, the proportion of the total shear resisted by the connectors n_w is also shown. It is to be expected that for some values of v and ψ, q_p is

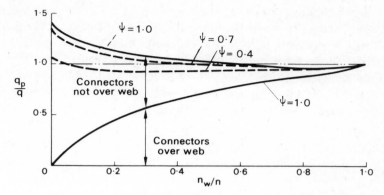

Fig. 9.6 Forces in shear connectors in composite plates

slightly less than q, because q is the full-interaction shear. Partial-interaction analyses always show some loss of interaction, however many connectors are used, due to the effects of slip.

9.4 Composite plates

In composite beam-and-slab bridges and in open-top box girders the steel flanges attached to the concrete slab are narrow, so that composite action in the direction at right angles to the span of the steel member can be neglected. Where the steel flange is wide, as in a closed-top box girder, composite action with the slab occurs in both longitudinal and transverse directions, and the slab and top flange together form a composite plate. The slab may be of uniform thickness (fig. 9.10), variable thickness (fig. 6.2), or may have a sudden change in thickness at the edge of the steel box (fig. 6.9), but it is unlikely to be precast because of the difficulty of providing uniform contact between the slab and the steel flange (section 6.2.1).

The composite plate has to be designed both for the overall compression or tension and in-plane shear given by the global analysis, and for any coincident local loading (e.g. HB wheel loads in highway bridges). In regions where steel stiffeners, stringers, crossframes or diaphragms are attached to the underside of the steel plate, local loading causes in-plane forces in the composite plate as well as bending and vertical shear.

To resist shear due to local loads, prevent separation and buckling of the steel plate, and reduce the risk of corrosion, shear connectors are provided over the whole area of the composite plate at not more than 600 mm spacing in each direction (plate 6).

Where a composite plate is used for only part of a bridge deck, the concrete slab is usually made the same thickness as in the remainder of the deck, which in Great Britain is rarely less than 200 mm.

Composite plates have been used for the whole area of the deck of several long-span bridges in continental Europe. In these, the concrete slab is much thinner, in relation to the steel plate, than in British practice. The Cologne-Deutz Bridge over the Rhine (1949) has a 120-mm slab composite with a top plate varying from 12 to 45 mm thick and Pont St Christophe at Lorient, France (1960) has a 60-mm slab on a plate between 8 and 20 mm thick. In both decks the shear connection consists of bent bars or flats welded to the plate, that also act as reinforcement for vertical shear; and the steel plate is supported on longitudinal stringers spaced at only 0.73 m (Cologne) or 1.0 m (Lorient). In the similar deck of total thickness 98 mm used in a suspension bridge at Bordeaux (1967) the stringers were at 2 m spacing.

The type of shear connection used in these decks becomes complex and expensive in regions where there is two-way bending of the composite plate, and does not seem suitable for regions where reversal of shear occurs. The most significant development of this type is at Tancarville Bridge (section 6.3.5), where stud connectors were used, at a spacing greater than the thickness of the composite plate. Results of tests on this type of plate are given in section 9.4.1.

In continuous box-girder bridges, composite plates have been used as the bottom (compression) flange adjacent to internal supports less often than one might expect, for there is little increase in the dead-load bending moment, and the concrete can be placed without difficulty. This method was used in the Pont d'Illarsaz, Switzerland (fig. 6.6), where the bottom of the box was filled with concrete to a depth of 0.7 m over each internal support, tapering to a minimum depth of 0.2 m at 12 m each side of the support. The design of such plates is simpler than that of deck plates, as no local loading need be considered, and is not discussed here. The use of composite plates as diaphragms over supports[24] also has obvious advantages.

9.4.1 Research on composite plates

A detailed investigation of the use of steel plates as reinforcement for concrete beams and slabs was completed in 1957.[124] Wire mesh was found to need too much welding to the steel plate to be effective as shear connection, and the welds in the fabric itself were found to be too weak. Welded stud connectors were then used, as it was found that small slips had no effect on the degree of composite action. Some beams failed in shear after excessive slip at overloaded connectors, and it was concluded that these should be spaced according to the shear force diagram. Design rules were given for shear connectors and for the reinforcement of negative moment regions.

An important difference between a simply supported composite plate and a composite beam is that in the plate, the base of the connector (where most of the shear is transferred) is in a region of high tensile strain. The shear connection is therefore less stiff, and a further rapid loss of stiffness in shear is to be expected in regions where the plate yields. Fortunately, the supported edges of composite plates in bridge decks usually lie in regions where the local moments are negative (so that the concrete at the base of the connectors is in compression), due either to continuity of the plate itself, or to the weight of a non-composite cantilever edge slab. (It is noted in passing that the sudden change of effective cross section in the transverse direction over an outer web (point B in fig. 9.4) will cause additional load on the shear connectors near the edge of the plate that should be considered in design.)

The relative weakness in vertical shear of composite plates spanning one way was observed in tests both by Gogoi[14b] and Maeda and Matsui[125] but the problem is less severe in bridge decks where the most severe vertical shear is that due to wheel loads, because these are carried by two-way bending, so the shear perimeter is relatively longer. Maeda and Matsui tested plates with steel on both faces of the concrete, and noted that closely spaced connectors were needed to stabilise the top plate at midspan. In a more recent set of tests[126] on such 'sandwich' plates, epoxy resin was used for the shear connection, and tensile failure of the lower plate or shear failure always occurred before yield of the top plate. Punching shear failures occurred at loads of 270 and 280 kN in specimens with a 90-mm concrete slab and two steel plates each 3.18 mm thick. It was found to be difficult to make a good joint between the slab and the steel top plate, and it was concluded that plates with steel on only the lower face show more promise.

This conclusion was in effect reached in France 18 years earlier. In 1958, two prototype composite plates for the Tancarville Bridge, each 6.15 by 5.32 m, were tested to failure[43] under four 'wheel' loads each applied through an oak block on an area 380 by 130 mm (fig. 9.7). The concrete was 95 mm thick, with a cylinder strength of 47 N/mm^2 when tested. The 10-mm steel plate formed the top flange of stringers at 2 m spacing. There were ten 16-mm Nelson stud connectors in each row of length 2 m, at spacings ranging from 155 to 280 mm. The spacing of rows was 110 mm in the first test and 150 mm in the second. The four equal loads were applied by jacks. Results are given as multiples of the design

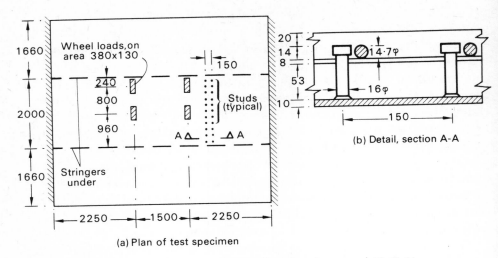

(a) Plan of test specimen

(b) Detail, section A-A

Fig. 9.7   Test on prototype composite plate for Tancarville Bridge

wheel load (W) of 65 kN.

In the first test, cracks became visible at $7W$, but these disappeared on unloading. Failure, by crushing of the slab, occurred at $10.6W$, when the cracks were about 1 mm wide. No deformation of the studs was observed. Results of the second test were similar except that failure occurred at $10.4W$; that is, at a total load of 2700 kN on the four wheels.

No attempt was made in these tests to simulate the global loading that may coexist with local wheel loads, but even so, the results show that the static strength of a composite plate about 100 mm thick should be adequate for British HA and HB loading.

The punching shear strength of this plate, treated as a reinforced concrete slab, is given by Part 4 of the Bridge Code as about 255 kN per wheel (cf 650 kN in the test). Another method of calculation is to assume that failure occurs when the mean interface shear across the critical perimeter for punching shear reaches the design strength of the shear connectors. The connector density in the second test was 33.3 per m², and values from Part 5 of the Bridge Code give an ultimate load of about 450 kN per wheel. This method is conservative in this case, because the shear connectors did not fail at a load of 650 kN.

Clarke and Morley[127] did static and fatigue tests on 12 one-way composite plates, studied the fatigue of connectors under rotating shear (section 10.5.1), determined the shear experienced by a connector due to the passage of a wheel load across a rectangular plate, and showed that when allowance is made for slip, the shears are lower than those given by full-interaction analysis. They did not consider cracking of concrete, so their conclusions are relevant to regions where there is global compression in the slab.

Allowance was made for cracking in elastic linear partial-interaction finite element analyses of two-way plates by Moffatt[128] and then Lou[129] developed an iterative method of analysis for the same problem. They found that the distribution of interface shear over the area of the plate is much altered by cracking and that the shear is in some areas higher than that given by full-interaction analysis. The maximum stress in the concrete in a simply-supported square plate under distributed load was found to occur midway between an edge and the adjacent quarter-point, and to be almost double the maximum stress given (at midspan) by full-interaction analysis; but, full-interaction theory was found to be suitable for calculating stresses in the steel plate. These analyses apply to slabs without membrane force due to global bending.

The authors have not found any research on the elastic behaviour of composite plates subjected to global tension, as in negative moment regions, and there are other unexplored problems relating to behaviour at the ultimate limit state. In all of the reported research, both experimental and analytical, the plates and loadings studied have been simpler than those that occur in practice. For example, combinations of global and local effects have rarely been considered and in test specimens, unstiffened steel plates have been used. Realistic laboratory research on this subject would be very expensive. There has been little, if any, monitoring of thin composite plates in actual bridges, but information is available on their behaviour in service.

Engineers responsible for the maintenance of the Tancarville Bridge, F. Hanus and B. Deroubaix, have kindly provided the following information. In 1977, the bridge had been in service for about eighteen years, and had been crossed by about four million heavy vehicles. The 12 000 m² of asphalt surfacing was replaced in 1976; until then, only 80 m² of deck had required minor repairs. During resurfacing, careful observation revealed fine cracks in the underlying slab. Its condition was satisfactory except over an area of 70 m², which was replaced. If the bridge were being rebuilt, the only change in the design of the composite plate that M. Hanus would recommend was the use of a more uniform mesh of slightly heavier reinforcement.

By 1985, the heavy-vehicle traffic had increased to about 900 000 per year, and no further repairs or resurfacing had been required.

9.4.2 Design of composite plates in bridge decks

Due to the lack of basic knowledge, discussed above, little detailed guidance on design of composite plates is given in the Bridge Code. At the serviceability limit state, elastic analysis is appropriate, and the principle of superposition can be used. The effects of the weight of wet concrete and of local wheel loading have to be combined with global tensile or compressive forces in the plate, which occur in the transverse as well as the longitudinal direction when composite cross-girders are used.

At the ultimate limit state, local and global effects can be considered separately. At the construction stage, stresses in the steel plate due to wet concrete and to global effects should be combined. Once the concrete has hardened, the midspan cross section may be compact. It can then be assumed that the concrete is carried by the composite section. The resulting stresses in the composite plate will be negligible in comparison with the global stresses. This simplification cannot be made at internal supports because the webs, and therefore the section, will probably be non-compact.

Some aspects of the design of composite plates are now discussed, with reference to three examples. Elastic analysis is used for local loads at the ultimate limit state, because no reference is made in the Bridge Code to other methods. Modes of failure of composite plates are considered in section 9.4.4.

The first example is relevant to box girders that are too wide for the top plate to span transversely. It is assumed to span longitudinally between crossframes at 3.0 m spacing (fig. 9.8(a)).

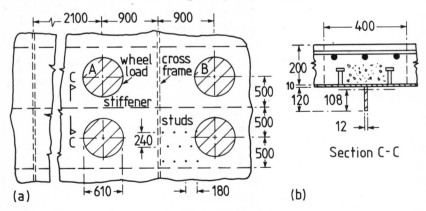

Fig. 9.8 Composite plate

THICKNESS OF STEEL FLANGE PLATE

Boxes can be designed with open tops, so the plate for a closed top can obviously be quite thin. The governing criteria are likely to be the deflection of the plate under the wet concrete or the local stresses at the shear-connector welds.

Stud connectors are almost always used for composite plates, as their properties are independent of the direction of the shear. To prevent fatigue failure, it is recommended in Part 5 of the Bridge Code that the diameter of studs attached to a flange plate in tension should not exceed one and a half times its thickness, nor twice the thickness of a plate not subject to tension. This rule is based on rather limited evidence from fatigue tests[14b] and is more conservative than the corresponding rule in the AISC specification[130] ($d \not> 2.5t$) because that is intended to ensure that a stud does not tear out from a thin plate before reaching its full shear capacity for static loading.

Studs used in bridges are usually of 19-mm or larger diameter, and this rule then requires plates in tension to be at least 13-mm thick. A 10-mm plate may be adequate for other purposes, and is used in the following example, together with 16-mm studs. The use of 13-mm studs would be more expensive due to the large number required.

LONGITUDINAL STIFFENERS

The steel plate cannot support the wet concrete over a span of 3 m without longitudinal stiffeners. Once the concrete has set they are not needed as stiffeners, and so can be quite small. A 200-mm slab of Grade 40 normal-density concrete will be used, as this is the minimum thickness likely to be practicable for the non-composite region on either side of the box girder. The deflection of a 10-mm one-way plate spanning 1.0 m due to its own weight and that of the slab is 0.8 mm for fixed ends, but 4.2 mm for simply-supported ends. The true deflection is unlikely to exceed twice the lower value (i.e. span/600), which is acceptable, so the stiffeners are placed at 1.0 m spacing. The maximum transverse bending stress in the plate is about 30 N/mm². This leaves an ample margin for the extra stress, due to heaping of concrete, personnel, etc., that occurs only before the concrete has hardened.

The effective area of plate associated with each stiffener to provide a steel T-section is given by Part 3 of the Bridge Code as $K_c bt$, where $K_c = 0.40$, $b = 1.0$ m, and $t = 10$ mm in this example. The detailed design of stiffeners is outside the scope of this book. It can be shown that a 120×12 mm flat is adequate here.

The T-section and its elastic neutral-axis are shown in fig. 9.8(b). It acts as a continuous beam with spans of 3.0 m. At the ultimate limit state, the longitudinal stresses at a crossframe due to the weight of the wet concrete, based on $wL^2/12$, are 73 N/mm² at the bottom of the stiffener and 15 N/mm² at the top of the plate. The webs of a large box are likely to be non-compact, even at midspan, so these stresses have to be combined with those due to global effects. Any overstress at the bottom of the stiffener can be redistributed to the slab.

PUNCHING SHEAR

It was found (section 8.3.1) that for a reinforced concrete deck slab spanning 3.0 m, punching shear due to a complete HB bogie (eight wheels) was more critical than that due to a single wheel. Here, it may be assumed that the stiffeners prevent shear failure over a large area and that the HA wheel load at the ultimate limit state (165 kN, including γ_{f3}) is the critical loading, as this is heavier than the factored wheel load (161 kN) for 45 units of HB loading.

Allowing for surfacing, the loaded area can be taken as a circle of diameter 0.4 m, and the shear perimeter as a circle of diameter 1.0 m. Calculations similar to those given in section 8.3.1 and the results of tests in punching shear on composite plates show that the strength of a 200-mm slab is ample for this loading.

LOCAL BENDING DUE TO WHEEL LOADS

Partial-interaction elastic analysis of composite plates is too complex for use in design, but there are several possible methods of full-interaction analysis for a typical deck panel of the type shown in fig. 9.8(a). It is assumed to be continuous over, and supported by, the crossframes at 3 m centres. It will be seen that the calculated local stresses are well below the limiting values.

A possible method is to assume that the composite plate spans transversely between the stiffeners, which themselves act as webs of slab strips 1.0 m wide that span longitudinally between crossframes. The transverse bending stresses in the plate due to positive moments caused by HA or HB wheel loads are very low (about 1 N/mm² compression in concrete and 6 N/mm² tension in the steel plate, in this example). The amount of top transverse reinforcement in the slab may be determined by the negative moments above stiffeners due to HB wheel loads located at points B in fig. 9.8(a). In this example, 16-mm bars at 200 mm spacing are stressed to about 100 N/mm², but the compressive stress in the plate is only about 10 N/mm², and it makes little difference whether concrete in compression is neglected or included.

The stresses in each slab strip are influenced quite strongly by the assumed effective breadth of its flange. The strip has a b/ℓ ratio of 0.5/3.0, or 0.17, and, for an internal span of a continuous composite beam of conventional proportions, with point load at midspan, Part 3 of the Bridge Code gives the effective-breadth ratio ψ at midspan as 0.36. This is certainly too low for a member of this cross section. It is taken as 0.40, so that the cross section of the slab strip is as shown in fig. 9.8(b). The second moment of area of the uncracked section, which is relevant to midspan of the girder, is about 96×10^6 mm^4 in 'steel' units, assuming a modular ratio of 7.4. This is less than that of a 1.0-m width of the unstiffened composite plate (170×10^6 mm^4), so it is simpler and at least as accurate to neglect the stiffeners, and analyse the panel as an elastic isotropic plate, for both sagging and hogging bending. The total amount of steel in a composite plate (about 6% in this example) is much higher than in a reinforced concrete slab, so that cracking of concrete has less influence on its flexural behaviour.

The top longitudinal reinforcement, in regions other than above an internal support of the girder, is likely to be governed by local hogging bending when a bogie of a 45-unit HB vehicle straddles a crossframe, as shown in fig. 9.8(a). The HB wheel load at the ultimate limit state is 146 kN (excluding γ_{f3}). Allowing for dispersal, the loaded area at the level of the neutral axis of the composite plate is a circle of diameter about 0.61 m. The longitudinal hogging bending moment above the crossframe is about 78 kNm per pair of wheels (e.g. A and B in the figure), and can be assumed to be resisted by a 1.0-m width of composite plate.

A more severe condition occurs near an internal support, where the slab resists global tension and will be cracked. The serviceability limit state is likely to govern because global and local effects have to be combined. The top longitudinal bars may be below the transverse bars, so the lever arm between them and the centre of the 10-mm plate is about 145 mm. The design local hogging moment is $78 \times 1.1/1.3 = 66$ kNm/m. This causes a tensile stress of 218 N/mm^2 in 20-mm bars at 150 mm spacing. The limiting stress is 345 N/mm^2 if the yield strength is 460 N/mm^2, so the margin left for global tension is 127 N/mm^2 in this example.

In a midspan region, where global compression can be assumed to prevent cracking, the top-fibre tensile stress in the concrete is about 7 N/mm^2, and the bottom fibre compressive stress in the stiffener, which will occur even though its presence has been ignored, is about 80 N/mm^2, far below the limiting stress of 355 N/mm^2.

HORIZONTAL SHEAR AND THE SPACING OF CONNECTORS

The shear connection can be designed for static loading as follows. One first provides studs at the maximum spacings that allow the plate to be treated as compact in accordance with Part 3 of the Bridge Code. The loads on these connectors due to global and local effects are then found for the serviceability limit state, and superimposed (by vectorial addition) as necessary. Any additional connectors that may be needed to reduce the maximum load per stud to the design value are then added. The checking of the shear connection for fatigue is discussed in section 10.5.

The maximum spacing of staggered rows of stud connectors allowed by Part 3 when the steel plate has $\sigma_{yc} = 355$ N/mm^2 is $18t_f$ longitudinally and $24t_f$ laterally, if the plate is to be treated as compact. Using these spacings, the density of connectors on the 10-mm plate used here is $(1/0.18) \times (1/0.24) = 23.1$ 16-mm studs per m^2. Studs of the standard height (75 mm) satisfy the condition that the spacing should not exceed four times the height (fig. 8.22).

The maximum shear per connector due to local loading will occur near an HA wheel load, which is 120 kN at the serviceability limit state. It can be calculated by the method given in the following amendment to Part 5, which is based on research at Imperial College.[131]

'The longitudinal shear forces due to local wheel loads in the regions of a composite plate supported by cross-members may be determined by considering the plate as an equivalent simply-supported beam spanning between these crossframes; the width of the equivalent beam, b, supporting the wheel load should be taken as:

$$b = u + 4x/3$$

where:

  u is the length of the wheel patch which is parallel to the cross-frame
  x is the distance from the centroid of the wheel patch to the nearest crossframe.'

Part 2 of the Bridge Code allows dispersal of wheel loads through the slab at a slope of 45° down to the neutral axis. In this slab, the neutral axis for sagging bending is 115 mm below the top surface, so $u = 0.4 + 2 \times 0.115 = 0.63$ m. For maximum shear, the wheel is located at $x = 0.315$ m from a crossframe, giving a vertical shear of $210 \times 2.685/3.0 = 107$ kN at the end of the equivalent beam that is supported by that frame. The breadth of this beam is:

$$b = 1.33 \times 0.315 + 0.63 = 1.05 \text{ m}$$

For the cracked reinforced slab in sagging bending, $Ay/I = 6.0$ m^{-1}, so that the longitudinal shear is:

$$q = 107 \times 6/1.05 = 611 \text{ kN/m}^2 \quad \text{(or 26.4 kN per stud)}$$

The design shear per stud at the serviceability limit state, assuming Grade 40 concrete, is given in Part 5 as $82/1.85 = 44.3$ kN, so that there remains, in this example, a margin, of $44.3 - 26.4 = 17.9$ kN per stud for the global shear.

BOTTOM REINFORCEMENT IN THE SLAB

There is no evidence that bottom reinforcement is needed for reasons of strength or for the prevention of local failure of the concrete near shear connectors. Where the deck slab is in overall tension due to global bending, it would be prudent to provide light reinforcement to control cracking, and so avoid the concentration of cracks along the lines of the shear connectors, which could lead to some loss of interaction.

UNSTIFFENED COMPOSITE PLATE

The steel plate in the preceding example had a span/depth ratio of 100 in the transverse direction; so, for small boxes, an unstiffened plate can be used, spanning between the webs. This was done in a viaduct at Poyle interchange between the M4 and M25 motorways (bridge 2.3 in table 6.2). Two decks 18.4 m wide are each composite with four steel boxes, which are generally 1.1 m wide. At midspan of the longest span (50.2 m), the top plate of each box is 20 mm thick, and is stiffened only by plate diaphragms at 3.75 m longitudinal spacing. In a bridge of this type, the deck slab and its reinforcement have to be designed to carry wheel loads over a transverse span (3.5 m here) between the webs of adjacent boxes that exceeds the width of the boxes. Above the boxes, it is not practicable to use a thinner slab or to reduce the reinforcement, so there is little or no advantage in designing the composite plate for local loading; the slab alone is strong enough already.

MEMBRANE TOP PLATE

Another concept, appropriate for medium-sized boxes, is to use a very thin top plate to carry the wet concrete by membrane tension, and to ignore the plate subsequently. This was done in two groups of bridges over the M6 motorway, with trapezoidal boxes and spans up to 36 m. The plates were 2.5 or 3.2 mm thick. They spanned transversely between webs up to 3.4 m apart, and longitudinally between stiffeners 2.2 m apart. Details of the bridges and of model tests to study the effects of the membrane system have been published.[132]

9.4.3 Loss of interaction in short-span composite members

The diameter of stud shear-connectors used in bridges ranges only from about 16 to 22 mm, and is virtually independent of the span of the member concerned. Thus the interface slip corresponding to working load for a shear connector is independent of the span.

In a simply-supported beam with symmetrical loading, the slips (s) at the two ends of the span are equal and opposite, so the mean rate of change of slip along the span, ds/dx, is inversely proportional to the span. For cross sections, the loss of interaction between steel and concrete depends not on the slip, but on the slip strain ds/dx (chapter 2 in Volume 1). This simple example shows what is true generally is that, for a given degree of shear connection, the shorter the span, the greater is the loss of interaction due to slip.

This result is of particular relevance to composite plates. For example, in their partial-interaction analysis of a member spanning 2.5 m, Clarke and Morley[127] found a 40% increase in longitudinal strain at midspan when allowance was made for slip. The loss of interaction under a point load also increases as the load approaches a simple support, as is evident from figs 25 and 27 of reference 133. Fortunately, it has been shown that design methods in which slip is neglected are adequate for composite plates. The relevant research, reported in the first edition of this book, is now available elsewhere.[134] As an example, the stiffened plate strip studied in section 9.4.2 was analysed. At the serviceability limit state, loss of interaction was found to increase stresses due to local loads by less than 7% in the steel stiffener and by less than 15% in the concrete slab. These increases are small in comparison with the conservative approximations made in calculating the local bending moments and full-interaction stresses.

9.4.4 Failure of composite plates

The design methods outlined above do not help the designer to envisage the circumstance (other than during construction) in which a composite plate might fail, or its mode of failure, for two reasons:

(1) Global and local effects are considered separately in design for the ultimate limit state.

(2) The composite section is usually non-compact, because of web slenderness, so that little account can be taken in design of inelastic behaviour in bending, even though the composite plate is compact.

Possible modes of failure under combined short-term global and local effects are now discussed, considering first a region of sagging global bending.

Within the lines of support of a plate panel, only two modes of failure to a single point load are possible: flexural failure by a mechanism of yield lines, and punching shear failure. Composite plates are normally only used in regions of a deck slab remote from a free edge (except near an end support, where longitudinal global compression is negligible), so that as soon as a mechanism begins to develop in a single panel, it becomes relieved of global compression, which now passes it by via the adjacent panels, by arching action in the plane of the deck. The flexural failure load is unlikely to be less than that given by simple yield-line theory, neglecting membrane effects. For the composite plate of fig. 9.8, this failure load is well over four times the design ultimate HB loading, whether one wheel or several wheels are considered.

The check on punching shear failure was done for loads at the ultimate limit state, neglecting the influence of global compression or tension. Both the calculations and available test data showed that existing designs have ample strength in punching shear. For thinner slabs, this strength must depend mainly on the detailing of the shear connection. The cross section shown to scale in fig. 9.7(b) was found to be very strong, but a composite plate with short and widely spaced connectors (fig. 9.9(a)) may have much less resistance to the mode of failure shown.

The addition of global compression is likely to increase punching shear strength, and so is not discussed. One might expect global tension to reduce the shear strength of the concrete but to increase the resistance of the steel plate to the mode of failure shown in fig. 9.9(a). No such tests on composite plates are known to the authors, but it has been shown[135] that biaxial membrane tension has little effect on the strength in punching shear of reinforced concrete slabs of the type used with composite beams.

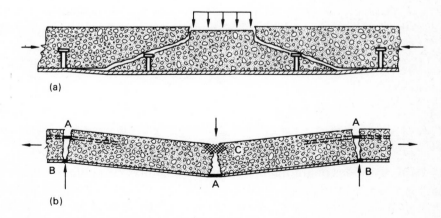

(a)

(b)

Fig. 9.9   Composite plate subjected to membrane forces

Any flexural failure of a composite plate subjected to global tension would have to be of the type shown in fig. 9.9(b), with extensive yielding in tension at points A and yielding of the plate in bending at points B, and with or without crushing of the concrete at C, depending on the magnitude of the global tension. For the full development of this mechanism, in-plane tensile strains would be large enough to relieve the plate panel of global tension, as was discussed above for global compression, so again the conclusion is reached that interaction between global and local effects does not reduce the strength of the composite plate.

LONGITUDINAL SHEAR

Shear forces at the steel-concrete interface due to local loading are only high in the vicinity of point loads, and so occur over a limited area. They do not influence the ultimate strength in shear of composite girders, and need be considered only at the serviceability limit state, as they may influence the design of the shear connection for fatigue (section 10.5).

9.5 Intersections of box girders with cross-girders

Several large highway bridges have been built with continuous main girders and closely spaced composite cross-girders (e.g. fig. 9.10 and plates 5 and 6). Where the cantilever end of a cross-girder intersects a

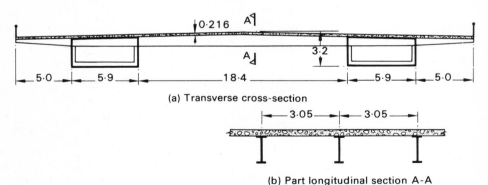

(a) Transverse cross-section

(b) Part longitudinal section A-A

Fig. 9.10   Cross sections of approach span, Avonmouth Bridge

main girder near an internal pier, the deck slab may be in biaxial tension throughout its thickness.

No account need be taken of biaxial interaction in designing the cross-girders or the main girder for bending moment and vertical shear. Longitudinal shears are overestimated because they are calculated neglecting the effects of cracking, and this may be assumed to more than compensate for any reduction in the stiffness or static strength of the shear connectors. The design of transverse reinforcement in the region of biaxial tension is discussed in section 8.3.2, the control of cracking in section 8.3.3, and fatigue of biaxially loaded shear connectors in sections 8.5.3 and 10.5.1. The design of the shear connectors for static loading is now considered.

One method in current use is to design separately the shear connections for the cross-girder and the box girder, and provide both sets of connectors. This is obviously conservative when stud connectors are used, because the vector sum of a longitudinal shear force and a transverse shear force is always less than the sum of their magnitudes.

The following method avoids the complexity of calculating the resultant shear force at many points on the composite plate near its intersection with a cross-girder, and takes advantage of the reduction in longitudinal force per connector given by equation (9.4). Initially, the connectors in area ABCD (fig. 9.11) are designed only for the transverse

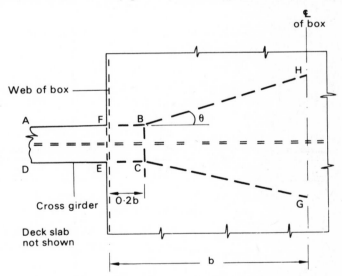

Fig. 9.11   Part plan of box girder and cross-girder

shear due to the bending of the cross-girder, and the connectors elsewhere are designed by the method explained in section 9.3.1 for longitudinal bending of a box girder. The total longitudinal force that would be resisted by box-girder connectors on area FBCE is then calculated from equation (9.4) and combined vectorially with the transverse force on that area. The 'cross-girder' connectors already designed are checked for this total force and augmented if necessary.

In area BCGH, the reduction in longitudinal force per connector, Q_x, given by equation (9.4) is substantial. If the most heavily loaded connectors (over the box-girder web) are designed for a load P_d per connector, which may be less than the design load, to provide a margin for local effects of wheel loads, then from equations (9.4) and (9.5) with $x/b = 0.2$:

$$Q_x = P_d(0.64K + 0.15)/(K + 0.15)$$

Using fig. 9.5 for K, we find that Q_x remains almost constant at $0.67P_d$ as n_w/n ranges from 0.1 to 0.5. The transverse force that can safely be applied to each connector to the right of BC therefore exceeds $(1 - 0.67^2)^{\frac{1}{2}}P_d$, which is $0.74P_d$, and those that are provided for longitudinal bending in the area BCGH (where θ is, say, 30°) will usually have sufficient transverse strength to resist, in addition, the cross-girder shear in this area.

Chapter Ten

Fatigue

10.1 Introduction

The principal problem arising from repeated loading that is found solely in composite bridges is the design of the shear connection for fatigue. This is the main subject of this chapter. The specified loadings are the same whether the superstructure is steel, concrete, or composite, and most other aspects of design for fatigue arise also in steel bridges. They are therefore discussed here only to the extent that is necessary to explain the types of loading and methods of analysis used when designing the shear connection for fatigue. A worked example is given for the effects of highway loading on a multi-span bridge (Appendix X). Methods of design for railway loading are similar, and are not considered further.

SHEAR CONNECTORS
The experimental data on the fatigue strength of shear connectors was reassessed for the Bridge Code. This work is summarised in Appendix B of this book. The new method of presentation used in Part 10[1] made possible a simple method of design for the shear connection, which is explained in section 10.4.

10.2 Loading for fatigue assessment of highway bridges

In the studies of highway loading made for Part 10, it was found[136] that the patterns of loading imposed by commercial vehicles could be represented with an accuracy sufficient for fatigue assessment, by the four-axle layout shown in fig. 10.1.

The datum weight of this 'standard fatigue vehicle' is 320 kN, divided equally between the sixteen wheels. The axle spacings are the same as for the shortest of the HB vehicles (fig. 7.7) but the wheel spacings along an axle are different, being based on typical heavy goods vehicles, which are

211

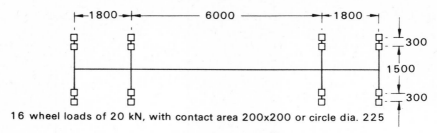

Fig. 10.1   Plan of load pattern for standard fatigue vehicle

narrower than the heaviest HB (abnormal) vehicles. For some types of bridge it is possible to obtain a global analysis for the effects of fatigue loading by scaling the results of an elastic analysis for HB loading, taking account of the different factors and, in a multi-beam bridge, of the different proportions of the total vehicle weight carried by each girder.

The relative frequencies, vehicle weights, and axle arrangements of the full range of commercial traffic on a busy arterial road in Great Britain can be represented by a 25-line spectrum, which has been deduced from traffic counts and weighbridge records. It includes vehicles ranging in weight from 370 tonnes to 3 tonnes, and in wheelbase from 48 m to 4 m, and is given in Part 10.

For the development of simplified design methods, this was condensed[136] to a 9-line spectrum (table 10.1) given in terms of the standard fatigue vehicle by weight factors K_w and frequency proportion factors K_n. Use of this spectrum gives approximately the same cumulative damage as the full spectrum for welded details (other than shear connectors, as explained below) when the relevant loop of the influence line for the detail considered has a base length of 25 m.

The vehicle weights given in these two spectra are used as design

Table 10.1   Loading spectrum for highway bridges, in terms of the standard 320-kN vehicle

Weight factor for standard fatigue vehicle, K_w	Proportion of total number of vehicles, K_n
6.75	0.000 01
5.03	0.000 02
4.09	0.000 03
2.10	0.000 44
1.06	0.104 5
0.82	0.105
0.72	0.09
0.43	0.32
0.20	0.38
	1.000 00

values. Thus the factors γ_{fL} and γ_{f3} are in effect both taken as 1.0. No explicit load factor appears in the calculations, for reasons discussed in section 7.1.4.

Particular types of loading event have no significance in design for fatigue unless they occur many hundreds of times. Most bridges are designed for a life of 120 years, so that any type of loading which occurs less than about once a month can be neglected. In highway bridges, for example, significant braking and accidental forces occur too rarely to cause noticeable fatigue damage. Furthermore, design is based on the number of repetitions of ranges of stress, as explained in Appendix B; and no account need be taken of the coexisting stresses due to any type of static loading. Thus, only the normal effects of passages of vehicles weighing about 30 kN or more need be considered in fatigue assessments.

10.3 Application of design loading

Calculations of fatigue damage in highway bridges become impracticable unless radical simplifications of observed traffic loadings are made. Any errors in predicted fatigue lives that arise from these are likely to be less than those due to an inevitable but more fundamental assumption that data based on traffic today truly represent what may be expected over the 120-year design life of the bridge.

NUMBER OF VEHICLES
The fatigue loading is specified as a number of 'loading events' per year, that depends on the type of road carried by the bridge. A loading event is

Table 10.2 Numbers of commercial vehicles per year for fatigue assessments

Category of road		No. of vehicles per lane per year, n_c (millions)		σ_{r1} from equation (10.15) (N/mm²)
Type	Carriageway	Each slow lane	Each adjacent lane	
1 Motorway	Dual 3-lane	2.0	1.5	33.6
2 Motorway All purpose	Dual 2-lane Dual 2 or 3 lane	1.5	1.0	34.8
3 All purpose	Single 3-lane Single 2-lane, 10 m wide	1.0	Not applicable	36.6
4 All purpose	Single 2-lane, 7.3 m wide	0.5		39.9

a single passage of the 320-kN standard vehicle along the full length of one traffic lane across the bridge, with no other vehicles on the structure. Design is based on the ranges of stress so caused.

Roads are divided into four categories. A condensed version of the specified numbers of loading events per year is given in table 10.2. There is no fatigue loading for the 'fast' lane of a 3-lane one-way carriageway because vehicles heavy enough to contribute to fatigue damage are not allowed to use this lane in Great Britain.

For a 120-year design life, and a girder receiving load from both a 'slow' and an 'adjacent' lane of a type 2 road, the total number of loading events is thus $2.5 \times 120 \times 10^6$, or 3×10^8. It will be shown later that significant fatigue damage is assumed to be caused only by vehicles with K_w exceeding about 4. From table 10.1, considering only those vehicles, the total number of loading events for the girder considered here is:

$$N_t = 3 \times 10^8 \times 6 \times 10^{-5} = 1.8 \times 10^4$$

This total is well below the endurance, about 10^7 cycles, above which test data on shear connectors are sparse.

PATH OF VEHICLES
Lanes are defined in Part 10 'as marked on the carriageway'. The mean path of the centres of all the vehicles is assumed to be parallel to the

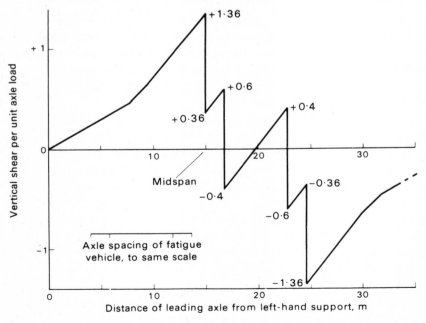

Fig. 10.2 Vertical shear at centre of 30-m span due to passage of fatigue vehicle

centreline of the lane and within 300 mm of it, and must be selected so as to cause the maximum range of stress in the detail being considered. Part 10 also gives a distribution of vehicle paths across the width of a lane that may be used in place of a single mean path. There is little advantage in doing so unless the slope of the transverse influence line near the centre of the lane is steep.

10.4 Design of shear connection for fatigue

Shear connectors (Class S) are one of the nine classes of welded detail covered in Part 10 of the Bridge Code. For most types of detail it is convenient to check the fatigue life of a given design, but for shear connectors near midspan, it is more convenient to calculate directly the spacing required than to check an assumed spacing.

It will now be shown how this could be done, in a very simple situation. The following theory is not a design method, for reasons discussed later, but should assist the reader to use the new methods with understanding. The loading spectrum given in table 10.1 will be used.

10.4.1 Direct calculation of connector spacing

The type of road and the layout of the structure are assumed to be known. The objective is to calculate the spacing p of shear connectors (having a known nominal static strength P_u) at a particular cross section of a girder, such that their design fatigue life is the specified life of 120 years. It is assumed that the live-load longitudinal shear is influenced by loading on one slow lane only.

The influence line for vertical shear at the section considered is first calculated by elastic theory. From this, the variation of vertical shear as the standard fatigue vehicle crosses the bridge is easily obtained. Figure 10.2 shows, as an example, the variation of vertical shear per unit axle load at the centre of a simple span of 30 m, with the distance of the leading axle from one support. If the girder is of uniform section, the diagram of longitudinal shear per unit length has exactly the same shape. The vertical shear diagram has one main cycle ($+1.36$ to -1.36) and three subsidiary cycles, due to the passage of the four axles of the fatigue vehicle across the section considered.

Multiple stress cycles are counted by the 'reservoir' method, which is explained in Part 10. It is an alternative and simpler presentation of the well-known 'rainflow' method, and gives the same result. There are four cycles in fig. 10.2, with the ranges shown in column 4 of table 10.3. The

Table 10.3   Relative importance of stress cycles due to the standard fatigue vehicle, at mid-span of a 30 m span

Cycle	Peak	Trough	Range, R	$(R/R_1)^8$	$(R/R_1)^5$
1	+1.36	−1.36	2.72	1.0	1.0
2	+0.40	−0.40	0.80	5.6×10^{-5}	2.2×10^{-3}
3	+0.60	+0.36	0.24	3.7×10^{-9}	5.3×10^{-6}
4	−0.36	−0.60	0.24	3.7×10^{-9}	5.3×10^{-6}

σ_r–N relation for shear connectors (equation B.2 in Appendix B) is:

$$N\sigma_r^8 = 2.08 \times 10^{22} \qquad (10.1)$$

where N is the design number of cycles (to failure) of constant-amplitude load that causes a stress range σ_r N/mm^2 on the throat area of a weld attaching a shear connector to a steel flange.

Since Part 10 was written, further research on the fatigue behaviour of stud shear connectors[137] has led to the conclusion that the value 8 for the exponent (m) in equation (10.1) is too high. The σ_r–N relation given in draft Eurocode 3[4], which, like equation (10.1), is based on a 2.3% probability of failure, is:

$$N\sigma_r^5 = 6.55 \times 10^{15}$$

This equation is compared with equation (10.1) in Appendix B. Figure B.1 shows that they give the same range, 147 N/mm^2, when $N = 95\,600$ cycles, and that at higher numbers of cycles (lower stress ranges), the Eurocode equation is the more conservative. For example, at $N = 10^8$ cycles, Part 10 gives $\sigma_r = 65$ N/mm^2 and draft EC3 gives $\sigma_r = 48$ N/mm^2, when account is taken of the different treatments of stress cycles of low amplitude (Appendix B) in the two codes.

If, as seems likely, the value of the exponent m is reduced from 8 to 5 when Part 10 is next revised it should be possible to align design methods for Class S welds more closely with those for the other classes, for which m lies in the range 3 to 4. The value 8 is so far above this range that, for example, the simplified method of Clause 8.3 of Part 10 is not available for Class S details. If $m = 5$, the errors in using it may be acceptable.

The rest of this chapter relates to the current (1980) version of Part 10, with $m = 8$, except that some comments relevant to $m = 5$ are made.

Returning to the example of fig. 10.2, equation (10.1) shows that cumulative damage is proportional to the eighth power of the range of stress (or of applied load). Relative values of the eighth powers of the four ranges in table 10.3 are given in column 5, and of the fifth powers in column 6. These show that the error in neglecting the three subsidiary cycles is less than 0.01% when $m = 8$, and rises only to 0.2% when $m = 5$.

Effects of individual axles are of course important in short-span members; these are discussed in section 10.5.

Let the range of longitudinal shear per unit length in the main cycle due to the passage of the fatigue vehicle with $K_W = 1$ be q_r, and the total specified number of loading events be N_t (found from table 10.2). For connectors at spacing p, the resulting range of load per connector is pq_r. It will now be shown how this quantity, also written as P_{r1}, can be calculated. It is a property of the specified load spectrum, the relation given in equation (10.1), and the type of connector, but is the same for all structures where only a single stress cycle has to be considered.

If σ_{r1} is defined as the range of mean shear stress at the weld throat due to one cycle of loading, then obviously:

$$pq_r = \sigma_{r1} A_w \tag{10.2}$$

where A_w is the effective area of the weld throat. For bar and channel connectors, this is calculated for the welds used, in accordance with rules given in Appendix B.

For stud connectors, it is shown in Appendix B that:

$$\sigma_{r1} = K_s(P_{r1}/P_u) \tag{10.3}$$

where:

$$P_{r1} = pq_r \tag{10.4}$$

and K_s is 425 N/mm^2 for normal-density concrete and 500 N/mm^2 for lightweight-aggregate concrete.

The remaining step is to calculate the range of stress σ_{r1} such that connectors at the spacing p given by equations (10.2) or (10.4) have the correct design life. It is now shown that σ_{r1} is a function only of the load spectrum and the total number of cycles, N_t.

Equation (10.1) can be written in the general form:

$$N\sigma_r{}^m = K \tag{10.5}$$

where m and K are known constants.

The load spectrum of table 10.1 consists of k frequency proportion factors, K_{n1} to K_{nk}, each associated with a weight factor K_w. The *spectral constant C* is defined by:

$$C = \sum_{j=1}^{j=k} K_{nj} K_{wj}{}^m \tag{10.6}$$

From table 10.1:

$$C = (10^{-5} \times 6.75^8) + (2 \times 10^{-5} \times 5.03^8) + (3 \times 10^{-5} \times 4.09^8) + \cdots$$
$$+ (0.38 \times 0.2^8)$$
$$= 54.00 \tag{10.7}$$

The first three terms of this expression for C are 43.1, 8.2 and 2.3, which have a total exceeding 99% of C. This shows that for shear connectors which are influenced by the passage of a complete vehicle rather than by axle loads, vehicles of design weight less than about 130 tonnes ($K_w = 4.09$) can be ignored. This result is sensitive to the value of m. When it is reduced to 5, the first three terms contribute only 52% of C, and even the last three terms contribute 5%.

For line j of the load spectrum, the range of stress is:

$$\sigma_{rj} = K_{wj}\sigma_{r1} \tag{10.8}$$

and the number of cycles is:

$$n_j = K_{nj}N_t \tag{10.9}$$

From equation (10.5) the fatigue life for a stress range σ_{rj} is:

$$N_j = K\sigma_{rj}^{-m} \tag{10.10}$$

The Miner's rule summation is:

$$\sum_{j=1}^{j=k} (n_j/N_j) = 1 \tag{10.11}$$

so from equations (10.8) to (10.11):

$$\frac{N_t\sigma_{r1}{}^m}{K} \sum_{j=1}^{j=k} (K_{nj}K_{wj}{}^m) = 1 \tag{10.12}$$

Substituting the spectral constant from equation (10.6):

$$\sigma_{r1} = (K/N_tC)^{1/m} \tag{10.13}$$

Thus the design range of stress at the weld throat, for all types of connectors, depends only on the properties K and m of the σ_r–N curve, the load spectrum considered, and the design life of the structure, which determine C and N_t.

TRAFFIC ON MORE THAN ONE LANE

The result given by equation (10.13) can be generalised to provide a direct design method for fatigue of shear connectors at locations where traffic on several lanes contributes to the cumulative damage.

At the location considered, let the ranges of longitudinal shear due to the loadings considered (with $K_w = 1$) be q_r, $\rho_1 q_r$, $\rho_2 q_r$, etc. It is convenient to arrange the loadings such that all the factors ρ are less than one. Let the spectral constants for these loadings be C_0, C_1, C_2, etc., and the specified numbers of cycles during the design life of the member be N_{t0}, N_{t1}, N_{t2}, etc. By the method used for equation (10.13) it can be shown that:

$$\sigma_{r1} = \left[\frac{K}{N_{t0}C_0 + N_{t1}C_1\rho_1{}^m + N_{t2}C_2\rho_2{}^m + \cdots} \right]^{1/m} \qquad (10.14)$$

where σ_{r1} is the range of stress due to the loading (with $K_w = 1.0$) that causes longitudinal shear q_r.

USE OF EQUATION (10.13) FOR TRAFFIC ON ONE LANE ONLY
Substituting the known values of K, C and m for shear connectors and the fatigue loading spectrum:

$$\sigma_{r1} = 374.3 N_t{}^{-0.125} \quad \text{N/mm}^2 \qquad (10.15)$$

Thus for a design life of 120 years and loading on one slow lane only of a dual 3-lane motorway:

$$\sigma_{r1} = 374.3(2 \times 10^6 \times 120)^{-0.125} = 33.6 \text{ N/mm}^2 \qquad (10.16)$$

Similar results for the other three categories of road are given in table 10.2.

The spacing, p, of bar or channel connectors of a given type (for which A_w is known) can now be calculated from equation (10.2). For studs, from equations (10.3) and (10.4):

$$p = \sigma_{r1} P_u / K_s q_r \qquad (10.17)$$

For example, for stud connectors with $P_u = 100$ kN in normal-density concrete, at a point in a girder where the passage of the 320-kN vehicle causes a range of longitudinal shear of 90 kN/m, in a bridge where equation (10.16) applies, the spacing of single connectors for a 120-year life is:

$$p = (33.6 \times 100)/(425 \times 90) = 0.088 \text{ m}$$

and in practice, groups of three at 0.26 m spacing might be used.

These simple results could be used in design only if the assumptions made above were valid. In fact, corrections have to be made for the following errors and oversimplifications.

(1) A single loading event on one lane may cause more than one cycle of stress in the detail considered.

(2) The stress cycle(s) associated with a loading event may be modified by the simultaneous passage of other vehicles (a) in that lane or (b) in other lanes.

(3) The stress cycle(s) may be influenced significantly by the differences between the axle spacings and loadings of real vehicles and those of the simplified loading spectrum and standard fatigue vehicle.

(4) The simplified spectrum given in table 10.1 was prepared for welded details for which the exponent m in equation (10.5) lies between 3 and 4; that is, for all classes except Class S, for which $m = 8$.

10.4.2 Fatigue assessment without calculation of cumulative damage

The simplest of the methods of assessment given in Part 10 of the Bridge Code consists of using corrected values of σ_{r1} (equation 10.13) that allow for the errors listed above. This can be done by replacing the four values of σ_{r1} given in table 10.2 by four curves of 'limiting stress range', denoted by σ_H. The curve for Class S details and a road of type 4 (table 10.2) is shown in fig. 10.3.

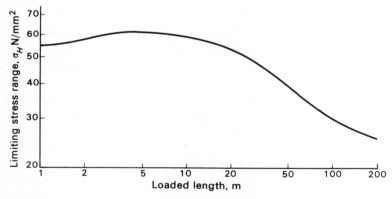

Fig. 10.3   Limiting stress ranges σ_H for class S details (two-lane all-purpose road, 7.3 m wide)

Error (1) above is important in short-span members, where stress cycles due to individual axle loads may govern design; error (2) increases at longer spans, where stress cycles due to combinations of vehicles may have a much greater range than that due to a single vehicle. This is why σ_H is a function of the 'loaded length', L. This is defined as 'the base

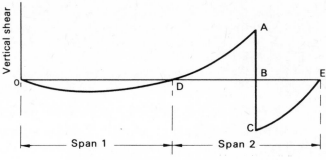

Fig. 10.4   Influence line for shear at point B

length of that portion of the point load influence line which contains the greatest ordinate'.

Figure 10.4 illustrates the interpretation of this when two ordinates are equal. It shows the influence line for vertical shear at point B in a two-span girder ODE, for which the two 'greatest ordinates', AB and BC, are equal. The relevant base lengths are DB and BE, and the longer of these (DB) is taken as the 'loaded length'.

When a member is influenced by loading on more than one lane, L is determined from the influence line that gives the greatest range of stress or stress resultant.

The effects on σ_H of errors (2) and (3) can be determined by trial calculations for various types of bridge member. Errors (1), (2) and (3) are responsible for the variation in σ_H at loaded lengths below about 15 m, which are relevant to fatigue assessment in cross-girders and composite plates. At longer spans, σ_H drops sharply (fig. 10.3), due mainly to error (2a), because additional vehicles increase the range of stress σ_r at a given location.

The curves of σ_H against L are of similar shape for all types of welded detail except Class S, where correction also has to be made for error (4). For a given loaded length, a higher proportion of the total cumulative damage is caused in Class S welds than in other classes by the heaviest (abnormal) vehicles in the full loading spectrum, because $m = 8$, rather than 3 or 4. The simplified spectrum, consisting of the standard fatigue vehicle with various load factors, cannot fully represent this difference, particularly for loaded lengths above 25 m. This is why the reduction in σ_H with loaded length is steeper for Class S welds than for the other classes.

A simple example of the use of this method is given in section X19. As the method cannot take account of specified features of deck layout and structure, it is inevitably conservative in particular cases. Trial calculations have shown that for Class S welds, the given values of σ_H can be up to 30% too low. In direct design, in which σ_{r1} in equation (10.2) or (10.3) is replaced by σ_H, it follows that the connector spacing so determined can in extreme cases be up to 30% lower than would be given by the more accurate method mentioned below.

10.4.3 Fatigue assessment with calculation of cumulative damage

This method, the 'vehicle spectrum method', consists of determining stress spectra for loading events in the various lanes that affect the detail considered, taking account of multiple cycles; combining these with the appropriate load spectrum; and calculating the cumulative damage by Miner's summation. It is applicable to shear connectors, but is not explained here, as it would be necessary to reproduce much of the explanatory material given in the appendices to Part 10 of the Bridge Code. The simpler method explained in section 10.4.2 will usually suffice in practice.

10.5 Local effects of repeated loading

As with global effects, only fatigue of the shear connection need be considered here, because all other local effects of repeated loading occur also in steel or reinforced concrete structures. The relevant properties of precast concrete permanent formwork are discussed in section 11.4.

10.5.1 Shear connectors on box girders

The passage of wheel loads may cause significant local shear forces on connectors on closed-top boxes. The information on these effects from research is sufficient to enable a design method to be given. In existing practice, the spacing of connectors is normally governed by the global longitudinal shear or by the empirical rules for maximum spacing, discussed in section 9.4.2. In calculations, connectors are often assumed to be effective only in the longitudinal direction of the member considered.

Particularly in regions of composite plates remote from the nearest steel web, the passage of a wheel load is believed to apply a multidirectional cycle of shear to each connector, of the type shown in fig. 10.5. These effects cannot easily be calculated, and accurate assessment of the fatigue life of connectors subjected to loading of this type is not yet possible, due to lack of knowledge of the following subjects.

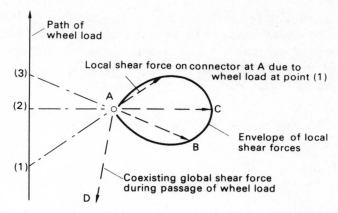

Fig. 10.5 Plan of path of wheel load and typical shear forces acting on connector at point A

(1) The reduction in the shear force applied to the connector, due to slip at the steel-concrete interface.

(2) The influence of the stiffness of the road surfacing on the amplitude of shear ranges of short duration.

(3) The interaction in time between cycles of global and local shear. This is influenced by the whole of the traffic on the span considered, and cannot be studied in detail until more statistical data on vehicle loadings and traffic flow become available. Local shears are further complicated by the effects of composite action with plate stiffeners and crossframes.

(4) The fatigue life of connectors subjected to load cycles of the type shown in fig. 10.5, with a coexisting global shear force. Fatigue tests on connectors under rotating shear have shown[110] that the equivalent range of shear can be taken as half the perimeter of the shear envelope (i.e. as the curved length ABC in fig. 10.5). A coexisting global shear in direction AC can probably be neglected, by analogy with the neglect of mean stress level when considering unidirectional shear, but little is known of the influence of global shears in other directions, such as AD.

In practice, designers are likely to use the simplified method given in Part 10 of the Bridge Code. For points on a composite plate forming part of a closed-top box, the influence lines for horizontal shear at the steel-concrete interface will usually give 'loaded lengths' between about 1.5 m and 3.5 m. In this range, fatigue damage is determined by the wheel loads from single axles or pairs of axles, not by total weights of vehicles. The values of σ_H (section 10.4.2) given in Part 10 for short spans should allow for this.

At the interface between steel and concrete, the horizontal shear forces very close to a point load can be assumed to act radially from the load, giving longitudinal and transverse shears of equal intensity for static loads, but in a bridge deck the ranges of shear are higher in the longitudinal direction for the following reasons.

(1) The motion of the vehicle causes a reversal of longitudinal shear, but a smaller variation of shear in the transverse direction.

(2) When allowance is made for the lateral dispersal of load through the surfacing and deck slab, the effective loaded area under a pair of wheels of the standard fatigue vehicle is about 800 mm wide but only 500 mm long. This, together with the variations in the width of vehicles and in their lateral position in a traffic lane, ensures that the transverse components of the local shears due to the different wheels on an axle tend to cancel out, whereas the longitudinal components are always additive.

Thus it will often be accurate enough to consider only the shear ranges in the direction of travel of the vehicles. Only the greatest ranges of shear need to be identified, because the slope of the σ_r–N line for shear connectors is so shallow. These ranges will not necessarily be due to the passage of the heaviest vehicles, as these tend to have wheels uniformly spaced along their axles, rather than in pairs.

In design to Part 10, the greatest range of shear will normally be that due to the heaviest specified axle load or pair of wheels. It is on the safe side, in relation to the fatigue of a specific connector on a composite plate, to assume that all such axles traverse the same path, that is, the one within 300 mm of the lane centreline that gives the greatest range of shear at that connector. An alternative method would be to take account of the distribution of lateral position of vehicles in a traffic lane, given in Part 10, and to calculate the cumulative damage. For shear in the longitudinal direction, there is little advantage in using this longer method.

10.5.2 Intersections of cross-beams with main beams

The ranges of longitudinal shear due to the passage of individual wheel loads close to shear connectors in plate girders and open-top box girders are usually well below the ranges due to global loads (section 10.4) and can then be neglected. The ranges of shear in the direction transverse to the span of the main beams are also low, and can be neglected, except near an intersection of a cross-beam with a main beam.

In the cross-beam, the greatest ranges of shear are likely to occur when axles or bogies cross the beam. The interaction in time between these events and the most adverse cycles of global shear in the main beam are

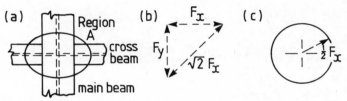

Fig. 10.6 Shear connection for biaxial fatigue loading

too little known and too complex to be allowed for in design.

It is usual to use stud connectors in these regions, and to check the studs on the cross-beam for the transverse shear ranges and those on the main beam for the longitudinal shear ranges, even though all the connectors near an intersection are subjected to biaxial fatigue loading.

The following simplified example shows that this method can be used with confidence. We consider a region (A in fig. 10.6(a)) within which the studs are required to resist N repetitions of both a transverse range of horizontal shear force F_x and a longitudinal range F_y, and assume that both forces vary sinusoidally with time, at the same angular frequency ω. If one force is much smaller than the other, it will cause stress ranges in the studs that lie below the non-propagating value, so the worst case is likely to be when F_x and F_y are equal. If the forces are in phase, they then combine as a single sinusoidal force in a direction of 45° to both beams, fig. 10.6(b), of range $\sqrt{2}\,F_x$. If σ_r is the design stress range for the studs, for N repetitions, then the number required is:

$$n_1 = 1.41 F_x/\sigma_r A_s$$

where A_s is the cross-sectional area of a stud.

If the forces are 90° out of phase:

$$F_{xt} = \tfrac{1}{2}F_x \sin \omega t$$
$$F_{yt} = \tfrac{1}{2}F_y \cos \omega t$$

where F_{xt} and F_{yt} are instantaneous values. In this case they combine as a rotating radial force of magnitude $\tfrac{1}{2}F_x$, as in fig. 10.6(c). From section 10.5.1, the equivalent range of shear is half the perimeter of the circle, which is $(\pi/2)F_x$, so the number of studs required is:

$$n_2 = 1.57 F_x//\sigma_r A_s$$

The design method recommended above consists of providing $F_x/\sigma_r A_s$ studs for the F_x force and an equal number for the F_y force, giving a total number:

$$n_3 = 2 F_x/\sigma_r A_s$$

which is 27% above the higher of the preceding results.

Chapter Eleven

Permanent Formwork in Bridge Decks

11.1 Introduction

The decision to use permanent formwork in a bridge deck will often have been made because of constraints imposed by the site; for example, where access for erection and/or dismantling of conventional shuttering and falsework is restricted. It is then likely that access for maintenance during the life of the bridge will also be difficult, so it is essential that permanent formwork is made from a material requiring little or no maintenance.

MATERIALS

The materials commonly used are glass fibre-reinforced cement (GRC) sheeting, sometimes with additional steel reinforcement, and precast concrete. Glass-reinforced plastic sheeting and profiled steel sheeting are used occasionally. Permanent formwork is competitive with multiple-use steel or timber formwork because of the savings of time, falsework and labour costs for dismantling that can be made.

STRUCTURAL PARTICIPATION

As with multiple-use formwork, the prime function of permanent formwork is to support the weight of the wet concrete and the temporary construction loads during casting of the concrete slab, without excessive deflection. Where any shear connection that exists between the shuttering and the mature concrete deck slab, due for example to natural bond, is regarded as incidental, the formwork is referred to as *structurally non-participating*, and is not considered to contribute to the resistance of the composite slab.

In contrast, when materials with a substantial tensile strength are used for permanent formwork and a positive mechanical shear connection can be developed between the formwork and the mature concrete slab, the formwork is referred to as *structurally participating*, and its contribution to the strength and stiffness of the deck is taken into account in design.

226

Precast concrete planks, either reinforced or prestressed, with reinforcement projecting into the *in situ* deck slab to form the shear connection (fig. 11.2) are an example.

Profiled steel sheeting, which is widely used in form-reinforced floors in buildings, is rarely used as structurally-participating formwork for bridge decks, for two reasons. For economy, it has to be very thin (less than 2 mm), and so is vulnerable to corrosion. Even when galvanised or resin coated, its maintenance over a design life of 120 years is costly and inconvenient, if not impracticable. Shear connectors are not easily welded to it, and the fatigue behaviour of some of the available types of mechanical interlock with concrete has not been established. Details of research on this subject are given in reference 138 and, with recommendations for design, in the first edition of this book.

Composite plates, considered in section 9.4, are of course another type of participating formwork, but are only economical when closed-top steel box girders are required for other reasons.

11.2 Loading

Permanent formwork for a bridge deck is usually unpropped during casting of the slab. It has to be designed to support its own weight, the weight of the wet concrete, and the temporary construction loading.

CONSTRUCTION LOADING

Part 2 of the Bridge Code gives limit state requirements for temporary loads (section 7.3.1). It does not specify their values, but requires that they be accurately assessed.

Designers thus have to make a realistic appraisal of the temporary loads that can occur before and during concreting, bearing in mind the level of control that is likely to be exercised on site. The *Code of practice for falsework*[139] encourages this approach, and gives guidance on the magnitude of temporary loads that can be expected. It recommends that a uniformly-distributed load of 1.5 kN/m² is normally sufficient to allow for construction operatives, impact and heaping of concrete during placing, and for small items of equipment and hand tools for immediate use, but warns that this may not be sufficient for excessive heaping or impact of concrete during placing.

This recommendation appears also in draft Eurocode 4[5] in the clauses on profiled steel sheeting. There is also a requirement that for spans of less than 3 m, a nominal line load of at least 2 kN/m parallel to the supporting beams and placed to cause maximum bending moment and/or shear, should be considered as an alternative to the distributed loading.

A more severe representation of the mounding of concrete was given by the loading proposed in 1970 by the Department of the Environment for use in research[138] on profiled steel sheeting. This consists of a load of 9 kN applied over a circular area of 0.6 m radius.

PARTIAL SAFETY FACTORS FOR LOADS

For the ultimate limit state, the partial safety factor γ_{fL} for temporary loading is given in Part 2 of the Bridge Code as 1.15. This is applicable to all *temporary* materials, plant and equipment, and personnel to be used during construction, but not to the weight of the wet concrete slab or the self-weight of the permanent formwork, for which the usual factors for dead load are used. Where there is uncertainty in the assessment of the nominal values of the construction loads, they should be increased by a factor sufficient to ensure that the loading is not underestimated..

The design loads and loading effects are calculated for the structure as it is at the time considered, by the methods that would be appropriate if it were a completed structure, with the factor γ_{f3} taken as 1.1. Where construction loads have a relieving effect on the structure they should either be considered as not acting or, where they act, the design load should be taken as the nominal load multiplied by a factor sufficiently less than unity to ensure that the relieving effect is not overestimated.

11.3 Non-participating permanent formwork

Glass fibre-reinforced cement (GRC) flat and ribbed sheets have been available since the early 1970s. The use of GRC in bridge decks is increasing because it has been shown[140, 141] that it can be made with lower permeability to water vapour and better freeze/thaw behaviour and resistance to carbonation than structural concrete.

The formulation can be varied to suit the requirements. A typical mix would have a water:cement ratio of 0.28 to 0.35, a ratio of fine sand to cement of 0.3 to 0.5, and a glass fibre content (by weight) of about 5%. A plasticiser may be used, and the fire resistance can be enhanced by including pulverised fuel ash and an air-entraining agent.

The formwork panels are made to measure from layers of GRC from 6 mm to about 15 mm thick. For spans below about 0.8 m, a single flat sheet can be used.

For a 200-mm deck slab, the longest practicable unpropped span is about 3 m. This requires a three-layer sandwich construction about

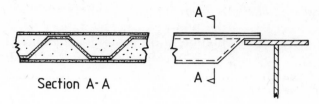

Fig. 11.1   Glass fibre-reinforced cement permanent formwork

230 mm deep with polystyrene cores (fig. 11.1). Design is governed by stress, not deflection. Limiting serviceability stresses lie in the range 5 to 9 N/mm². A designers' handbook is available.[142]

The panels span between the top flanges of the steel beams, and are bedded on mastic beads, mortar fillets, or foam strips. If used for cantilever edges, they require support from falsework at their outer ends.

The good protective properties of GRC are beginning to be recognised. The Norwegian Roads Directorate has accepted[141] that when GRC formwork at least 10 mm thick is used, the concrete cover to bottom main-reinforcing bars can be reduced to 10 mm. At present (1985), no relaxation in the rules for concrete cover or crack-width control have beeen made in the UK.

The rather high depth-to-span ratio of all-GRC panels can be reduced by the use of embedded steel reinforcement. One type of reinforced panel consists of a flat soffit sheet of GRC 6 mm thick, with upstanding ribs up to 60 mm high, in which steel flat bars are embedded. These panels can be tailor-made, with spans of up to about 3 m, and can be pre-cambered to counteract the significant deflection due to the wet concrete. It is not known for how long the thin skin of GRC can protect the embedded steel, so the panels are designed as non-participating.

11.4 Structurally participating permanent formwork

The Department of Transport allows the formwork for the decks of composite bridges to be designed as participating only when specific prior approval has been given, and it advises[2] that particular attention be paid to the following aspects of the design:

(1) fatigue behaviour
(2) durability
(3) bond between permanent formwork and the concrete slab under both long term and impact loading
(4) corrosion protection.

No detailed guidance is given in Parts 3 or 5 of the Bridge Code on the design of profiled steel sheeting as participating formwork. It is rarely used in bridges, for the reasons given in section 11.1.

11.4.1 Precast concrete formwork

Design methods for precast concrete planks and slabs that act with an *in situ* concrete deck to form a composite slab are given in the clauses on composite concrete construction in Part 4 of the Bridge Code. For the construction stage, the formwork is designed as a normal reinforced or prestressed precast element for the loads described in section 11.2.

After construction, the entire load may be assumed at the ultimate limit state to be resisted by the combined cross section, provided that the interface between the precast and *in situ* concrete is designed for the corresponding longitudinal shear. The relevant method is similar to that for transverse reinforcement near shear connectors (section 8.6), except that the first term of equation (8.58), $k_1 s L_s$, is replaced by $v_1 L_s$, where v_1 is the ultimate longitudinal shear stress in the concrete on the shear plane considered. For normal-density concrete, $k_1 s$ has the constant value 0.9 N/mm^2 in Part 5, but in Part 4, v_1 depends both on the cube strength of the concrete and on the type of contact surface between the precast and *in situ* concrete. Its range is from 0.3 N/mm^2 for a 'rough as cast' surface of Grade 20 concrete to 0.8 N/mm^2 for an exposed-aggregate or sand-blasted surface of Grade 40 concrete.

BEHAVIOUR UNDER REPEATED LOADING

Extensive experience of composite concrete bridge decks has shown that concrete-to-concrete interfaces designed for static longitudinal shear are well able to resist the effects of repeated wheel loadings.

One widely-used type of precast unit, 'Omnia' bridge decking (plate 12), has been subjected to full-scale fatigue tests[143] which were done on negative (hogging) moment regions of parts of continuous composite bridge decks, as shown in fig. 11.2.

The Omnia units, which spanned 3.0 m in these tests, have two distinct sets of reinforcement. There is a light welded lattice consisting of one top bar and pairs of inclined links, which are anchored within the precast unit by being welded to 5-mm longitudinal bars. The lattice is assumed to be effective only during construction. The additional unwelded longitudinal bars within the unit are specified by the designer of the deck, and treated as part of its permanent reinforcement. Thus, no account need be taken of the welding when designing the slab for fatigue.

In the tests, fatigue failure occurred in the 5-mm reinforcement near

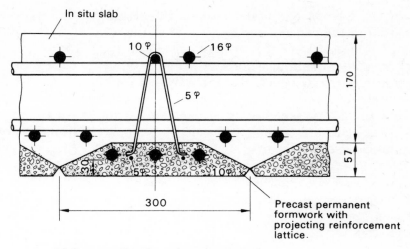

(a) Cross-section through composite slab

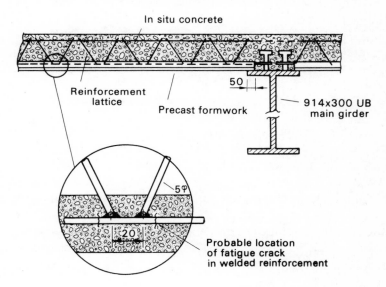

(b) Longitudinal section through composite slab

Fig. 11.2 'Omnia' precast concrete permanent formwork

a weld (fig. 11.2) and directly under a wheel load, but not until after the composite slab had been subjected to 1.1 million applications each of 9, 12, 15, 18 and 21 ton wheel loads and 700 000 cycles of a 24-ton wheel load. Even after the failure of the 5-mm diameter bars, the design ultimate flexural strength of the composite slab was achieved. The results for the fatigue resistance of the welded reinforcement lie close to the relationship proposed by Burton and Hognestad[144] for tack-welded reinforcement in concrete beams.

PRECAST FORMWORK IN A BRIDGE AT BERNE

Precast prestressed concrete slabs only 60 mm thick were used as permanent formwork for a composite beam-and-slab highway bridge of 34 m span and 25° skew that was completed in Berne, Switzerland, in 1976. The slabs are about 2.7 m wide, and support 220 mm of *in situ* deck concrete over a span of about 2.55 m (fig. 11.3). There are seven

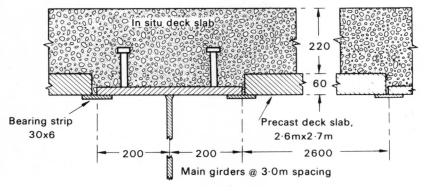

Fig. 11.3   Use of precast slabs as permanent formwork

main girders at 3 m spacing, and the slabs rest on strips welded to the tips of their flanges. All the steelwork is weathering steel. There is no interlock between the abutting edges of adjacent slabs, but the difference between the deflections of two adjacent slabs at midspan does not exceed 2 mm at the joints that were accessible for inspection. Gaps at the edges of the slabs were sealed with plastic foam strip.

Approximate calculations show that a slab of these dimensions would be strong enough to carry the wet concrete and a characteristic superimposed load of 1.5 kN/m², but not a load of 9 kN distributed over a circular area of 0.6 m radius.

Chapter Twelve

Cased Beams and Filler-beam Construction

12.1 Scope

Concrete-encased composite beams have been used occasionally in bridge decks, either to eliminate periodic repainting[145] or for reasons of appearance. The Preflex process (section 13.3) has provided a convenient method of encasing the bottom flanges of the steel members before erection, but the encasement of the webs and top flanges is usually expensive in relation to the benefits obtained. The use of weathering steel provides an alternative and sometimes cheaper solution to the maintenance problem, and the Preflex process is now less widely available, so it seems unlikely that cased beams will find much application in bridges in the future.

A filler-beam deck is essentially a concrete slab in which the longitudinal reinforcement consists of rolled steel I-sections. Such bridges can be designed and built quickly, and have been used occasionally in situations where lack of time or of alternative materials have made them economic.

There has been little research on either type of structure, so that their design has been mainly empirical. For these reasons the design methods given in Part 5 of the Bridge Code are limited in scope and, where test data are lacking, are necessarily conservative. Provision is made for simply-supported and continuous beams fully encased in normal-density concrete with cover nowhere less than 50 mm, and for simply-supported filler-beam decks in which the soffits of the beams may be encased or exposed.

12.2 Distribution of bending moments and vertical shear forces

Provision is made only for elastic methods of analysis, for these reasons.

(1) There are few test data on the ultimate static strength or the fatigue

life of cased beams of the proportions used in bridges, or of filler-beam decks.

(2) Little is known about the extent to which longitudinal shear forces at the ends of a span, due to shrinkage and cycles of temperature, combined with those due to repeated loading, can gradually cause deterioration of bond strength between concrete and the flat steel surfaces of rolled sections or plate girders.

(3) In continuous cased beams, yielding of the compression flange of the steel member near an internal support could cause spalling of the surrounding concrete. Redistribution of moments at the ultimate limit state is not recommended because no simple method is available for ensuring that premature spalling does not occur.

No detailed guidance is given on the global analysis of beam-and-slab decks using cased beams. The extent of cracking and its effect on relative stiffness of members is similar to that in reinforced concrete construction; the torsional stiffness of an encased I-section is little more than that of a reinforced concrete member of the same overall dimensions; and shear lag in concrete flanges is essentially the same as in uncased structures. For these reasons the methods of analysis used for other types of beam-and-slab deck are generally applicable.

12.2.1 Global analysis of filler-beam decks

Such structures can be analysed for non-uniform loading as orthotropic plates (section 6.4.2), but as the chief attraction of these decks is their simplicity, a much simpler method of global analysis is needed. The method given in Part 5 of the Bridge Code is now explained.

It is based on figs 4.29 to 4.31 in reference 64. These give the ratio of the maximum positive transverse moment caused by abnormal (HB) loading, M_{yb}, to the maximum longitudinal moment caused by standard or increased standard (HA) highway loading, M_{xa}, for isotropic slab decks, rectangular in plan, of breadths from 10.7 to 25.6 m and spans from 6.1 to 18.3 m.

A study was made of the applicability of these results to filler-beam decks, which are not isotropic because of the presence of the steel beams. These substantially increase the longitudinal flexural rigidity per unit width of deck (i in the notation of reference 64), but not the transverse flexural rigidity, j. The flexural parameter in the distribution analysis (θ) is proportional to $(i/j)^{\frac{1}{4}}$, and it was found that the assumption $i = j$ gives a value of θ less than 10% below the correct value. The torsional parameter α is proportional to $(i_o + j_o)/(ij)^{\frac{1}{4}}$, where i_o and j_o are the torsional rigidities. A few typical calculations showed that the contribution of the steel beams to i_o is less than 1%, if the increase in their torsional rigidity due to the prevention of warping by the surrounding concrete is neglected. Tests on encased I-beams in torsion have shown that the variation of i_o due to cracking of concrete is much greater than this. Neglecting the steel beams leads to an error in α of up to 20%.

Study of the design curves in reference 64 showed that these errors in θ and α had a relatively small influence on the maximum transverse bending moment, and that the error was on the safe side. This was confirmed by analysis of a typical deck of breadth 14 m and span 9 m, which showed that the maximum transverse moment due to 45 units of HB loading was 6% lower than that given by the curves for isotropic decks.

For spans between 6 m and 18 m, the nominal HA and HB loads given in Part 2 of the Bridge Code are essentially the same as those used by Rowe,[64] except that the increased HA loading for spans between 6.1 and 12.2 m is not given. After correction to take account of this change, figs 4.29 to 4.31 of reference 64 are as shown in fig. 12.1. The simplified design rule given in the Bridge Code (in terms of nominal loads) is:

$$M_{yb} = (0.95 - 0.04L)M_{xa} \qquad (12.1)$$

where L is the span in metres. Figure 12.1 shows that this is conservative

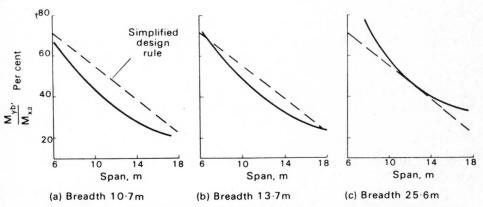

Fig. 12.1   Transverse moments in filler-beam decks due to 45 units of HB loading

but quite accurate for decks of breadth 10.7 m and 13.7 m, but is unsafe for decks of breadth 25.6 m. Its applicability has therefore been limited to decks of span between 6 and 18 m, and of breadth not exceeding 14 m. It may be used for decks with an angle of skew up to 20°, because below this limit distribution analysis gives satisfactory results[146] although the theory was developed for right decks.

Within these limits, it may be assumed that transverse moments due to 45 units of HB loading always exceed those due to HA loading. The calculation of the longitudinal moments M_{xa} has been simplified by allowing the deck to be analysed as a set of separate longitudinal strips each of width not exceeding the design width of a traffic lane. For a simple two-lane bridge with footways, only four strips would be needed

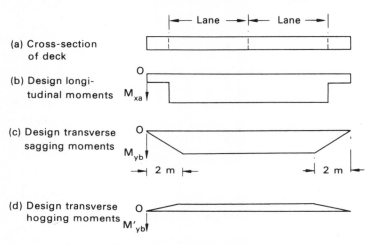

Fig. 12.2 Bending moments for a two-lane filler-beam deck

(fig. 12.2a) and the variation of the HA longitudinal moments across a typical section would then be as in fig. 12.2(b). Equation (12.1) is applicable only in regions not less than 2 m from a free edge and it is assumed that M_{yb} varies linearly from its value at 2 m from an edge to zero at the edge, so the variation of M_{yb} across the deck is found to be as in fig. 12.2(c). These assumptions are based on reference 64 as is the further rule that the negative transverse moment at any point may be taken as $0.1M_{yb}$ (fig. 12.2(d)).

It is important to remember that equation (12.1) is given in terms of bending moments due to *nominal* HA and HB loads. When used with moments due to design loads (M_y and M_x, say) it has the form:

$$M_y = (0.95 - 0.04L)M_x\alpha \qquad (12.2)$$

where α is the ratio (γ_{fb}/γ_{fa}) and γ_{fb} and γ_{fa} are the appropriate partial safety factors for HB and HA loading, respectively (tables 7.1 and 7.2). If equation (12.1) is used for design transverse moments the result is always on the safe side.

If the bridge is being designed for HA loading only, the bending moments M_x are the design longitudinal moments. The transverse moments M_y given by this simplified method may then be rather conservative, and some economy may result from the use of more accurate values determined, for example, by analysis of the deck as an orthotropic plate.

If the bridge is being designed for both HA and HB loading, it will usually be found that the design longitudinal moments are not the values M_x, because those due to HB loading will be higher.

TRANSVERSE BENDING AND VERTICAL SHEAR
The preceding method is based on the assumption that the filler joists are sufficiently closely spaced for their effects to be 'smoothed' across the breadth of the deck. If their bottom flanges are exposed it is inconvenient to provide much bottom transverse reinforcement, and if the deck is designed for HA loading only, the amount required to resist transverse sagging moments may be very low. There would then be little to prevent a vertical shear failure between widely-spaced steel girders (fig. 12.3). For

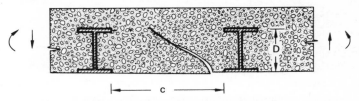

Fig. 12.3   Shear failure in filler-beam deck

these reasons, the simplified method includes the rule that the dimension c in fig. 12.3 may not exceed two-thirds of the depth D of the girders. There is also a rule for minimum top transverse reinforcement, to resist the splitting effect of longitudinal shear at the steel-concrete interface and local negative moments due to wheel loads.

12.3 Analysis of cross sections for longitudinal bending and shear

The methods of Part 5 of the Bridge Code, as amended by the Department of Transport,[2] can be summarised as follows.

(1) The composite section is classified as compact if the steel section satisfies the criteria given in Part 3, treating any steel flange or web without shear connection as if it were uncased.

(2) The methods of analysis for bending are as for uncased beams, but including concrete encasement in compression and longitudinal reinforce-

ment in tension as part of the cross section. In filler-beam decks, shear lag may be neglected at the serviceability limit state.

(3) Vertical shear is resisted by the steel section alone. This is obviously conservative, particularly for filler-beam decks, but there is little demand for a more accurate method.

No guidance is given on design of encased bottom flanges to resist lateral-torsional buckling, probably because of the lack of relevant research, so they are treated for this purpose as if uncased.

12.4 Longitudinal shear in cased beams

The limit-state design of cased beams without shear connectors must be based on rational criteria for both the serviceability and ultimate limit states. The following explanation of the methods given in Part 5 of the Bridge Code is based mainly on studies for the design of cased beams in buildings[147] (where they are more common than in bridges) and on reports of two series of tests.[148,149]

INTERPRETATION OF RESULTS OF TESTS

The behaviour in longitudinal shear of a cased beam under increasing load has several phases. Initially there is *adhesion* between concrete and steel, without slip. The maximum interface shear stress, the *bond strength*, f_b, corresponds to the very small slip of about 0.1 mm. There is substantial redistribution of interface shear stress before the bond strength is reached. It is convenient to relate it to a calculated *local bond stress* $f_{b\ell}$. This is the mean shear stress round the perimeter of the encased steel section, and can be calculated in design from the rate of change along the member of the longitudinal force in the steel (or the concrete). In most tests on cased beams, there have been two lengths of beam (the shear spans) in which the vertical shear (and hence the longitudinal shear, in the elastic range) is constant. The mean local bond stress along such a length is the *anchorage bond stress*, f_{ba}, and it is this which can reliably be found from tests. Because $f_{b\ell}$ and f_{ba} are approximately equal in the test specimens, it is assumed that the test value of f_{ba} so found is a suitable limiting value for $f_{b\ell}$ in a beam in which longitudinal shear varies along the span.

'Bond failure' is an imprecise term implying that the slip is greater than that associated with the bond strength. At bond failure, there is usually a slight increase in the deflection of the beam, and some longitudinal cracking of concrete where the cover is small or, in a T-beam, at the web-flange junction. Bond failure, as here defined, can be considered as a serviceability limit state.

After bond failure, there is progressive slip, cracking, and deflection, and shear is transferred by friction. The *frictional shear stress* depends on the constraints which prevent separation of the surfaces. It can be lower than the bond strength, but it increases with the number or size of stirrups, the thickness of the concrete cover, and in regions where vertical loads are applied to the top of the concrete slab. A beam in which the total frictional shear is less than that needed to develop the plastic moment of resistance of the composite member is similar to a beam with incomplete shear connection.[113] Such beams occur in buildings, and a partial-interaction design method for them at the ultimate limit state is given in the draft Eurocode 4.[5]

The design method of Part 5 is based on the two requirements suggested by Hawkins.[149] These are that the load for bond failure should exceed the load at first yield, and that the available constraints after bond failure should be sufficient to ensure a frictional shear stress large enough for the ultimate load to exceed the load for bond failure (as otherwise there would be no warning of impending failure).

CALCULATION OF LONGITUDINAL SHEAR

In uncased beams, longitudinal shear is calculated assuming the concrete to be uncracked and unreinforced because this gives a result on the safe side in negative moment regions, where it is not possible to predict accurately the extent of cracking. Concrete in tension rarely occurs in positive moment regions of uncased members, other than composite plates.

For cased beams, in which the neutral axis can be assumed to lie within the steel section, the longitudinal shear in midspan regions is slightly reduced by cracking. This is shown in fig. 12.4. The longitudinal shear per unit length per unit vertical shear is given by elastic theory as $A\bar{y}/I$, where A is the excluded area, $\bar{y}$ is the distance of its centroid from the neutral axis, and I is the relevant second moment of area of the section. The variation of $A\bar{y}/I$ across the depth of the section is given for the encased section shown, in which the thicknesses of the steel plates are: top flange, 20 mm; web, 12 mm; and bottom flange, 40 mm. The slab reinforcement is included only in the calculation for the cracked section with negative moments. The curve for the uncracked section is applicable in both positive- and negative-moment regions, and gives a value at the interface (point A) higher than that for the cracked section (B). The difference is small, and there is certain to be some cracking in this type of section, so for uniformity with the method for uncased beams Part 5 of

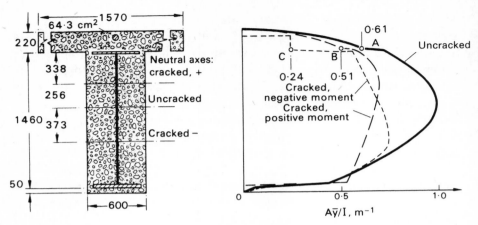

Fig. 12.4 Influence of cracking on calculated longitudinal shear in encased girder

the Bridge Code requires that the *cracked* cross section be used in midspan regions.

In negative-moment regions of cased beams, the shear based on the uncracked section (point A) is significantly higher than that for the cracked section (point C), so again the design rule can be the same as for uncased beams. To sum up:

> *Longitudinal shear and bond stresses calculated by the elastic theory are always based on the cracked section in positive-moment regions and the uncracked section in negative-moment regions of continuous members.*

Shear lag can be neglected at the ultimate limit state. Earlier comments on this subject (section 8.4.1) are applicable also to encased beams.

DESIGN SHEAR STRESSES IN BOND AND FRICTION

Mean concrete cube strengths for the tests on cased beams considered above ranged from 25 to 32 N/mm^2. The data are not sufficient to justify use of design values related to cube strength. A beam with 25 mm cover tested by Hawkins[149] had a mean bond strength of 0.59 N/mm^2 at 79% of the yield load. In all three of the beams with 50 mm cover, yield preceded bond failure, which occurred at 0.60, 1.03 and 0.81 N/mm^2. Several tests by other workers have given higher results. The design local bond stress in CP 117: Part 2 was 0.41 N/mm^2. This was increased to 0.5 N/mm^2 at the serviceability limit state for Part 5 of the Bridge Code, but account has been taken of possible weakness due to poor compaction of concrete by excluding the soffits of the steel flanges from the effective perimeter of the steel section.

The higher value of local bond stress given for filler beams (0.7 N/mm^2) takes account of the greater degree of containment of the steel I-section, and the conservative nature of the method of calculation when applied to a cross section with a large area of concrete in tension.

At the ultimate limit state, provision is made in Part 5 of the Bridge Code only for longitudinal shear failure on surfaces of type 5-5 in figs 8.24(d) and 12.5 because in tests failures have occurred on this surface rather than round the perimeter of the steel section. The method of calculation of the shear strength of this surface is the same as for other shear surfaces (section 8.6), except that in calculating L_s, account is taken of only two-thirds of the length of the steel-concrete interface, because the friction that can be developed on this plane is likely to be less than the shear strength of concrete. The design frictional shear stress implicit in this method is 0.6 N/mm^2 (two-thirds of the coefficient 0.9 in equation (8.81)), which is below the few values that can be deduced from tests (e.g. 1.38 N/mm^2 in reference 148).

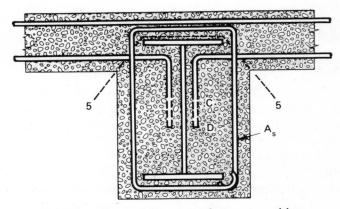

Fig. 12.5   Shear reinforcement for an encased beam

Where the steel section projects above the soffit of the slab it is convenient to use a combination of stirrups and bottom slab reinforcement to provide the required area A_e of shear reinforcement, as given by equation (8.81).

It can easily be shown by typical calculations that for shallow cased beams the limiting bond stress of 0.5 N/mm^2 at the serviceability limit state will ensure that relatively light shear reinforcement is sufficient for surface 5-5 at the ultimate limit state; but for deep cased beams, surface 5-5 may require heavy reinforcement for shear. No shear tests on cased beams of this type are known to the authors, and it is possible that other types of shear failure could occur if the restraint offered by the casing is weak. A significant proportion of the reinforcement A_e should therefore be in the form of closed stirrups and any slab reinforcement that is below the level of the top steel flange should be anchored by extending it along the face of the web either horizontally or vertically (CD in fig. 12.5).

MINIMUM STIRRUPS IN CASED BEAMS

In Part 5 of the Bridge Code, the rule for minimum stirrups is similar to that for minimum bottom transverse reinforcement in haunched beams (section 8.6), but with the length of the shear plane taken as that of the relevant plane of type 5-5 in fig. 12.5.

This requirement for minimum stirrups exceeds by about 60% that in Part 4 of the Bridge Code for stirrups in reinforced concrete beams. The increase has been made to improve ultimate-load behaviour in longitudinal shear and, in the absence of sufficient test data, to substantiate a less conservative method.

The maximum spacing of stirrups is given as 600 mm, to be consistent with the rule for spacing of transverse reinforcement in the slab.

12.5 Effects of temperature and shrinkage in cased beams

The distribution of temperature in a composite bridge deck with cased beams is likely to be similar to that in a reinforced concrete structure of similar shape, as the more rapid thermal response of steel becomes much less significant when it is insulated from the environment by a layer of concrete. In other respects the calculations of primary and secondary moments and vertical shears due to both temperature and shrinkage are as for a similar uncased structure. The primary longitudinal force due to shrinkage will be larger, but as the corresponding shear at an end support acts in the opposite direction to the shear due to imposed load, it need not be considered in design. Due to the encasement of the web, the line of action of the longitudinal force in the concrete is closer to the centroid of the composite section, so the shrinkage curvature and hence the secondary stresses due to shrinkage will be lower in the cased structure.

Shear connection by bond is very stiff, so that the longitudinal shear force at the free ends of a cased beam due to temperature effects (Appendix A) could cause local bond failure. There is no simple method of checking this, so Part 5 of the Bridge Code requires shear connectors to be provided for this force, as for uncased members.

Although the shapes of shear/slip curves for shear connectors and for bond followed by friction are quite different (fig. 12.6), a high proportion of the bond strength remains at slips of 0.2 to 0.3 mm, at which studs and

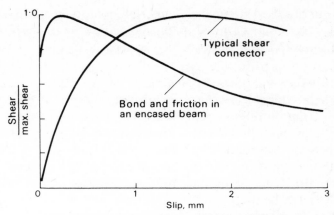

Fig. 12.6 Curve relating longitudinal shear to slip

other shear connectors reach their design load at the serviceability limit state. This shows that shear connectors and bond-friction can be used together in cased beams, at least to the limited extent that is proposed here. Their general use in parallel, to resist imposed loads, has not been allowed, due to the lack of test data (particularly fatigue tests) on cased beams designed in this way.

In filler-beam construction, there is no requirement in Part 5 of the Bridge Code for temperature and shrinkage effects to be considered, because there is no evidence from structures in service to suggest that this is necessary. It is assumed that the normal detailing rules and construction processes for reinforced concrete will be followed, to avoid cracking due to shrinkage. In trial calculations on filler decks using a wide range of universal beam sections at various spacings, it was found that design for bending at midspan was always governed by combination 1 loads at the ultimate limit state. When the top of the slab is at a higher temperature than the soffit, longitudinal shear due to temperature acts in the same direction as that due to imposed load, but it is reduced by the effects of shrinkage. Trial calculations showed that the net effects of temperature and shrinkage could safely be neglected in simply-supported filler decks.

12.6 Control of cracking

There has been no research on control of cracking in cased beams and filler beams, as the economic benefit from such work is likely to be small. For consistency with reinforced concrete design, the equation for tension stiffening given in Part 5 of 1979 has been replaced[2] by the one in Part 4 of 1984, equation (8.2), with the consequence that no account is now taken of any contribution from the encased steel tension flange to the control of crack spacing, and hence to tension stiffening.

It is necessary to provide longitudinal bars in the side faces of deep cased beams to control flexural cracking. A simple rule is given in Part 5 for bar size and spacing in the sides of cased beams of overall depth exceeding 750 mm. It relates the bar diameter and vertical spacing to the yield stress f_y of the bars and the breadth (i.e. overall thickness) of the casing at the level considered. This rule is the same as that in CP 110[189] and is based on research on reinforced concrete. If a more accurate method is needed, the equations for crack width in uncased beams (section 8.3.3) can be used.

Filler-beam decks with full encasement of the steel sections are treated for longitudinal bending in the same way as cased beams.

In the transverse direction, a filler-beam deck is essentially a reinforced concrete slab. It is assumed in Part 5 of the Bridge Code that bottom transverse reinforcement will be provided in accordance with the method of Part 4 for the control of cracking in slabs. In fully-encased decks these bars should pass under the bottom flanges of the steel beams to hold the bottom cover firmly in place but, where the bottom flanges are exposed, they have to be threaded through holes in the steel webs. The flanges are likely to be left exposed only when the risk of corrosion is low, and the Code requires that both surfaces of such flanges should be fully protected against corrosion.

Chapter Thirteen

Prestressing of Composite Beams

13.1 Introduction

The more widespread use of prestressed composite beams in continental Europe than in Britain or North America appears to be due to the different approaches that have been followed in these countries to the problem of controlling flexural cracking in concrete bridge decks. These approaches were discussed in section 6.2.3, and are now briefly re-stated. In Britain and North America it has long been the practice to permit limited tensile cracking in concrete bridges, other than in class I prestressed concrete members, provided that the deck has a waterproof membrane. There has also been extensive research[155, 156] on the prediction of crack widths in reinforced concrete members, much of which has been incorporated into recent codes of practice. In contrast to this, the practice in many European countries is to prevent tensile cracking by prestressing the concrete.

The authors are not aware of any serviceability failures due to excessive flexural cracking in bridge decks designed to British specifications that incorporate crack control clauses, although there is ample evidence of this type of failure in concrete bridges designed to earlier standards that did not include any check on crack widths. Prestressing does not significantly increase the ultimate strength of a composite beam for the reasons explained in section 15.1, so that if British experience with non-prestressed decks continues to be favourable, then the use of prestressing in Europe is likely to decline.

The following account of methods for prestressing composite beams by jacking the steelwork is based mainly on European experience.[14d] The associated losses of prestress are discussed in section 6.2.3.

13.2 Methods of prestressing

Part 5 of the Bridge Code lists five methods by which prestressing can be achieved in a composite beam: but, there are others. The system chosen will depend on the particular advantages sought by the designer. The five methods are:

(1) A system whereby a bending moment is applied to the steel section, such that the stresses produced are of the same sign as those produced by the full design loading. The tension flange of the steel beam is then encased in concrete and the applied moment released when the concrete has attained sufficient strength, thereby introducing precompression in the concrete flange.

(2) A system of jacking to alter the relative levels of supports of a continuous member after part or the whole of the concrete slab has been cast and matured.

(3) Prestressing the concrete slab, or sections of the slab by tendons or jacking whilst it is independent of the steel beam, and subsequently connecting them.

(4) Prestressing the steel beam by tendons prior to concreting. The tendons may or may not be released after the concrete has matured.

(5) Prestressing the entire composite section by tendons or jacking.

The Preflex system (section 13.3) is an example of method (1). It is virtually the only type of prestressed composite beam found in Britain. Once competitive for bridges with spans in the range 25 to 40 m, it has now been superseded by standard section-pretensioned prestressed concrete beams, which have two of the principal advantages claimed by the Preflex system viz, an increased span:depth ratio, and a maintenance-free high-quality concrete casing to the steel beam. Because of this limited use of prestressed composite beams in Britain, no detailed recommendations for their design are given in Part 5 of the Bridge Code.

Of the more generally applicable methods, those systems that introduce prestressing as part of the erection procedure have generally been found in European countries to be more economic than when tendons are employed.[14d] The reason is probably that in most major bridges, the erection procedures are complex, whether the bridge is prestressed or not, so jacking does not then cause a significant increase in cost.

A common method of introducing precompression in the concrete slab of a composite beam is by raising the steelwork above its level in the finished bridge, prior to casting the deck slab. After the slab (or part of it) has been cast and has matured, the composite section is lowered to its final level. Two versions of this process are shown in figs 13.1 and 13.2.

(1) Slab cast whilst the steel beams are prestressed by jacking from temporary falsework within the span.

(2) Falsework removed after the slab has matured.

Fig. 13.1 Prestressing simply-supported spans from temporary supports within the span

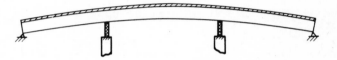

(a) Slab cast after jacking the steel beam upwards.

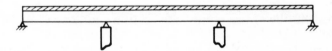

(b) Composite section lowered onto its final supports after the slab has matured.

Fig. 13.2 Prestressing continuous spans by jacking from supports

In continuous multi-span bridge decks, the distance through which the steel beams must be raised at the supports in the latter method becomes excessive. Roik[14d] gives a solution to this problem using a continuous bridge deck of eight 50-metre spans as an example. To obtain a useful amount of precompression in the concrete slab, the steelwork would have to be raised initially by 6 m at the central support, as shown in fig. 13.3(a). The same result can be achieved by introducing hinges as shown in fig. 13.3(b) and cambering each section of the steel beam so that the curvature is as in fig. 13.3(a). The concrete deck slab is then cast in stages. Once the concrete has attained sufficient strength, the deck is prestressed by applying moments at the hinge positions. This can be done using jacks placed in gaps left in the slab at the hinge positions (fig. 13.4) and using bottom-flange splice plates to transmit the corresponding tensile force. After jacking, the gap between the ends of the concrete deck slab is filled with concrete, leaving the jacks in pockets until the infill has matured. The temporary web cover-plates are then replaced by a full-strength splice.

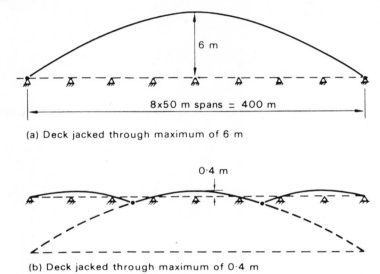

(a) Deck jacked through maximum of 6 m

(b) Deck jacked through maximum of 0·4 m

Fig. 13.3   Prestressing by induced cambering

The prestress could equally well have been introduced by jacking upwards from temporary falsework towers at the hinge positions. In both cases, the final level of precompression throughout the length of the deck slab, the reactions at the supports, and the stresses in the steelwork will be the same as if the precambering shown in fig. 13.3(a) had been used.

The amount of initial precompression that can be developed in the concrete slab is limited only by the compressive strength of the concrete,

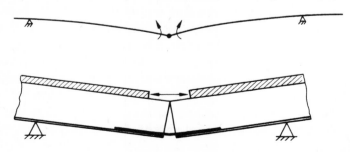

Fig. 13.4   Prestressing moment applied at a hinge

provided the stability of the steel beam during prestressing is ensured, but losses of prestress can be high (section 6.2.3).

When the deck is prestressed by means of tendons placed in the concrete slab, it is normal practice to prestress only the regions near internal supports (fig. 13.5(a)), since imposed loading will cause compression elsewhere. External tendons (fig. 13.5(b)) have occasionally been used, and can be useful for strengthening existing bridges.

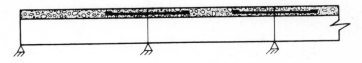

(a) Prestressing by tendons in the concrete deck slab

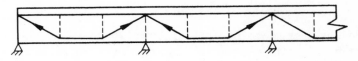

(b) Prestressing by external tendons in truss action

Fig. 13.5 Prestressing by tendons

The methods so far discussed assume that the concrete slab is cast *in situ* and acts compositely with the steel girder before the slab is prestressed. If it is precast, the slab (or parts of it) can be prestressed before composite action is developed, provided a satisfactory shear connection can be achieved. The problems of detailing precast slabs and making the shear connections are discussed in sections 6.2.1 and 8.5.3.

13.3 The Preflex beam

The Preflex beam was first conceived in 1949 by Lipski, a Belgian consulting engineer, although it is reported[14h] that attempts to reduce tensile cracking in the concrete encasement of cased steel beams by preloading the steel beam had been made independently in Britain as early as 1939 by James Drake (then Borough Engineer of Blackpool). The design rules for Preflex beams in use in the 1970s[157] differed little from those put forward by Baes and Lipski[158] between 1951 and 1954. Reference 157 has now been withdrawn, as the beams are no longer marketed in the UK. The following outline of their use is restricted to simply-supported beams.

The Preflex system consists essentially of preloading a steel beam, usually by two loads applied at quarter span, in such a way that it is bent in the same sense as it would be under its design working load. The tension flange is then encased in a high-strength concrete and the point loads are removed after the concrete has gained sufficient strength. Their removal induces a precompression in this concrete, which reduces with time due to the effects of shrinkage and creep. The *in situ* concrete top flange and the encasement to the web are added after the Preflex beam is in position in the bridge. It is normal practice to precamber the steel beam upwards so that, under dead load, the bridge deck presents a hogging profile.

The advantages to be gained by preflexing a steel beam are as follows.

(1) An increase in the stiffness and a corresponding decrease in deflection over the range of working load compared to a cased or uncased composite beam of the same size.

(2) Permanent encasement of the structural steelwork by a high-strength concrete in which there are fewer and narrower cracks at working load than in a cased beam of the same size.

(3) An increase in the span:depth ratio compared with an uncased composite beam designed for the same loading, when design is based on elastic theory at working load levels.

Preflexion does not increase the ultimate flexural strength of the beam, since cracking of the concrete surrounding the tension flange before collapse 'releases' the restraint afforded by the concrete encasement.

The Department of the Environment's design method[157] for Preflex beams in highway bridges was based on elastic theory with limiting stresses related to working loads, although there is limited evidence[159] to support the use of simplified rectangular-stress-block theory (as described for compact uncased beams) to predict the ultimate flexural strength of Preflex beams with shear connectors on the top flange.

Table 13.1 Nominal static strengths of horseshoe-type shear connectors

Connector material	Nominal static strengths (kN) for concrete strengths f_{cu} (N/mm²)			
	20	30	40	50
Mild steel, Grade 43 to BS 4360	130	150	170	170

LONGITUDINAL SHEAR

It used to be standard practice to provide horseshoe shear connectors on the tension flange of the steel beam. Where they are used, the number of connectors should be sufficient to transmit the longitudinal shear force between the steel flange and the precast casing on release of the preflex loads. As for other types of shear connector, the design static strength of horseshoe connectors at the serviceability limit state can be taken as $0.55P_u$. Values of P_u for horseshoe connectors of the type and size shown in fig. 13.6 are given in table 13.1 for different grades of concrete. They were determined by applying the 'lowest of three results' procedure described in section 7.2.3 (p. 67) to the results of push-out tests[160] carried out in 1964. Horseshoe connectors are not used in conventional composite beams because they provide no resistance to uplift.

The longitudinal shear force per unit length is calculated by elastic theory neglecting concrete in tension, as for cased beams (section 12.4). To reduce the risk of fatigue failure occurring in the welds of connectors attached to the tension flange, reference 157 recommended that no connectors should be attached between the loading points for preflexion (i.e. over the region of maximum tensile stress in the flange). Early repeated loading tests on U-shaped connectors[14h] showed that fatigue failure was usually initiated in the weld at the end of the connector. In horseshoe connectors the ends of the welds are staggered (fig. 13.6). This avoids having more than one concentration of stress at any single cross section.

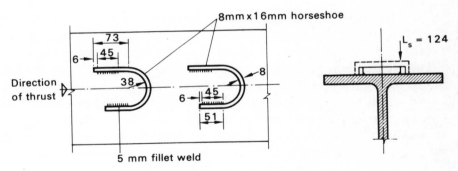

Fig. 13.6 Horseshoe-type shear connectors

The only fatigue tests on horseshoe connectors known to the authors are those of Menzies[161] which were discussed in the first edition of this volume.

The design of the shear connection between the *in situ* concrete top flange and the steel beam is as described for cased beams in chapter 9, except that in highway bridges and footbridges, bond to the steel beam should be assumed to be developed only over both sides of the web and the upper surface of the top flange, since the precast concrete casing to the bottom flange may be cracked under maximum service loading.

LOSS OF PRESTRESS

For simply-supported Preflex beams, reference 157 gave a detailed method for estimating the losses of prestress due to the shrinkage and creep of concrete that takes place between release of the preflexion loads and the casting of the *in situ* top flange; alternatively, it allowed the loss (from values immediately after release) to be taken as 50%. The origin of these procedures is not known. Measurements carried out by the Transport and Road Research Laboratory[162] on five Preflex beams used in a road bridge in Wales found an average loss of prestress of 10% prior to release and a further 30% over the next 15 months, after which the creep and shrinkage in the precompressed flange were substantially complete.

Baes and Lipski's method for calculating losses in Preflex beams was based on work by Froelich (1950) and Dischinger (1939). Other methods for calculating the effects of creep and shrinkage in prestressed composite beams include those of Bruggeling (1964) and, more recently, Haensel.[98]

Chapter Fourteen

Composite Columns

14.1 Introduction

The use of composite columns in bridge structures is less frequent than in buildings, where concrete encasement is also often used as a convenient method of providing fire protection to the stanchions in steel-framed buildings. In Britain, the most widely reported[145,163] use of composite columns in bridgeworks is in the four-level interchange on the M5 motorway at Almondsbury, near Bristol, where pin-ended concrete-filled steel tubes were used to advantage in a situation where the cross sections of reinforced concrete columns would have been unacceptably large.

At that time, there was little information available to the practising engineer on the design of composite columns, but a general ultimate strength theory to describe their behaviour had recently been completed by Bondale at Imperial College.[164] This analysis was used to check the initial design of the columns at Almondsbury. Tests were also carried out at the Building Research Station on half-scale columns to support the results from the theoretical analyses.

Research on composite columns was continued at Imperial College by Basu, who published in 1967 a theoretical analysis[165] for the ultimate strength of pin-ended concrete-encased steel sections and concrete-filled rectangular hollow sections (RHS), subject to any combination of end load and end moments about one or both of the two principal axes. Basu used his analysis to study systematically a wide range of columns having different cross sections and slendernesses, under combinations of end load and end moments, and found that the behaviour of such columns could be adequately predicted by four parameters for which he derived algebraic expressions. The design method developed was published by Basu and Sommerville in 1969.[166] Subsequently, numerical methods were developed[167,168] for the full-range inelastic analysis of composite columns with elastic end restraints, subjected to any combination of axial load and biaxial end moments, and were shown to agree well with the results of tests.[169]

14.2 The ultimate-strength design method of the Bridge Code

The design method for composite columns in Part 5 of the Bridge Code is basically the Basu and Sommerville method, which is described in Volume 1. However, it has been modified where design studies[170-172] have shown the original method to be unduly conservative, or where simplifications have been found possible.

The method assumes that the problem of an elastically restrained column subjected to any combination of end loads and moments can be reduced to that of an equivalent pin-ended column. The design of this length of pin-ended column is considered in section 14.3, and the replacement of the restrained column by an equivalent pin-ended column in section 14.4.

Only single-storey braced and unbraced frames are considered in Part 5 of the Bridge Code, and it is assumed that members framing into one another are rigidly connected. Steel cross sections have to be symmetrical about both principal axes. The concrete section and its reinforcement are not required to have the same centroid or be doubly symmetric, although this was assumed when the design methods were developed, because a small degree of asymmetry is acceptable.

The design resistances for columns given in Part 5 *do not include* γ_{f3}. As in Part 4, it is assumed that this factor (1.1 at the ultimate limit state) is applied to the loading. This differs from Part 3, in which expressions for the resistance of steel compression members do include γ_{f3}.

14.3 Pin-ended columns

The loading on the column is defined by the axial load, N, and end moments M and βM as shown in fig. 14.1, where β is the ratio of the smaller to the larger end moment in a particular plane of bending. In this chapter, as in Part 5 of the Bridge Code, the subscript x or y is used in conjunction with M (and other symbols) to denote bending about the major or minor axis, respectively. Where no suffix is used, it may be assumed that the parameter is that relevant to the axis being considered at the time. Except where significant transverse loads are present, the effects of transverse shear forces have been found to be negligible in tests and so need not be considered further.

14.3.1 Properties of the cross section

Three of the four main parameters required for the prediction of the

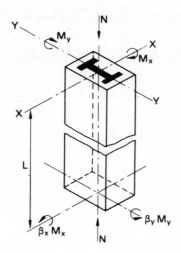

Fig. 14.1 Positive directions of end loads and moments acting on a composite column

failure load of an isolated pin-ended column are properties of its cross section. They are the squash load N_u, the concrete contribution factor α_c, and the ultimate moment of resistance in pure bending, M_u. The expressions for them are applicable to all composite sections that fall within the scope defined in section 14.2, other than circular hollow sections (CHS), which are treated in section 14.3.5.

The *squash load* is the ultimate axial load that a short column can sustain, assuming that the steel is yielding in compression and the concrete is crushing. Its design value is given in Part 5,[1] as amended,[2] by:

$$N_u = 0.95 A_s \sigma_y + 0.87 A_r f_y + 0.45 A_c f_{cu} \qquad (14.1)$$

where A_s, A_r and A_c are the cross-sectional areas of structural steel, reinforcement, and concrete, respectively, and σ_y, f_y and f_{cu} are the nominal or characteristic strengths of the materials.

The last two terms in equation (14.1) are as in Part 4 for reinforced concrete columns. The limiting stress in concrete, $0.67 f_{cu}/\gamma_m$, with $\gamma_m = 1.5$, is the same as for elastic-plastic analysis of composite beams (section 8.4.3). The 'steel' term incorporates $\gamma_m = 1.05$, which is consistent with the treatment of short axially loaded compression members in Part 3.

The *concrete contribution factor* is the ratio of the axial load carried by the concrete in a short column to its squash load. It is given by:

$$\alpha_c = 0.45 A_c f_{cu}/N_u \qquad (14.2)$$

The scope of the method is restricted by limits on α_c given in Part 5:

For concrete-encased sections $0.15 \leqslant \alpha_c \leqslant 0.8$
For concrete-filled hollow sections $0.10 \leqslant \alpha_c \leqslant 0.8$

These limits correspond to the range of cross sections studied by Basu and Sommerville[166] and are not restrictive in practice.

The *ultimate moment of resistance* of the composite section has to be determined for the axis of bending considered. The methods used are the same as for the plastic moment of resistance of a composite beam (section 8.4.3). The relevant assumptions are stated in Appendix C of Part 5, together with equations for calculating M_u for encased sections and concrete-filled rectangular hollow sections. The theory is given in chapter 5 of Volume 1.

For concrete-filled circular hollow sections, the principles and assumptions are the same, but the expressions for M_u are more complex. A design chart for M_u is therefore given in Part 5. It is reproduced in Appendix C (fig. C.5), where its construction and use are explained.

14.3.2 Axially loaded columns

The failure load of a concentrically loaded column decreases as its slenderness increases. For a perfectly elastic, perfectly straight pin-ended steel strut, the axial failure load, N_a, is equal to the *Euler load*, N_E, so the ratio of the axial failure load to the squash load is given by:

$$\frac{N_a}{N_u} = K_1 = \frac{\pi^2 E_s I_s}{L^2 A_s \sigma_y} = \frac{\pi^2 E_s}{\sigma_y (L/r)^2} \qquad (14.3)$$

where r is the relevant radius of gyration of the section.

In practice, initial imperfections and material yielding cause the column to fail about its minor axis, unless constrained to do otherwise at a load lower than the Euler load, but the failure load is still primarily a function of the column slenderness, so that design rules such as the well known Perry-Robertson curve shown in fig. 14.2 have been used instead.

For composite columns of rectangular section, Basu and Sommerville adopted a curve for K_1 that was a lower bound to computed results for axially loaded columns obtained using Basu's theoretical analysis[165] and an expression for the equivalent radius of gyration of a composite column. The curve for K_1 lies close to the Perry-Robertson curve for mild steel struts (fig. 14.2) except at high slenderness ratios. Design studies[170,171] revealed that in some situations, Basu's K_1 curve predicted failure loads for encased sections that were lower than that of the bare steel strut. To overcome this problem, Virdi and Dowling[172] related the

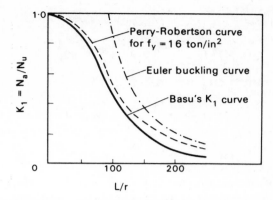

Fig. 14.2 Comparison of various basic strut curves for axially-loaded columns

failure load of an axially loaded composite column to the strut curve for the bare steel section, taking account of the contribution of the concrete to the strength and stiffness of the composite section by a new definition of slenderness, λ.

The *slenderness parameter*, λ, is defined in Part 5 for a given plane of bending by:

$$\lambda = \ell_e/L_E \qquad (14.4)$$

where ℓ_e is the effective length of the column considered, and L_E is the length of the column for which its Euler load, N_E, and squash load, N_u, are equal. As N_u (equation (14.1)) includes values of γ_m, the same partial safety factors were applied[2] to the flexural rigidities to obtain an equivalent 'design' value of the Euler load:

$$E_dI_d = 0.95E_sI_s + 0.87E_rI_r + 0.45E_cI_c \qquad (14.5)$$

whence:

$$N_E = \pi^2 E_dI_d/L_E^2 \qquad (14.6)$$

Putting $N_E = N_u$ in equation (14.6) and substituting in equation (14.4) gives

$$\lambda = (\ell_e/\pi)(N_u/E_dI_d)^{\frac{1}{2}} \qquad (14.7)$$

This slenderness parameter can be related to the slenderness ℓ_e/r that is used for all-steel columns, as follows. If the column consists only of steel, $I_r = I_c = 0$, and $I_s = A_s r^2$. From equations (14.1) and (14.5):

$$N_u = 0.95A_s\sigma_y$$

and:

$$E_dI_d = 0.95E_sA_s r^2$$

Then from equation (14.7):

$$\lambda = \frac{\ell_e}{\pi}\left(\frac{\sigma_y}{E_s r^2}\right)^{\frac{1}{2}} = \frac{\ell_e}{r}\left(\frac{\sigma_y}{355}\right)^{\frac{1}{2}}\left(\frac{355}{205\,000\pi^2}\right)^{\frac{1}{2}}$$

where σ_y is in N/mm^2 units. Rearranging:

$$(\ell_e/r)(\sigma_y/355)^{\frac{1}{2}} = 75.5\lambda \qquad (14.8)$$

THE EUROPEAN COLUMN CURVES

The behaviour of an axially loaded steel strut cannot be accurately predicted by a single strut curve, as it is influenced by the geometry of the cross section, the magnitude and distribution of the residual stresses in the cross section, and initial imperfections in straightness that arise from the methods of production and fabrication. The European Convention for Constructional Steelwork developed four curves of Perry-Robertson type, that relate the design ultimate axial stress in the steel section (σ_c in Part 3) to the slenderness of the column length. A fifth curve was added later, and all five are given in draft Eurocode 3,[4] labelled a_0, a, b, c and d. The slenderness is represented by the parameter λ defined above (though in EC3 the symbol $\bar{\lambda}$ is used), and the curves differ only in the imperfection parameter used. This parameter is similar to that for lateral buckling, which is explained in section 8.4.4. All the curves have a plateau ($K_1 = 1$) where $\lambda < 0.2$. From equation (14.8), this corresponds to $(\ell_e/r)(\sigma_y/355)^{\frac{1}{2}} < 15$, and it appears in Part 3 in this form.

The design method given in Part 3 of the Bridge Code is almost identical with that of Eurocode 3, but the presentation is different. The column curves relate σ_c/σ_y to $(\ell_e/r)(\sigma_y/355)^{\frac{1}{2}}$, and there are four of them, labelled A, B, C and D. Curve A lies between the Eurocode curves a_0 and a; curves B and b coincide; and curves C and D lie slightly above curves c and d. Consequently the rules given in Part 3 that determine which curve is used differ slightly from those in EC3. Only the former are referred to here. The curve to be used depends on the method of fabrication of the column length, and on the ratio of the relevant radius of gyration of its cross section, r, to half of its overall depth in the plane of bending, y. For rolled steel I-sections, the relationship is as follows.

All sections with flange thickness > 40 mm	Curve D
Other sections: $r/y \leqslant 0.45$	Curve C
$0.5 \leqslant r/y \leqslant 0.6$	Curve B
$r/y \geqslant 0.7$	Curve A

The column curves and these rules are given in a different form in the 1979 edition of Part 5, but have now been deleted.[2] For a composite column, one now calculates λ from equations (14.1), (14.5) and (14.7), and converts it to an equivalent steel slenderness using equation (14.8). The ratio σ_c/σ_y obtained from the curve in Part 3 is taken as K_1, and from equation (14.3):

$$N_a = (\sigma_c/\sigma_y)N_u \tag{14.9}$$

To validate the method of relating the strength of a composite column to that of the steel section used in it, Virdi and Dowling[172] compared the results of computed failure loads for a wide range of axially loaded composite columns with all the available test results on composite columns, and with the results predicted by the European curves for bare steel struts using different values of E_c in equation (14.5). They found that there was good agreement between theory and tests when E_c was taken as $1000f_u$. For design, when f_u is taken as $0.67f_{cu}/\gamma_m$ and γ_m is equal to 1.5 at the ultimate limit state, the expression for E_c becomes:

$$E_c = 450f_{cu} \tag{14.10}$$

This expression is given in Part 5 of the Bridge Code, for use only in calculating the slenderness parameter, λ. Elsewhere in the analysis of composite columns (e.g. calculation of relative stiffness of members) the appropriate value of E_c given in Part 4 is applicable. These range from $600f_{cu}$ to $1250f_{cu}$, depending on the cube strength.

Virdi's computed results for composite columns designed to curve c using a value of $E_c = 1000f_u$ are shown plotted against the European curve in fig. 14.3. In general there is good agreement, but over the high

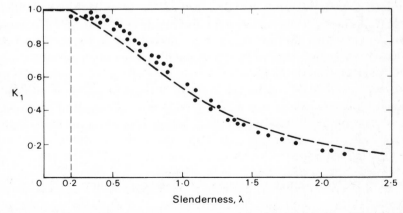

Fig. 14.3 Comparison between computed results for rectangular columns and column curve c

slenderness range ($\lambda > 1.0$), the computed results lie just below the design curve adopted. Figure 14.4 shows the results of all the available relevant tests on axially loaded composite columns plotted against the same curve. Most of the test results lie above it, but the majority relate to columns of low or medium slenderness ($\lambda < 1.0$). The lack of test results for slender columns where the computed results lie below the design curve is the reason why limits on the slenderness of columns are imposed in Part 5 of the Bridge Code. They are not restrictive in practice.

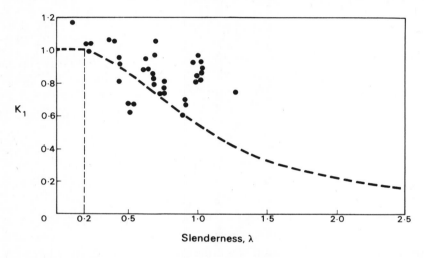

Fig. 14.4   Comparison between test results for rectangular columns and column curve c

ALLOWANCE FOR CONSTRUCTION TOLERANCES IN AXIALLY LOADED COLUMNS
In practice, eccentricities of loading can arise in columns which would otherwise be designed as axially loaded, due for example to tolerances in the position of bearings. These result in failure of the column at a slightly lower load than predicted for a concentrically loaded column. To allow for this, Part 5 of the Bridge Code requires a designer to assume the end load to be applied at a minor-axis eccentricity of not less than 0.03 times the least lateral dimension of the composite column. The same concept is used for reinforced concrete columns in Part 4 of the Bridge Code, but there the allowance is 0.05 times the smallest column dimension, but not more than 20 mm. The reasons for the use of a smaller allowance in Part 5 are now explained.

As for reinforced concrete columns in Part 4, the design methods for a composite column depend on whether the length considered is *short* or *slender*. It is short if both ℓ_x/h and ℓ_y/b are less than or equal to 12, where ℓ_x and ℓ_y are the effective lengths for bending about the two principal axes, and h and b are the greatest and least lateral dimensions of the concrete in the cross section.

For short columns, it is convenient to allow for nominal eccentricity of loading by making a simple percentage reduction in the axial load capacity rather than using the full design procedure for a column subjected to end moments. The use of an eccentricity ratio of 0.05 has been shown[173] to reduce the axial load capacity of encased H-sections by as much as 22% under the most adverse combination of conditions. To limit the reduction to 15%, the value used for reinforced concrete columns, the eccentricity ratio must not exceed 0.03. This change in eccentricity allowance is strictly only necessary for encased sections, since for filled tubes (CHS and RHS), the reduction in load capacity is less. Part 5 of the Bridge Code does not differentiate between different types of composite columns in this respect, but it leaves a designer free to calculate more accurately the effect of the nominal eccentricity, if it is to his advantage.

It is reasonable to reduce the eccentricity allowance for composite columns in bridges for the following reasons.

(1) Column cross sections in bridges are likely to be larger than in buildings, but absolute constructional tolerances are not.

(2) The use of bearings in bridges (but not in buildings) reduces the uncertainty of the line of action of the applied loading. For example, in a composite column whose minimum dimension is 500 mm, the tolerance assumed is ± 15 mm.

For slender columns, the use of a simple percentage reduction is too conservative, so slender columns which are nominally axially loaded, are treated in Part 5 as columns bent in single curvature about their minor axes, with the end moments taken as the product of the design end load at the ultimate limit state and the nominal eccentricity, $0.03b$.

14.3.3 Columns with end moments

The failure load of a pin-ended column subject to any combination of end load, N, and uniaxial end moments, M and βM, is defined by a *load-moment interaction diagram* as shown in fig. 14.5. The interaction curve can be approximately described by coefficients K_1, K_2 and K_3 which were expressed by Basu and Sommerville[166] as functions of the column properties α_c, β (fig. 14.1), ℓ_e/r and M/M_u (as explained in Volume 1).

The expressions for K_2 and K_3 in Part 5 of the Bridge Code are similar to those given by Basu and Sommerville but the slenderness ratio ℓ_e/r has been replaced by the slenderness parameter, λ. From fig. 14.5, it can be seen that the ratio M/M_u can exceed unity at low axial loads. No account of this is taken in the Bridge Code because its availability depends on the loading history of the column.

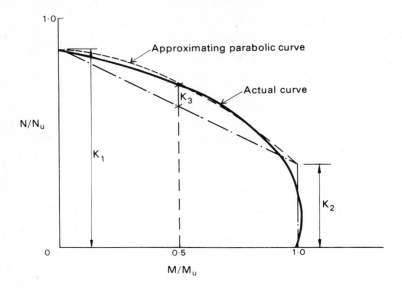

Fig. 14.5 Approximation to uniaxial interaction curve

In Volume 1, two methods are given for checking whether a column length is strong enough to resist a known axial load N and end moments M and βM. Let the greater of these moments be given by point E on fig. 14.6. The first method assumes that the loading path is OA, with failure at point B, and consists of checking that $N_2 \geqslant N$. The second method assumes that the loading path is OEA, with failure at C, and consists of checking that $N_1 \geqslant N$. For uniaxial bending, both methods give the same result, because if $N_1 > N$, then $N_2 > N$, but for columns in biaxial bending, the first method introduces a conservative error. The second method is used in Part 5 for this reason, and because it gives the simpler design equation (Volume 1).

14.3.4 Biaxial interaction

Where moments are applied simultaneously about the two principal axes of a column, failure will occur in a biaxial mode and is defined by an interaction surface of the type shown in fig. 14.7. This mode of failure is not restricted to columns loaded biaxially but also occurs in columns bent about their major axis which are unrestrained against buckling about the minor axis, due to initial imperfections in the straightness of the steel member.

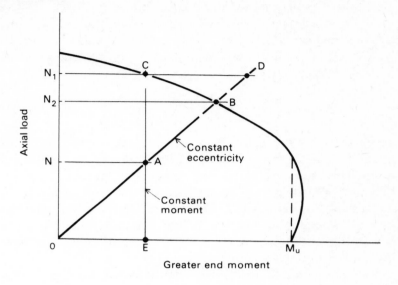

Fig. 14.6 Interpretation of load-moment interaction diagram

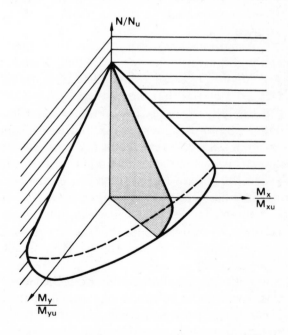

Fig. 14.7 Interaction surface for columns failing biaxially

No simple method of defining the biaxial interaction surface has been found. Part 5 of the Bridge Code gives an interaction formula proposed by Virdi and Dowling[167] for the biaxial failure load, N_{xy}, which is in terms of the uniaxial failure load of the column for the two principal axes each considered separately, and the strength of the column under axial loading assuming minor axis buckling to be prevented. The equation given is:

$$\frac{1}{N_{xy}} = \frac{1}{N_x} + \frac{1}{N_y} - \frac{1}{N_{ax}} \qquad (14.11)$$

It appears to be the same as that previously used by Basu and Sommerville but the terms N_x and N_y are defined in a slightly different way, for the reasons explained in chapter 5 (Volume 1).

Both tests and theory[172] confirmed that the original formula could safely be used throughout the range of practical column slenderness to predict the strength of columns failing biaxially, but found that the results for slender columns were unduly conservative. The new definitions will always give numerically higher values of N_x and N_y than when the Basu and Sommerville definitions are used, and thus will predict a higher biaxial failure load in the interaction formula (equation 14.11). Subsequent work has shown[174] that the revised formula is still conservative in comparison with a better method that uses simple bilinear interaction curves.

14.3.5 Triaxial containment in concrete-filled circular hollow sections (CHS)

Short columns of this type under predominantly axial loading have a considerably higher compressive resistance than is predicted using the normal design strengths for concrete and steel. This is evident from fig. 14.8, where the results of tests on axially-loaded concrete-filled CHS are compared with the appropriate European column curve.

This additional strength occurs because the increase in compressive strength of the concrete core, which is restrained laterally by the surrounding steel tube, outweighs the reduction in the yield strength of the steel in vertical compression due to the circumferential tension needed to contain the concrete.

This containment effect is not present in concrete-filled RHS, except in the corner regions, because the hoop tension cannot be developed in the same way. In concrete-filled CHS, the effects of containment reduce as bending moments are applied because the mean compressive strain in the concrete (and the associated lateral expansion) is then reduced. It also

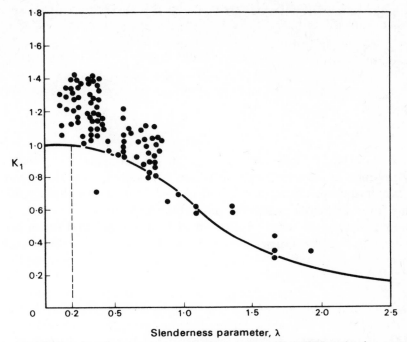

Fig. 14.8 Comparison between test results for concrete-filled CHS and column curve a
neglecting the effect of triaxial containment

diminishes with increasing slenderness of the column, because bowing
prior to failure again increases the bending moment and reduces the
mean compressive strain in the concrete.

In Part 5 of the Bridge Code, allowance is made for this additional
strength only in concrete-filled CHS that are 'axially' loaded. This term
includes columns with end moments due solely to eccentricities arising
from construction tolerances (section 14.3.2). The adjustment is made by
replacing equation (14.1) for the squash load, N_u, by:

$$N_u = 0.95 A_s f_{yc} + 0.45 A_c f_{cc} \qquad (14.12)$$

where f_{yc} is a reduced yield strength for the steel (given as f'_y in Part 5),
and f_{cc} is an increased cube strength for the concrete. Derivations of the
expressions for f_{yc} and f_{cc} are now given.

Sen[175] proposed that the squash load of a concrete-filled CHS should
be calculated from:

$$N_u = A_s(f_y/\beta) + A_c[\sigma_u + 2\mu\gamma(t/D)(f_y/\beta)] \qquad (14.13)$$

where σ_u is the ultimate strength of the concrete in direct compression,
and β, μ and γ are constants for which values were deduced from tests.

For consistency with the design methods for other types of composite column, f_y must be replaced by the design strength $0.95\sigma_y$, and σ_u by $0.45f_{cu}$. Equations (14.12) and (14.13) are then the same if f_{yc} and f_{cc} are defined by:

$$f_{yc} = \sigma_y/\beta \qquad (14.14)$$

and:

$$f_{cc} = f_{cu} + 4\mu\gamma(t/D)(\sigma_y/\beta) \qquad (14.15)$$

where t is the wall thickness and D the outside diameter of the steel tube.

Virdi and Dowling[172] considered that containment should be neglected in columns with $\lambda > 1$, for failure is then governed by Euler buckling rather than material yield. For most practical columns, $\lambda = 1$ corresponds to ℓ/D ratios between 24 and 29, but for simplicity they set the limit at $\ell/D = 25$, and proposed linear expressions for μ and γ as follows:

$$\mu = 0.25[25 - (\ell/D)] \qquad (14.16)$$

and:

$$\gamma = 0.02[25 - (\ell/D)]$$

$$(14.17)$$

β was given by Sen as:

$$\beta = (1 + \gamma + \gamma^2)^{\frac{1}{2}} \qquad (14.18)$$

Equations (14.14) and (14.15) are given in Part 5 of the Bridge Code in the form:

$$f_{yc} = C_2 f_y \quad \text{and} \quad f_{cc} = f_{cu} + C_1(t/D)f_y \qquad (14.19)$$

and the constants C_1 and C_2, calculated from equations (14.14) to (14.19), are given in tabular form.

It follows that the concrete contribution factor, α_c for such columns is given by:

$$\alpha_c = \frac{0.45A_c f_{cc}}{N_u} \qquad (14.20)$$

where N_u is calculated from equation (14.12). This value of N_u is then also used in calculating L_E.

Figure 14.9 shows the improvement obtained when the test results from fig. 14.8 are replotted against the European curve, taking account of triaxial containment by the method of Part 5.

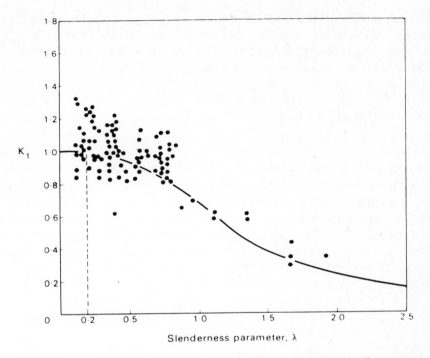

Fig. 14.9 Comparison between test results for concrete-filled CHS and column curve a including triaxial effects

14.3.6 Long-term loading

Under sustained loading, creep of the concrete causes an increase in the lateral deflection of a composite column and a reduction in its strength and stiffness. Theoretical analysis[165] suggests that the maximum adverse effect occurs in encased columns of intermediate slenderness when the steel cross section is light and the concrete contribution factor, α_c exceeds about 0.3. Basu allowed for the effect of shrinkage and creep in his theoretical analysis by increasing the instantaneous strain level in the concrete by a factor of two. In the design method developed from this[166] Basu and Sommerville gave magnification factors to be applied to that part of the loading which was considered to be sustained (Volume 1). These factors were functions of the column slenderness, the area of concrete and its disposition in the cross section.

As yet, only a few tests have been carried out on composite columns under long-term loading[176] and it is not possible to make simple recommendations. Even if this were possible, for the designer there remains the problem of deciding what proportion of the loading is sustained. In a bridge structure this is easier to define than in a building, but it is still an area of uncertainty. For these reasons, load magnification factors such as those given by Basu and Sommerville have not been used in Part 5 of the Bridge Code. Instead, the design compressive strengths of concrete used in calculating the axial and flexural strengths of the column have been reduced.

For concrete in direct compression, it is normally assumed that the best estimate of the ultimate strength is the cylinder strength, which Basu and Sommerville took as $0.83f_{cu}$ for calculating the squash load. For calculating the ultimate moment of resistance, they assumed that the concrete was able to sustain a uniform stress over its compression zone of $0.67f_{cu}$. The two factors 0.83 and 0.67 are consistent with the values used in the earlier Code of Practice CP 114[177] for load-factor design of reinforced concrete.

In limit-state philosophy, design strengths are obtained by dividing the characteristic (or nominal) strengths by γ_m. At the ultimate limit state, $\gamma_m = 1.5$ for concrete, which gives design strengths in axial compression and in bending of $0.55f_{cu}$ and $0.45f_{cu}$, respectively.

In Part 5 of the Bridge Code, these values have been reduced to $0.45f_{cu}$ and $0.4f_{cu}$ to allow for the effects of creep. They are then consistent with the design strengths used for reinforced concrete columns in Part 4 of the Bridge Code. No other allowance is made for creep in columns.

14.3.7 Comparisons between design methods for pin-ended composite columns

The validity of using the European curves for bare steel struts in conjunction with a new definition of slenderness to predict the failure load of composite columns has been demonstrated for axially-loaded columns by Virdi and Dowling[172] by comparing the design curves with results computed from a theoretical ultimate-strength analysis for a wide range of composite columns and with over one hundred results obtained from composite columns tested to failure.

In Part 5 of the Bridge Code, the method is applied not only to axially-loaded columns but also to columns with any combination of end moments about their two principal axes. In order to assess the likely differences between available design methods for composite columns subject to end loads and moments the failure loads calculated by the design method in Part 5 of the Bridge Code were compared[171] with those predicted by the original Basu and Sommerville method[166] for ten different combinations of axial load and end moments, in each of twelve different composite cross sections and three slenderness ratios. In total 720 failure loads were calculated for the two methods. The cross sections considered represented a wide range of encased steel sections and concrete-filled CHS and RHS. In addition, where the design resistances predicted by the two methods were significantly different, comparisons were also made for the axial loading case only with the failure loads predicted by the BS 449 cased-strut method (chapter 5, Volume 1) and the failure load of the bare steel strut as predicted by BS 449[178] and the appropriate European curve.

For the purpose of these studies it was assumed that the ratio of short-term to total loading was 0.4 for calculating the load magnification factors in the Basu and Sommerville method, and the 'permissible' working loads predicted by BS 449 were increased by a factor of 1.5 to provide an equivalent basis for comparison with the failure loads predicted by limit-state design methods.

The many differences between these methods have different effects on the calculated resistances of columns with different types of cross section, so the results of this study are complex. They are given on pp. 315–21 of the first edition of this volume. The work confirmed the validity of the methods of Part 5, which give a far more uniform margin of safety than do the earlier methods.

14.4 Elastically restrained columns

The preceding section dealt with the design of isolated pin-ended columns subject to any combination of end load and end moments acting about one or both principal axes. In practice other forms of end restraint are commonly encountered, for example, where a column forms part of a frame. Frames of more than a single storey are unlikely to occur in bridge structures and for this reason, only single-storey frames are considered in Part 5 of the Bridge Code. It is assumed that the frame has rigid joints and the beams remain elastic.

14.4.1 Requirements of a design method for restrained columns

Any valid design method for restrained composite columns must be capable of taking account of the following.

(1) The reduction in member stiffness due to inelasticity and axial compression.

(2) Redistribution of the bending moments due to the effect of axial loads and changes in the geometry of the structure as it is loaded (including both joint displacement in members free to sway and bowing

within the length of a member).

(3) The effect of end restraint afforded to the column from members framing into its ends.

(4) Initial imperfections in the steel member.

Rigorous analyses capable of taking account of all these effects[167, 168] are too complex for use in design.

The semi-empirical method given in Part 5 of the Bridge Code assumes that the restrained column can be replaced by an equivalent pin-ended column loaded with the same end loads and moments as the restrained column (except as noted below for unbraced columns), the failure load then being found by the method for pin-ended columns as described in section 14.3. The steps in this procedure are now described, to show how allowance is made for each of the effects (1) to (4).

14.4.2 Semi-empirical design method

The bending moments, shear and axial force acting on the restrained column (whether it forms part of a frame or not) are determined by a normal elastic analysis, neglecting the effects of axial load on member stiffness and on the deflected shape of the structure.

Deflections of members free to sway are, of course, included in the analysis, as in a normal moment distribution, but the effects of axial loads on the deflected shape (second order effects) are neglected.

The relative stiffnesses (K, equal to I/L) of the column and its restraining members are based on the gross transformed composite cross section (i.e. assuming the concrete to be uncracked), with L taken as the distance between centres of end restraints, and, an appropriate modular ratio is used, based on values for the moduli of elasticity of steel and concrete given in Parts 3 and 4, respectively, of the Bridge Code.

EFFECTIVE LENGTH

This is the length of an equivalent pin-ended column that would have the same elastic critical load as the restrained column. It depends on the stiffnesses of the restraints.

In BS 5400, as in previous British codes of practice, only idealised forms of end restraint are given; for example, 'pinned' or 'effectively restrained'. This is precise enough for isolated columns, but for columns forming part of a frame, a more accurate method of determining the effect of end restraint afforded to the column is necessary, particularly in frames which are free to sway, where the maximum effective-length ratio

(ℓ/L) of 2.0 given in Part 5 is by no means an upper limit. Wood has published charts[179] which enable the effective length of braced and unbraced columns with any form of elastic end restraint to be determined accurately. These are based on the deteriorated elastic critical load of the restrained column in a limited substitute frame, which is easily calculated from stability functions. Effective length charts suitable for single-storey frames are given in figs 14.10 and 14.11.

The charts can also be used for isolated columns where, for example, it is required to take account of the stiffness of a foundation base, by

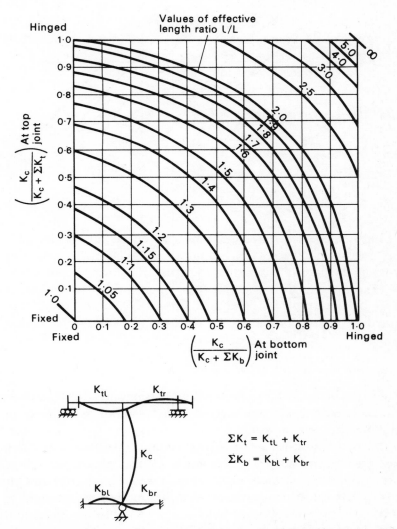

$$\Sigma K_t = K_{tl} + K_{tr}$$
$$\Sigma K_b = K_{bl} + K_{br}$$

Fig. 14.10 Effective length of columns restrained by beams but with unrestrained sway

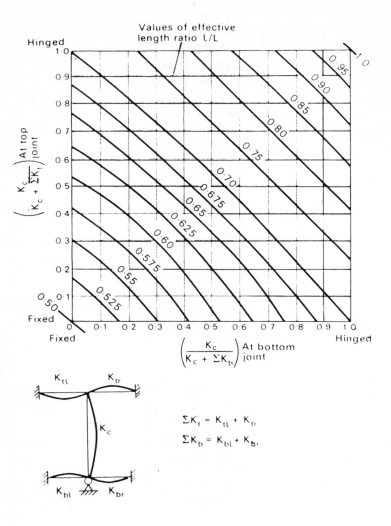

Fig. 14.11 Effective length of columns restrained by beams and with sway prevented

inserting imaginary beams to form a limited frame, as shown in fig. 14.12. For the isolated column cantilevered from a spread base shown in fig. 14.12(a) many designers would assume that the lower end of the column was effectively restrained against rotation and translation by the base, and use an effective length ratio of two. Others may suspect that the rotational restraint afforded to the column was less effective and use an effective length ratio greater than two: but how much greater? If the foundation stiffness is assumed to be equal to that of the column, then from fig. 14.10, with $\Sigma K_{bb} = K_c$ and $\Sigma K_{tb} = 0$:

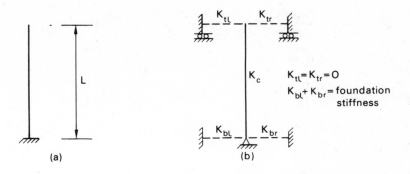

Fig. 14.12   Representation of isolated cantilever column by equivalent limited frame

$$\frac{K_c}{K_c + \Sigma K_{bb}} = \tfrac{1}{2} \qquad \frac{K_c}{K_c + \Sigma K_{tb}} = 1$$

Hence, $\ell/L = 2.5$. Clearly the problem of determining the stiffness of the foundation (or restraining members) still exists, but even an order of magnitude estimation will give a safer prediction of ℓ/L than the assumption that the base is rigid.

Figures 14.10 and 14.11 can be used to determine minimum stiffnesses of restraining members. For example, the table of effective-length ratios in Part 5 gives $\ell/L = 0.7$ for a braced column 'effectively restrained against rotation at both ends'. Figure 14.11 shows that this value is conservative provided that the stiffnesses of the end restraints (ΣK_t and ΣK_b) are at least equal to the stiffness of the column length (K_c).

BENDING MOMENTS APPLIED TO THE EQUIVALENT PIN-ENDED COLUMN
It is assumed that the pin-ended column of length ℓ is loaded with the same end loads and moments as the restrained column, except that where the column (or frame) is not restrained against sway, the pin-ended column is always assumed to be in single-curvature bending, with the smaller end moment taken as the greater of the calculated moment and 75% of the larger end moment. This adjustment is a device to allow for the increase in end moments due to sidesway. It does not allow for the possibility of 'unwrapping' failure, which can occur in columns bent in double curvature ($\beta \simeq -1$), when the end load is increased beyond the capacity of the column without end moments. Unwrapping is allowed for in Part 5 in the expressions for K_2 and K_3 (fig. 14.5), which correspond to the use of the curve for $\beta = -0.5$ (fig. 14.13) for all values of β between -0.5 and -1.0.

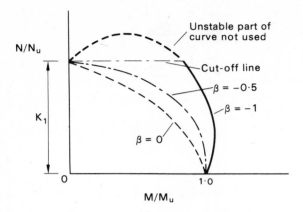

Fig. 14.13 Typical load-moment interaction curves for various combinations of end moments

TRANSVERSE LOADS

Transverse loads (for example, due to the impact of a vehicle) should be included in the elastic analysis of the restrained column if their effect could be to produce a more severe loading condition. For columns in unbraced frames, this includes loads applied at beam level.

When the resultant moment, M_R, anywhere within the column length, due to the combined effects of transverse and other loads, exceeds half of the modulus of the algebraic sum of the end moments, the alternative loading condition of single curvature bending must also be considered, with both end moments taken as M_R. The sign convention assumed here is that single curvature bending produces end moments of the same sign.

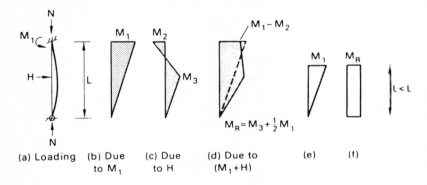

Fig. 14.14   Treatment of transverse loads on columns bent in single curvature

This requirement is given in Part 5 to ensure that premature flexural failure within the column length will not occur. The following example illustrates its application. In fig. 14.14(a), a column forming part of a braced frame pinned at its base is subjected to an end load N, a moment M_1 at its upper end, and a transverse load H acting at mid-height. The elastic distributions of bending moment due to M_1 and H are shown in figs 14.14(b) and (c), respectively, and their combined effect is shown in fig. 14.14(d). The equivalent pin-ended column will have an effective length ℓ less than the actual column length L. Its resistance will be checked for the axial load N combined with the following bending-moment distributions.

(1) The moments with H absent, shown in fig. 14.14(e), since the effect of H is to reduce the moment at the upper end of the column.

(2) If $M_R > \frac{1}{2}|M_1 - M_2|$, the single-curvature distribution shown as fig. 14.14(f).

JUSTIFICATION OF THE METHOD

This process of inelastic design of a pin-ended column subjected to end loads and moments found by first-order elastic analysis of a frame or restrained column, cannot be expected to give the bending moments that would actually occur. However, taken as a 'package deal' with the modified Basu and Sommerville method, it has been shown[168, 172] to be satisfactory for columns braced against sidesway. The evidence is summarised on pp. 332–7 of the first edition of this volume.

The more difficult problem of the column in a single-storey frame free to sway is adequately covered by the various conservative modifications to the method given in Part 5 of the Bridge Code. The most likely source of error is overestimation of the stiffnesses of the other components of the frame, leading to too low an effective-length ratio for the column length.

Chapter Fifteen

Workmanship and Construction

15.1 General

The procedures to be followed and the precautions to be taken in the construction of composite bridges are, in general, a combination of those required for the erection of steelwork and for reinforced concrete construction. Only those of special importance for composite structures are mentioned here.

The proposed method and sequence of construction for the super-structure are essential inputs to the design process, and should be clearly set out in the contract documents. They should, where possible, allow the contractor some flexibility. If he subsequently proposes changes, their consequences for the behaviour of the structure must be evaluated by the designer, and the changes approved. The agreed procedures should be described in detail on the final drawings and instructions sent to the site. This subject is further discussed in section 15.2.

The effectiveness of shear connectors is sensitive to poor workmanship in two respects: poor compaction of the few cubic centimetres of concrete at the base of each connector through which most of the load is transferred (sections 7.2.3 and 8.5.3); and incorrect welding procedures (section 15.3).

15.2 Method and sequence of construction

Methods of construction for composite superstructures were described in sections 6.2.1 and 6.2.5. The influence of some of the simpler procedures on design of composite beams is now considered. The same principles can be applied to the more complex situations that can arise in practice.

PROPPED AND UNPROPPED CONSTRUCTION

Fully propped construction, which enables the whole of the dead load to be carried by the composite section, is practicable (but not necessarily economical) for structures such as low-level viaducts. The method is more sensitive to settlement of the foundations for the props than when it is used for beams in buildings, because of the relatively greater stiffness of the steel beams. Settlement led to some loss of composite action during the propped construction of Moat Street flyover, Coventry.[28]

Most composite bridge decks are built by first erecting the whole of the steelwork, and then constructing the deck slab in stages, without propping. When the deck is precast, the units are usually erected working from one end of the span. *In situ* concrete is sometimes cast in that sequence, but for the longer bridges it is usual to cast midspan regions first, and lengths over internal supports last, to minimise the longitudinal tensile strain in the slab. The reverse procedure is adopted for continuous decks with longitudinal prestress; the lengths of slab over the internal supports are cast and prestressed first, since prestress is more effective when applied to a more flexible structure (i.e. steel only at midspan).

To illustrate the influence of the method of construction on design, we consider a span in which the whole of the deck slab is cast before any of it is old enough for composite action to have developed. If unpropped, the steel members must be designed for the whole of the weight of the wet concrete, and for the formwork, construction plant, and personnel. This load case occurs only at this stage, because the formwork, etc., are later removed. Design for this condition is often governed by lateral-torsional buckling, so that the steel flange requires lateral restraint. In theory, this could be provided by the formwork, but no simple solution to the practical problems has been found, and temporary or permanent steel bracing is usually used. This is a potential area of development for GRC permanent formwork (section 11.3).

When the composite section is non-compact, so that elastic analysis is used, account has to be taken in the design of the completed structure of the effects of sequence of construction, as explained in section 8.4.3. The effects of a change from propped to unpropped construction are illustrated in fig. 15.1, which shows the midspan cross section of a simply-supported composite beam of span 30 m, and the global effects of combination 1 design loads at the ultimate limit state. In fig. 15.1(b) the horizontal scale is proportional to longitudinal strain, but the numbers are stresses in N/mm^2.

For propped construction, compressive stress never governs in a steel top flange. It may do so in the slab, so the reduction due to the use of unpropped construction (20% in this case) is significant. The increase in

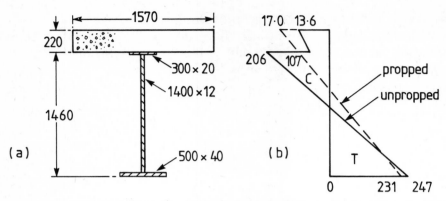

Fig. 15.1   Influence of method of construction on longitudinal bending stresses

the compressive stress in the steel top flange is 93%, which will influence the extent to which it needs lateral bracing during construction. There is also a small increase (7%) in the tensile stress in the steel bottom flange, which will influence design if that flange governs.

The design moment for combined bending and shear, as given by equation (8.13), increases by 10% in this example when propped construction is used. This should not influence design, because the vertical shear in a midspan region is low.

For a beam of compact section, the difference between the elastic bending-stress distributions for propped and for unpropped construction are as they would be for a non-compact section. But, if failure by lateral buckling is prevented, the ultimate moment of resistance is independent of the method of construction, because of inelastic redistribution of stress within the section. This is why load applied to the steel member can be treated, in the design of compact sections, as if it were applied to the composite member.

SEQUENTIAL CASTING OF THE DECK SLAB

Except in the smallest of bridge decks, the concrete slab will usually have to be cast in stages. The sequence of casting is likely to be determined by the need of the designer to stabilise compression flanges of steel beams at an early stage of construction, or to limit tensile strain in the slab over internal supports; or by the sequence in which the steelwork is erected. The effects of casting the deck slab for a 37-m span in five stages are analysed in Appendix X.

Ideally, sections of the slab should only be cast when the concrete in previous sections is either very young (i.e. still in a plastic state), or sufficiently mature to resist the stresses imparted to it by the shear connectors without permanent damage. It is to reduce the risk of permanent damage that Part 5 of the Bridge Code recommends that loading of the composite section should be avoided until the concrete has attained a cube strength of 20 N/mm^2. The reasons for the choice of this limit are now discussed.

(1) There has been no systematic research into the strengths of shear connectors embedded in concrete with cube strengths significantly less than 20 N/mm^2, but as the connector modulus (load per unit slip) is likely to be lower when the concrete is weak, there will be a greater loss of interaction, so that the error in using full-interaction calculations is increased.

(2) Some allowance must be made for the reduction in strength and stiffness of concrete slabs loaded at an early age. Even if the lower value of E_c is allowed for simply by using an increased modular ratio in elastic calculations, the amount of calculation increases several fold, and the validity of such an approach at very low concrete strengths is questionable. When the cube strength of the concrete is above 20 N/mm^2 it may be sufficiently accurate to rely on the original design calculations.

In reality, there is a continuous growth of composite action with time; with modern cements and admixtures, a cube strength of 20 N/mm^2 can be reached in little over 24 hours. Loading of connectors in young concrete in fact occurs near the free ends of every length of slab that is cast, due to the effects of shrinkage. There is no evidence that this causes damage.

When account has to be taken in design of the loading of a composite section at an early age, the limiting compressive stress in the concrete and the nominal strengths of shear connectors should be related to the cube strength at the time considered. Figure 15.2 shows the nominal static strengths of 19-mm diameter stud connectors plotted against concrete cube strength, as given in Part 5 of the Bridge Code. It is likely that the results of push-out tests using weaker concrete would give a curved line such as OA, because in this range the strength of the connector is governed by that of the surrounding concrete rather than by the properties of the steel stud. For design, a straight line from A towards O should be safe, but no account should be taken of composite action where the concrete strength is very low (say less than 5 N/mm^2) for the reasons previously discussed.

During the construction of a deck slab that is cast in stages, and for a few weeks after its completion, there are significant variations in both the elastic modulus and the cube strength of the concrete over the area of the deck. For the analysis of cross sections it is a simple matter to use values

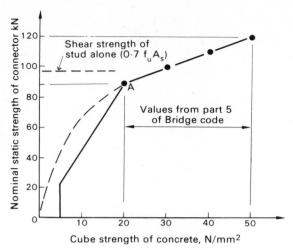

Fig. 15.2 Nominal static strengths of 19-mm stud connectors

of E_c and f_{cu} appropriate to the region considered, but for global analyses for loads that occur during construction, simplification is essential. Where the whole of the concrete in the structure considered is more than 28 days old, the value of E_c associated with the specified 28-day cube strength (f_{cu}) would normally be used in the analysis. Part 5 of the Bridge Code allows this uniform value to be used also when younger concrete is present, provided that its strength at the time considered is at least $0.75f_{cu}$. The reason is that when f_{cu} is at least 30 N/mm^2 (as is usual in bridges), the variation of E_c corresponding to a range of $0.25f_{cu}$ of concrete strength is little more than the variation which may be expected between concrete of the same specified strength (but different actual strengths) in a given structure. This conclusion was confirmed in design studies[85] on continuous composite bridges in which the slab was cast in stages.

Where accurate control of profile is critical during construction, the effects of creep should be considered when calculating deflections. Another potential problem with sequential construction is the restraint to early thermal movement in newly cast slabs afforded by sections of slab previously cast. If excessive cracking is to be avoided, care must be taken in detailing the distribution reinforcement.[180, 181]

15.3 Testing of shear-connector welds

Requirements for the testing of the welds for shear connectors are given in BS 5400: Part 6.[1] Those for fillet welds (as used in bar-and-hoop connectors, for example) are as for fillet welds generally. Specific references are made to stud shear connectors in respect of procedure trials and production tests. Procedure trials are needed because stud connectors are welded by an automatic process in which the stud is plunged into a pool of molten weld metal. The soundness of the weld depends on using the correct combination of current and weld time. These have to be determined because they can vary with the welding equipment, power supply, and length of leads used.

Procedure trials are done before production welding begins, as may be required by the engineer. Each trial consists of welding six studs, using the procedures and samples of the materials to be used on the job. Three of the welded studs are sectioned and subjected to metallographic examination and hardness tests on the weld metal and heat-affected zone. The other three studs must be bent until the lateral movement of the head reaches half the height of the stud (i.e. through about 30°), and then bent straight again without failure of the weld.

Site testing should begin with a visual inspection. The weld to a stud connector should form a complete collar around the shank, be free from cracks and excessive splashes of weld material, and have a 'steel blue' appearance. Two site tests are specified.[1] One is to strike the side of the head of the stud with a 2 kg hammer. The stud should emit a clear ringing note; otherwise, there may be a crack in the weld. The second test is to strike the side of the head with a 6 kg hammer until the head is displaced laterally through one-quarter of the height of the stud. The weld must not show signs of cracking or lack of fusion. The stud is left bent, because re-bending can cause excessive strain hardening. Bent studs can be seen in plates 6 and 11.

The second test corresponds to bending the shank of the stud through about 15°, which is the site test given in draft Eurocode 4.[5]

Other types of test have been used. On the Tay Road Bridge, the stud connectors were tested[24] by jacking apart adjacent connectors until the shear load on the studs corresponded to their design load. Excessive bending of the studs was avoided by applying the load near the base of the connector. Proof loading of the welded connector in direct tension has occasionally been specified, but has the disadvantage that the stud is tested in a different mode from that in which it acts in the structure.

Appendix A

Stresses due to Temperature and Shrinkage

Expressions for longitudinal stresses in an uncracked composite cross section due to the effects of temperature and shrinkage are derived in general form, suitable for computation. Simpler expressions are then obtained for shrinkage of an unhaunched slab. The reinforcement in the slab is neglected. The expression for the transfer length ℓ_s given in Part 5 of the Bridge Code is derived. Finally, the effects of slip, shear lag, reinforcement, and cracking are discussed. The treatment of shrinkage effects given in section 8.7.1 should be read first.

In this subject, great care is needed with sign conventions. Here, the usual conventions of structural mechanics are used: tensile force, stress and strain and sagging bending moments are taken as positive. The subject is sometimes referred to as TSC (Temperature and Shrinkage as modified by Creep).

PRIMARY LONGITUDINAL STRESSES

The cross section to be analysed (fig. A.1(a)) is divided into n horizontal slices by lines at depths y_1 to y_n from its upper surface. Each slice must be of a single material and of constant breadth, except that tapering haunches can be represented by a single slice of mean breadth. To allow for a discontinuity of strain at the steel-concrete interface (necessary for shrinkage calculations), an extra slice of zero area can be included at this level (i.e. $y_2 = y_3$ in fig. A.1(b)). To reproduce temperature distributions of the type specified in Part 2 of the Bridge Code (fig. 7.8), it is necessary to use two slices each for the deck slab and the steel web. One is needed for each steel flange, giving a total of six slices for an unhaunched section.

Temperature distributions through the depth of a cross section, of the type shown in figs 7.8 and 7.9, can be represented by linear variation between temperature changes T_r at the boundaries of the slices. These changes are defined as the rises of temperature above the uniform temperature at which thermal stresses are assumed to be zero. The free

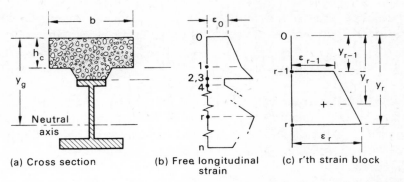

(a) Cross section (b) Free longitudinal strain (c) r'th strain block

Fig. A.1   Free strain in composite section due to temperature and shrinkage

thermal strains at the boundaries of the slices are given typically by:

$$\varepsilon_r = \beta T_r \qquad (A.1)$$

where β is the appropriate coefficient of thermal expansion.

The following theory is applicable also to a free shrinkage, ε_{sc}, of concrete, for which the *numerical* value will be negative. For slices as shown in fig. A.1(b):

$$\varepsilon_0 = \varepsilon_1 = \varepsilon_2 = \varepsilon_{sc} \qquad (A.2)$$

and ε_3 to ε_n are zero when shrinkage is considered.

For each slice, the mean strain $\bar{\varepsilon}_r$ and the depth of the centroid of the strain block, $\bar{y}_r$, are found from:

$$\bar{\varepsilon}_r = (\varepsilon_{r-1} + \varepsilon_r)/2 \qquad (A.3)$$

and:

$$\bar{y}_r = [\varepsilon_{r-1}(2y_{r-1} + y_r) + \varepsilon_r(y_{r-1} + 2y_r)]/6\bar{\varepsilon}_r \qquad (A.4)$$

The cross-sectional area of each slice, A_r, is taken as the actual area for slices in steel; and the transformed effective area for slices in concrete:

$$A_r = h_r b_{rp}/\alpha_e \qquad (A.5)$$

where:

$$h_r = y_r - y_{r-1} \qquad (A.6)$$

and α_e is the modular ratio E_s/E'_c, where E'_c takes account of creep (section 7.3.5), and b_{rp} is the effective breadth of the slice for primary effects. For temperature effects this breadth is usually the actual breadth, but for the effects of shrinkage prior to the completion of the deck, a smaller value is often appropriate, as explained later.

The free change of length in each slice is that which would be caused by a force F_r applied at the centroid of its strain block, where:

$$F_r = E_s \bar{\varepsilon}_r A_r \qquad \text{(A.7)}$$

Initially, these changes of length are assumed to be completely prevented by forces $-F_r$ applied at the centroid of each strain block (fig. A.2(a)). Thus if $\bar{\varepsilon}_r$ is a tensile strain, the rth slice is restrained by a compressive force of magnitude F_r. The extensions of the slices, all being zero, are compatible.

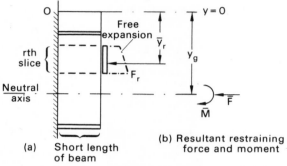

Fig. A.2   Primary effects of temperature or shrinkage

Let the n restraining forces be statically equivalent to a total force $\bar{F}$ at depth y_g and a bending moment $\bar{M}$, acting in the directions shown in fig. A.2(b). Taking moments about the level $y = 0$:

$$\bar{M} + \bar{F} y_g = \sum_1^n (F_r \bar{y}_r) \qquad \text{(A.8)}$$

where:

$$\bar{F} = \sum_1^n F_r \qquad \text{(A.9)}$$

and y_g is the depth of the centroid of the composite section. The position of this centroid should be calculated making the same assumptions about creep and effective breadth as when calculating the forces F_r, so that:

$$y_g = \sum_1^n (A_r y_{gr})/A \qquad \text{(A.10)}$$

where:

$$A = \sum_1^n A_r \qquad \text{(A.11)}$$

and y_{gr} is the depth of the centroid of each slice of material, and so may differ from $\bar{y}_r$.

If the free strains consist only of a constant strain $\bar{\varepsilon}_1$ in a slice consisting of the deck slab, the distributions of stress and strain at this stage are as shown in Fig. A.3(b) and (c).

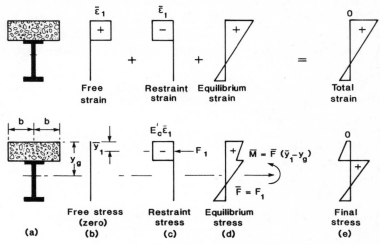

Fig. A.3   Uniform free strain in deck slab only

Equilibrium is now restored, without upsetting compatibility, by applying a force $\bar{F}$ and a moment $\bar{M}$ at the centroid of the composite section (fig. A.3(d)) acting in the opposite directions to those shown in fig. A.2(b). The total *strains* in the cross section (ε_t, say) are due to this force and moment alone, which are assumed to act on the effective cross section of the member. In 'steel' units, this has second moment of area:

$$I = \sum_{1}^{n} [(A_r h_r^2)/12 + A_r(y_{gr} - y_g)^2] \tag{A.12}$$

Thus the strain at the lower edge of the rth slice is:

$$\varepsilon_{tr} = \bar{F}/E_s A + \bar{M}(y_r - y_g)/E_s I \tag{A.13}$$

The total stresses are those due to $\bar{F}$ and $\bar{M}$ acting on the composite member plus those due to the restraining forces initially applied to the slices (fig. A.3(c)). These latter stresses at level y_r are $-E_s\varepsilon_r$ in steel and $-E'_c\varepsilon_r$ in concrete. From equation (A.13) the total longitudinal stress in the steel at any level y below the top surface of the member is:

$$f_s = -E_s\varepsilon_y + \bar{F}/A + \bar{M}(y - y_g)/I \tag{A.14}$$

where ε_y is the free strain at the level considered. Similarly, the stress in the concrete at any level y is:

$$f_c = [-E_s \varepsilon_y + \bar{F}/A + \bar{M}(y - y_g)/I]/\alpha_e \qquad (A.15)$$

These stresses are tensile if positive.

The curvature of the member (sagging positive) is given by the usual expression:

$$\phi = 1/R = \bar{M}/E_s I \qquad (A.16)$$

Equations (A.14) to (A.16) are of general applicability. Some applications are now given.

LONGITUDINAL FORCE IN AN UNHAUNCHED CONCRETE FLANGE

This is the force for which the shear connection near each free end of the member must be checked. From equation (A.15), the longitudinal force (tension if positive) is:

$$Q = \sum_{1}^{n_f} [A_r \{-E_s \bar{\varepsilon}_r + \bar{F}/A + \bar{M}(y_r - y_g)/I\}] \qquad (A.17)$$

where n_f is the number of slices into which the concrete flange is divided. This result is valid for both temperature and shrinkage calculations.

SHRINKAGE STRESSES IN AN UNHAUNCHED FLANGE

Further simplification is now possible. The slab can be represented by one slice; from equation (A.2), $\varepsilon_0 = \varepsilon_1 = \varepsilon_{sc}$ and ε_2 to ε_n are zero. From equations (A.3) to (A.6), $\bar{\varepsilon}_1 = \varepsilon_{sc}$, $y_1 = h_1 = h_c$, $\bar{y}_1 = h_c/2$ and $A_1 = b_{1p} h_c / \alpha_e$.

From equations (A.7) and (A.9):

$$\bar{F} = F_1 = E_s \varepsilon_{sc} A_c \psi_p / \alpha_e \qquad (A.18)$$

where A_c is the actual area of the concrete slab, bh_c, and ψ_p is the effective-breadth ratio, b_{1p}/b.

From equation (A.8):

$$\bar{M} = -\bar{F}(y_g - h_c/2) \qquad (A.19)$$

From equation (A.17), the longitudinal force in the concrete slab is:

$$Q = (A_c \psi_p / \alpha_e)[-E_s \varepsilon_{sc} + \bar{F}/A + \bar{M}(h_c/2 - y_g)/I] \qquad (A.20)$$

and the curvature is given by equations (A.16) and (A.19).

TRANSFER LENGTH FOR LONGITUDINAL FORCE

As explained in section 8.7.1, the longitudinal distribution of the shear force at the steel-concrete interface that transfers the force Q to the steel beam (curve OB in fig. 8.29) can be found only by a theory that takes account of slip. The expression for the equivalent transfer length ℓ_s (fig. 8.29) given in Part 5 of the Bridge Code can be derived using linear partial-interaction theory, as follows.

It is shown in chapter 2 (Volume 1) that for a simply-supported beam of span L with uniformly-distributed load w and free shrinkage strain ε_{sc}, the interface slip is given by:

$$s = \beta wx - [(\beta w + \varepsilon_{sc})/\alpha]\, \text{sech}(\alpha L/2)\, \sinh \alpha x \qquad (A.21)$$

where α and β are functions of the properties of the cross section and of the shear connection.

If shear connectors of stiffness k (load per unit slip) are provided at longitudinal spacing p, the longitudinal shear flow per unit length at the interface is:

$$q = ks/p$$

whence from equation (A.21) with $w = 0$:

$$q = -(k\varepsilon_{sc}/\alpha p)\, \text{sech}(\alpha L/2)\, \sinh \alpha x \qquad (A.22)$$

This is the curve OB in fig. (8.29). The negative sign arises from the sign convention used, and can be ignored here. The maximum shear flows occur at the ends of the beam ($x = \pm L/2$) and are:

$$q_{max} = \pm(k\varepsilon_{sc}/\alpha p)\, \tanh(\alpha L/2) \simeq \pm k\varepsilon_{sc}/\alpha p \qquad (A.23)$$

since in practice, $\alpha L/2$ usually exceeds 2, so $\tanh(\alpha L/2) \simeq 1$; q_{max} is then independent of the span.

The total longitudinal force in the concrete slab at midspan is:

$$Q = \int_0^{L/2} q\, dx = (k\varepsilon_{cs}/\alpha^2 p)[1 - \text{sech}(\alpha L/2)] \qquad (A.24)$$

whence:

$$Q \simeq k\varepsilon_{cs}/\alpha^2 p \qquad (A.25)$$

since $\text{sech}(\alpha L/2) \ll 1$ when $\alpha L/2 > 2$. This assumption is equivalent to neglecting the effects of slip at midspan, so that this result for Q is identical with equation (A.20), which was obtained by full-interaction theory. The algebraic transformation from equation (A.20) to equation (A.25) is tedious.

From the definition of the transfer length ℓ_s (section 8.7.1):

$$Q = q_{max}\ell_s/2 \qquad (A.26)$$

Eliminating q_{max} and α from equations (A.23), (A.25) and (A.26) gives:

$$\ell_s = 2(Qp/k\varepsilon_{sc})^{\frac{1}{2}} \qquad \text{(8.63) bis}$$

as in chapter 8 and Part 5 of the Bridge Code. This result can be used for longitudinal forces due to temperature effects by replacing ε_{sc} in equation (8.63) by ε_d, the difference between the free strains at the centroids of the slab and the steel member.

Values of p/k are given in Part 5, and are discussed in section 8.7.1.

REDUCTION IN THE EFFECTS OF SHRINKAGE DUE TO SHEAR LAG

The choice of the ratio ψ_p in equations (A.18) and (A.20) is now considered. Its influence on the longitudinal force in the slab is seen more clearly by eliminating $\bar{M}$ and $\bar{F}$ from equations (A.18) to (A.20); putting $y_g - h_c/2 = d$; and defining an equivalent transformed area of the slab, A_{eq}, given by:

$$A_{eq} = A_c \psi_p / \alpha_e$$

This gives:

$$Q = A_{eq} E_s \varepsilon_{sc} [-1 \stackrel{+}{-} A_{eq}\{(1/A) + (d^2/I)\}] \qquad \text{(A.27)}$$

As expected, Q is proportional to ε_{sc}, but its variation with A_{eq} is non-linear.

When a length of slab is cast between existing lengths, shear lag is negligible, and $\psi_p = 1$. When it is cast alone (CD in fig. A.4), or against an

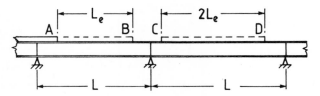

Fig. A.4 Influence of sequence of casting on shrinkage effects

existing length at one end only, ψ_p depends on the breadth of the slab, b, and on the length from the free end of the slab to the nearest cross section at which shear lag cannot occur, L_e say. This section is at mid-length of a slab cast alone (from symmetry), or can be assumed to be at an existing concrete surface (A in fig. A.4), so that values of L_e are as shown. They are not influenced at all by the span, L.

The variation of the force Q due to the primary effects of shrinkage along a length such as AB has been studied by linear-elastic finite element analysis[182] including the effects of slip of the shear connection. Typical results are compared in fig. A.5 with the variation given by the

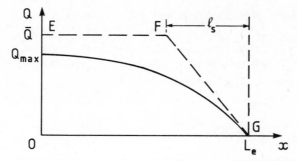

Fig. A.5   Influence of shear lag and slip on shrinkage effects

methods of the Bridge Code when shear lag is neglected, line EFG. The force $\bar{Q}$ is as given by equation (A.27) with $\psi_p = 1$, and ℓ_s is from equation (8.63). The ratio $Q_{max}/\bar{Q}$ depends on b/L_e, the proportions of the cross section, and the stiffness of the shear connection, and, in these tests, ranged from 0.6 to 1.0 for the members analysed.

In simply-supported beams, no advantage is gained by taking account of this reduction in Q; for simplicity, ψ_p should be taken as 1.0.

In continuous structures, the secondary effects depend on integrals along members of the primary shrinkage curvature ϕ, (equation (A.16)), which depends on Q and is proportional to ψ_p. For example, if the deck slab for a beam with two equal spans L is cast at one time, $L_e = L$, and the secondary moments and shear are proportional to:

$$\int_0^{L_e} \phi(L_e - x)\,\mathrm{d}x$$

This case was also studied[182] and the reductions in secondary effects were found to range from 5 to 50%.

It was found that a simple and sufficiently accurate approximation to these results is obtained by taking ψ_p as the effective-breadth ratio ψ given in Part 3 of BS 5400 for a quarter-span of a simply-supported beam, which is the upper of the two curves labelled Q in fig. 8.8. In that figure, b is the half-breadth of the slab and ℓ is the span. When b is the whole breadth (as here) and L_e is as shown in fig. A.4, the x-axis of fig. 8.8 should be labelled $b/4L_e$, not b/ℓ.

The reduction in shrinkage effects given by this method is significant only when the steel webs are widely spaced. For example, in the Avonmouth Bridge, there are four webs for a deck of breadth 40 m (fig. 9.10), so $b \simeq 10$ m. If an isolated 40-m width of slab only 10 m long is cast in a midspan region, $L_e = 5$ m, $b/4L_e = 0.50$, and $\psi_p = 0.36$. It would be reasonable to neglect completely the longitudinal effects of shrinkage that occurred while this slab remained isolated, but not the transverse effects, because ψ_p for the cross-girders is close to 1.0.

Shear lag can often be ignored. In situations where it is significant, its treatment using effective breadths is different in each of the three stages of the calculation.

(1) When calculating $\bar{F}$, $\bar{M}$, Q, and the primary curvature, use the effective-breadth ratio for primary effects, ψ_p, found as explained above and based on the length of slab L_e.

For temperature effects, the relevant value of L_e is half the distance between the free ends of the slab in the completed structure, irrespective of how the slab was cast. The ratio ψ_p is always close to 1.0, so shear lag should be neglected.

(2) When calculating secondary bending moments and shears from primary curvatures, ignore shear lag, as is usual in global analyses.

(3) When calculating stresses due to the secondary moments (for checks at SLS), use the same section moduli as for other dead-load effects. These are found using the effective-breadth ratios, ψ, based on span lengths (section 8.4.1).

INFLUENCE OF CRACKING OF CONCRETE ON THE EFFECTS OF SHRINKAGE
The theory for the primary curvature of a member due to shrinkage took no account of the cracking of concrete: yet, at internal supports both primary and secondary effects of shrinkage cause tensile strain in the concrete slab which is likely to be cracked by the effects of other types of loading, even if not by shrinkage.

The primary sagging curvature due to shrinkage does not disappear as soon as the slab cracks, because shrinkage strains still exist in the concrete between the cracks. But the curvature does diminish as the cracks widen and become more numerous[112] and when the calculated tensile stress in the reinforcement reaches yield, the remaining shrinkage curvature is negligible. Primary stresses due to shrinkage should therefore be omitted when checking hogging moment regions at ULS. Under the action of repeated loading, the curvature probably disappears at lower stress levels, by analogy with the current assumption about tension stiffening (section 8.3.3), but there are no relevant test data.

This reduction in primary curvature reduces secondary effects only slightly, because typically it occurs over a very small proportion of the span. It is thus not over-conservative to design for secondary effects calculated neglecting cracking.

Appendix B

Fatigue Strength of Shear Connectors

Rules for the evaluation of fatigue damage to shear connectors are given in Part 10 of the Bridge Code, and are discussed in chapter 10. The data on fatigue resistance were obtained as explained below. The design resistance of stud connectors is shown to differ significantly from the more recent data given in draft Eurocode 3.[4]

The method of CP 117: Part 2[77] for the design of shear connectors for repeated loading differed in several ways from the design methods for welded details given in BS 153.[18] In drafting Part 10, it was found to be possible to align these procedures, but shear connectors had to be placed in a separate class from other welded details, because their σ_r–N curve has a different slope.

σ_r-N CURVE FOR SHEAR CONNECTORS

In CP 117: Part 2, design fatigue strengths are given as percentages of ultimate capacity for stud connectors, but as nominal shear stresses on the weld throat for other types, and the data are presented in terms of the maximum and minimum stresses (or loads) caused by each cycle of loading. It was found that these same data could be presented more simply in terms of σ_r, the range of stress caused by a loading cycle, because when the maximum stress in each cycle is not excessive, fatigue damage depends almost entirely on range of stress. In limit-state design, maximum stress is controlled by the design for static loading, so it is possible to use range only when designing for fatigue.

Design data for welded details are given in part 10 of the Bridge Code as algebraic expressions relating the stress range σ_r to the number of cycles N of that range that cause fatigue failure. In graphical form, these are known as σ_r–N curves, and are conveniently plotted using log–log scales. Welds are designed for a 2.3% probability of failure, so the σ_r–N curves that are used give mean fatigue lives reduced by two standard deviations. In principle, this method has been used for shear-connector welds, but it was necessary to make certain assumptions, explained below, in deriving design values from the available test results.

The most extensive set of data relates to $\frac{3}{4}$ in. (19 mm) diameter headed studs, but the tests were done on specimens set in concretes with cube strengths ranging from 23 to 70 N/mm². Allowance was made for the influence of the strength of the concrete by analysing all results of constant-amplitude fatigue tests in terms of the ratio of the range of the applied shear per connector to the nominal static ultimate strengths per connector as given in Part 5 of the Bridge Code. In other words, it was assumed that the strength of the concrete influences the range of load for fatigue failure of the weld in the same way that it influences static strength. This assumption is consistent with the results of Menzies' tests[76] but has no thoretical basis.

Regression analyses were done on the results of 67 fatigue tests on headed studs, from three sources.[14b, 76, 117] These were subdivided into those with a ratio of maximum to minimum load (R) of -1, and those with $R \geqslant 0$. It was found that the mean endurance (number of cycles to failure) of the first group was about three times that of the second group. Design is therefore based on the 'mean less two standard deviations' line for the tests with $R \geqslant 0$, which was found to be:

$$Nr^8 = 19.54 \tag{B.1}$$

where r is the ratio of load range to nominal static strength, and N is the number of cycles to failure.

The test results for channel and bar connectors were analysed in terms of the range of shear stress on the weld throat, and lines were drawn at the lower limit of the band of scatter on the σ_r–N diagram, as there were insufficient data to justify the use of regression analyses. All the results for channels were for $R = -1$. These were corrected for use when $R \geqslant 1$ by dividing all endurances by three (by analogy with the results for studs), leading to the design σ_r–N line shown as ABD in fig. B.1:

$$N\sigma_r^8 = 2.08 \times 10^{22} \tag{B.2}$$

where σ_r is the range of shear stress in N/mm² units.

It is convenient to use equation (B.2) for studs as well as for channel connectors. From equations (B.1) and (B.2), $\sigma_r = 425r$, so this can be done if the range of shear stress for a stud is calculated from:

$$\sigma_r = 425r = 425(P_r/P_u) \tag{B.3}$$

where P_r is the range of load on a stud with nominal static strength P_u. Equation (B.3), given in Part 10, is consistent with tables 3 and 4 of CP 117: Part 2.

The design σ_r–N line for bar connectors was found to be:

$$N\sigma_r{}^8 = 3.494 \times 10^{20} \tag{B.4}$$

This can be written:

$$N(\sigma_r/0.6)^8 = 2.08 \times 10^{22}$$

so that equation (B.2) can be made applicable to the design of welds for bar connectors by requiring the effective leg length of the weld to be taken as not greater than 0.6 times the actual length. This was done in Part 10, and enabled all shear-connector welds to be placed in one class, Class S.

The slope of the σ_r–N curve for connectors ($m = 8$) is shallower than those of the curves for other welded details in Part 10 (e.g. the curves for Class C ($m = 3.5$) and Class W ($m = 3$), shown in fig. B.1). Subsequent

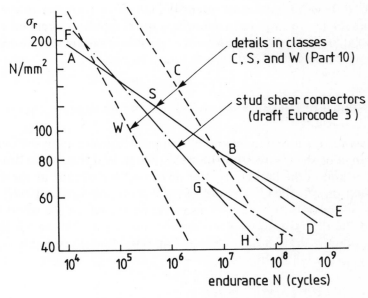

Fig. B.1 σ_r–N curves for shear connectors

evidence has shown (section 10.4.1) that $m = 8$ for shear connectors is too high. The σ_r–N curve in draft Eurocode 3 uses $m = 5$, and is shown as FGH in fig. B.1. Even the revised curve is significantly shallower than the curves for all-steel welded details. The reason is thought to be the influence of the concrete on the results of fatigue tests of the push-out type. It has the consequence that in highway bridges, the heaviest vehicles (which cause low-N, high-σ_r cycles) have a relatively greater influence on the fatigue design of shear connectors than of other types of welded detail.

STRESS CYCLES OF LOW AMPLITUDE

Equation (B.2) is based on fatigue tests on shear connectors in which failure occurred at less than 3×10^7 cycles, for which $\sigma_r = 72$ N/mm^2. Design for stress cycles of lower amplitude is now discussed.

It is well known from constant-amplitude fatigue tests on all-steel specimens that σ_r–N curves flatten out when N exceeds about 10^7 cycles, because cycles of stress below a certain level ('low-stress cycles') do not propagate fatigue damage.

For spectrum loading of the type found in highway bridges, behaviour is more complex, because gradual propagation of an initial defect by early high-σ_r cycles successively lowers the threshold value of σ_r below which no propagation occurs. It would therefore be unsafe to terminate a σ_r–N curve by a horizontal cut-off at 10^7 cycles, but a reduction in slope can be made. By analogy with the behaviour of all-steel specimens, it was assumed that the σ_r–N curve for shear connectors when N exceeds 10^7 cycles can be taken (in N/mm^2 units) as:

$$N\sigma_r^{10} = 1.40 \times 10^{26}$$

This is shown as line BE in fig. B.1. It can be written:

$$N(\sigma_r/82)^2\sigma_r^8 = 2.08 \times 10^{22} \tag{B.5}$$

Comparison of equations (B.2) and (B.5) shows that the line ABD can be used for cycles of all ranges of stress if the number of repetitions of each stress range σ_r less than 82 N/mm^2 is reduced in the proportion $(\sigma_r/82)^2$. This method is given in Part 10. For shear connectors, the effect of the reduction is negligible, and it is simpler to use line ABD with the full number of repetitions. The corresponding reduction given in Eurocode 3 is shown in fig. B.1 as line GJ.

LIGHTWEIGHT–AGGREGATE CONCRETE

In a limited number of tests, Menzies found[76] the fatigue strengths of studs embedded in 'Lytag' concrete to be about 90% of those of similar connectors in normal-density concrete of the same strength. In Part 10 of the Bridge Code, this ratio is taken as 85%, the same as for static strengths.

MAXIMUM SIZE OF WELDS TO SHEAR CONNECTORS

The rules given in Clause 5.3.3.1 of Part 5 relate to the maximum size of weld that may be *used* to attach a shear connector to a steel flange. Their purpose is to ensure that fatigue failure does not occur in the flange. They should not be confused with the rules given in Part 10 which refer to the maximum size of weld that may be *assumed* when calculating the stress range due to a cycle of loading. Their purpose is to limit the maximum range of load (or spectrum of loads) that can be assumed in design to be imposed on a connector, and so to prevent fatigue failure in regions of the connector other than the weld itself.

For both bar and channel connectors, the leg length may not be taken as greater than half the thickness of the steel flange. This is also the limiting size given in Part 5.

For bar connectors, the assumed weld size may not exceed 0.6 times the actual size (as explained above), nor one-eighth of the sum of the height and breadth of the bar. The second limit controls the compressive stress in the concrete at the face of the bar. The breadth is included in order to limit the load that could be applied to a high narrow bar, which would be less stiff than the bars of square cross section normally used.

For channel connectors, the assumed weld size may not exceed the actual size of the weld, nor the thickness of the web of the channel. The second limit is to prevent failure of the web.

Appendix C

Ultimate Moment of Resistance of a Concrete-filled Steel Circular Hollow Section

As explained in section 14.3, the expressions given in Part 5 of the Bridge Code for the ultimate moment of resistance, M_u, include γ_m (taken as 1.05 for structural steel and 1.50 for concrete in compression) but not γ_{f3}. No account is taken of any reinforcement (for simplicity), and the other assumptions are those used when calculating M_u for a beam of compact section (section 8.4.3):

(1) steel is stressed to its design yield strength f_d ($=0.95\sigma_y$) in tension or compression

(2) the concrete on the compression side of the plastic neutral axis is stressed to its design strength for bending compression, $0.4f_{cu}$

(3) the tensile strength of concrete is neglected.

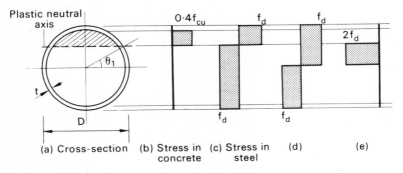

(a) Cross-section (b) Stress in concrete (c) Stress in steel (d) (e)

Fig. C.1   Stress distributions in concrete-filled CHS at M_u

Figures C.1(b) and (c) show the stress distribution across the section. The stress blocks shown in (d) and (e) are together equivalent to (c). The stress distribution shown in fig. C.1(d) represents the state of stress in the steel tube when it is at its plastic moment of resistance,

From figs C.1(d) and C.2, for the steel tube alone:

296

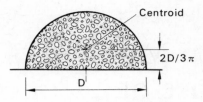

Fig. C.2   Centroid of semi-circle

$$M_p = 2f_d[(2D/3\pi)(\pi D^2/8) - (2/3\pi)(D - 2t)\{\pi(D - 2t)^2/8\}]$$

$$= f_d t^3[(D/t)^2 - 2D/t + 1.33] \simeq f_d t^3[(D/t) - 1]^2 \qquad (C.1)$$

Therefore the plastic section modulus for the steel section, S, is given by:

$$S = t^3[(D/t) - 1]^2 \qquad (C.2)$$

From figs C.1(e) and C.3, for an elemental section of the tube:

$$dN = f_d t[(D - t)] \, d\theta$$

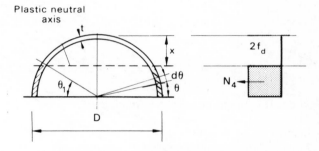

Fig. C.3   Tensile force in steel tube

and the moment of force dN about the plastic neutral axis is:

$$[(D - t)/2](\sin \theta_1 - \sin \theta) \, dN$$

By integration, the total force N_4 and its moment M_4 are:

$$N_4 = 2f_d t(D - t)\theta_1 \qquad (C.3)$$

$$M_4 = f_d t(D - t)^2(\theta_1 \sin \theta_1 + \cos \theta_1 - 1) \qquad (C.4)$$

For the concrete in compression, from figs C.1(b) and C.4, the force in an elementary strip of depth dy is:

$$dN = 0.8f_{cu}R \cos \theta \, dy$$

and its moment about the plastic neutral axis is

$$dM = 0.8f_{cu}R \cos \theta(y - y_1) \, dy$$

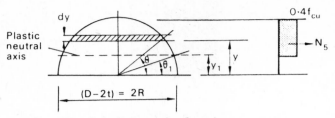

Fig. C.4   Compressive force in concrete

where R, y and y_1 are defined in fig. C.4. By integration, the total force N_s and its moment M_5 are:

$$N_5 = 0.05 f_{cu}(D - 2t)^2 (\pi - \sin 2\theta_1 - 2\theta_1) \tag{C.5}$$

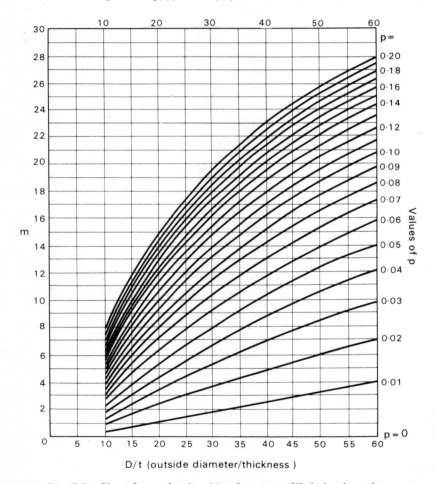

Fig. C.5   Chart for evaluating M_u of concrete-filled circular tubes

$$M_5 = 0.1f_{cu}(D - 2t)^3\{(\cos^3 \theta_1)/3 - 0.25 \sin \theta_1(\pi - \sin 2\theta_1 - 2\theta_1)\} \quad (C.6)$$

For equilibrium, $N_4 = N_5$, so from (C.3) and (C.5):

$$16(D - t)t\theta_1 = p(D - 2t)^2(\pi - \sin 2\theta_1 - 2\theta_1) \quad (C.7)$$

where:

$$P = 0.4f_{cu}/f_d \quad (C.8)$$

The ultimate moment of resistance of the composite section is:

$$M_u = M_p + M_4 + M_5$$

which is:

$$M_u = f_d S(1 + 0.01m) \quad (C.9)$$

where:

$$m = (100/S)[t(D - t)^2(\theta_1 \sin \theta_1 + \cos \theta_1 - 1)$$
$$+ 0.25p(D - 2t)^3\{(\cos^3 \theta_1)/3 - 0.25 \sin \theta_1(\pi - \sin 2\theta_1 - 2\theta_1)\}]$$
$$(C.10)$$

The chart for evaluating m (fig. C.5, given also in Appendix C of Part 5 of the Bridge Code) is constructed by solving equation (C.7) for θ_1 for assumed values of p and D/t, and then using equations (C.2) and (C.10). Equations (C.2) and (C.9) are given in Part 5 of the Bridge Code.

For example, for $p = 0.03$ and $D/t = 40$, trial-and-error solution of equation (C.7) gives $\theta_1 = 9.5°$. From equation (C.2):

$$S = (D/40)^3 \times 39^2 = D^3/42.08$$

Substituting in (C.10), $m = 7.2$. Figure C.5 gives the same value. From equation (C.9):

$$M_u = f_d D^3(1 + 0.07)/42.08 = 0.025f_d D^3$$

Appendix X: Worked Example

Three-span Continuous Composite Plate Girder Bridge Deck

X.1 Introduction to worked example

In the following sections, calculations are given for the design of the outer girder in the centre span of a three-span continuous composite plate girder bridge deck. Typical calculations for the design of the deck slab are also given. The design procedures are not fully explained but should be easily understood by referring to the relevant clauses of BS 5400 or to the relevant parts of the main text.

The design procedures comply with the requirements of BS 5400 Parts 1 to 5, as amended by the Department of Transport[2] for highway structures in Britain.

The example includes preliminary calculations that would be carried out to determine suitable section sizes before a detailed global analysis can be undertaken.

In practice initial section sizes are often estimated from previous designs with similar span dimensions, girder spacings and live loading. Whatever method is used, it should be noted that in continuous uncased composite bridge decks, cracking of the deck slab in negative moment regions reduces the stiffness of the section and modifies the moment distribution to a greater extent than in a continuous prestressed or reinforced concrete deck (section 8.2).

In this example, the term 'design loads' is always used to mean the nominal loads multiplied by γ_{fL}. For steel elements designed to Part 3 of the Bridge Code, these design loads must be less than or equal to the design resistance given in Part 3. For concrete elements designed to Part 4 and for shear connectors and transverse reinforcement designed to Part 5 of the Bridge Code, the design loads are multiplied by γ_{f3} to obtain the design load effects, which are then compared with the design resistances given in Part 4 or Part 5. The reasons for this are explained in section 7.1.2.

X.2 Abbreviations, symbols and units

Abbreviations used in the worked examples are explained either below or, in a few instances, where they occur. The symbols are as defined and used in the main text, and usually are consistent with those used in BS 5400.

BM	bending moment
HA	type HA loading for highway structures in Britain
HB	type HB loading for highway structures in Britain
HAK	knife edge load component of HA loading
HAU	uniformly distributed component of type HA loading
LS	longitudinal shear
sdl	superimposed dead load
SLS	serviceability limit state
S-1, S-2, etc.	serviceability limit state, load combination 1, 2, etc.
SW	self weight
TSC	temperature, and shrinkage as modified by creep
ULS	ultimate limit state
U-1, U-2, etc.	ultimate limit state, load combinations 1, 2, etc.
udl	uniformly distributed load
VS	vertical shear
T20	deformed reinforcing bars with nominal yield stress 460 N/mm² and of diameter 20 mm (and similarly for other sizes)
3/6.5.4	Clause 6.5.4 of BS 5400: Part 3 (and similarly for other clauses)

UNITS

For the concise presentation of calculations, the ideal set of units is such that most numerical values lie between 0.01 and 1000. For bridge structures, this is not possible using preferred SI units only. The units used here give most numbers of convenient size, and reduce to a minimum the occasions on which powers of ten enter into calculations.

The basic units of force and length are the kilonewton and the metre. For cross sections, dimensions and areas are in millimetre units, leading to second moments of area given as $10^{-6}I$, in mm^4 units. Divison of $10^{-6}I$ by a depth $\bar{y}$ in mm gives section moduli, in the form $10^{-6}Z$, in mm^3 units. These are used because a bending moment in kNm divided by $10^{-6}Z$ mm^3 gives directly a stress in N/mm^2 (or MN/m^2 or MPa, as preferred). For longitudinal shear, the factor $A\bar{y}/I$ is given in m^{-1} so that multiplication by vertical shear in kN gives longitudinal shear in kN/m, and hence spacing in metres of connectors whose strength is given in kN.

All dimensions in diagrams are in metres or millimetres. The unit is not stated because it will be obvious which is being used.

CONVENTION OF SIGN

As is usual in engineering practice, unsigned quantities are sometimes used where only the magnitude is of interest, or where the direction is obvious from the context. Signed quantities are used where necessary to avoid ambiguity, and then sagging moments, clockwise shear, and tensile stresses and strains are taken as positive.

X.3 General arrangement of bridge deck and design criteria

X.3.1 Geometry

Span arrangement: three spans, 25 m – 37 m – 25 m, continuous.

Skew: 10 degrees.

Cross section: as shown in fig. X.3.1 the deck carries a dual three-lane motorway and has an overall width of 36 metres. Each 14.9 m carriageway comprises a 3.3 m hard shoulder, three traffic lanes marked out over a width of 11 m and a 0.6 m median strip. Services are carried in the 0.6 m wide footways and in the 4 m central reservation.

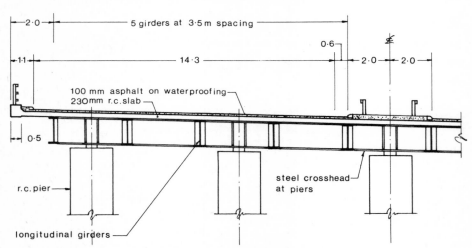

Fig. X.3.1   Cross section of bridge deck

STEELWORK DETAILS

Ten longitudinal plate girders framing into composite steel cross-girders at intermediate piers. Main girders spliced at a point 6.5 m either side of internal supports.

X.3.2 Nominal loadings

Dead: unit weight of steel, 77.0 kN/m^3.
unit weight of reinforced concrete, 25.0 kN/m^3.

Formwork and construction loads: 0.5 kN/m^2.

Superimposed dead:
asphalt surfacing and waterproof membrane, 21.0 kN/m^3.
infill concrete in footways and reservation, 23.0 kN/m^3.
parapets, 0.5 kN/m each.

Live loading: the more severe of either:
(i) HA loading as specified in BS 5400 Part 2 and amended by Department of Transport Interim Revised Loading Specification and Technical Memorandum BD 24/84.[186]
or (ii) 45 units HB loading combined with HA loading, as specified in BS 5400 Part 2 as amended by Department of Transport Technical Memorandum BD 24/84.

Wind and temperature: as specified in BS 5400 Part 2 for an inland site, 15 miles west of central London at 100 m above sea level for a design life of 120 years.

Shrinkage (modified by creep): as specified in Clause 5.4.3, Part 5 for a normal open-air environment.

X.3.3 Materials and strengths

Structural steel: Grade 50 C, BS 4360[72]   3/6.5.4
                $E_s = 205$ kN/mm^2 3/6.6
                $v = 0.3$
                $\sigma_y = 355$ N/mm^2 (irrespective of thickness)   3/6.2

Concrete: Grade 40, $f_{cu} = 40$ N/mm^2
             $E_c = 31$ kN/mm^2 4/4.3.2.1
             $v_c = 0.2$
             free shrinkage, $\varepsilon_{cs} = -200 \times 10^{-6}$ 5/5.4.3

Reinforcement: Grade 460, to BS 4449 or BS 4461
             $f_{ry} = 460$ N/mm^2 4/5.1.4.3
             $E_r = 200$ kN/mm^2 4/4.3.2.2

Table X.3.1 Design resistance of materials

Material/loading	Nominal strength†	Serviceability Limit State*		Ultimate Limit State			
		Design resistance†	BS 5400 ref.	Design resistance	BS 5400 ref.	γ_m	γ_{f3}
Steel flange in tension	$\sigma_{yt} = 355$	$\sigma_{yt}/\gamma_m\gamma_{f3} = 355$	3/4.2.2	$\sigma_{yt}/\gamma_m\gamma_{f3} = 307$	3/4.3.2, 4.3.3 3/9.9.1.2, 9.9.1.3	1.05	1.1
Steel flange in compression							
– compact	$\sigma_{yc} = 355$	$\sigma_{yc}/\gamma_m\gamma_{f3} = 355$	3/9.9.8	$\sigma_{\ell c}/\gamma_m\gamma_{f3}$ with $\sigma_{\ell c} = \sigma_{\ell i}$	3/9.8.2, 9.9.1.2	1.20**	1.1
– non compact	$\sigma_{yc} = 355$	No check necessary	3/4.2.2	$\sigma_{\ell c}/\gamma_m\gamma_{f3}$ with σ equal to lesser of σ_{yc} or $D\sigma_{\ell i}/2\gamma_t$	3/9.8.3, 9.9.1.3	1.20**	1.1
Webs in shear	$\tau_v = \sigma_{yw}/\sqrt{3}$ $= 205$	No check necessary	—	$\tau_\ell/\gamma_m\gamma_{f3}$	3/9.9.2.2	1.05	1.1
Webs in combined bending and shear	—	No check necessary	—	As determined from 3/9.9.3.1, 9.9.5.3		—	1.1
Concrete in compression	$f_{cu} = 40$	$0.5f_{cu} = 20$	4/table 2	$0.67\, f_{cu}/\gamma_m = 17.87$ or $0.4f_{cu} = 16$	4/5.3.2.1	1.5	1.1
Reinforcement in tension	$f_{ry} = 460$	$0.75f_{ry} = 345$	4/table 2	$f_{ry}/\gamma_m = 400$	4/4.3.2.2	1.15	1.1
Shear connectors (static)	$P_u = 139$ kN	$0.54P_u = 75$ kN	5/5.3.2.5	ULS need only be considered when required by 5/5.3.3.5		—	—

† Nominal strengths and design resistances are in N/mm², except where stated
* At SLS, $\gamma_m = 1.0$, $\gamma_{f3} = 1.0$
** $\gamma_m = 1.20$ for compression flanges acting compositely with concrete deck

Shear connectors: 22 mm dia. × 100 mm welded studs to Part 5

$$P_u = 139 \text{ kN} \qquad \text{5/table 7}$$

Coefficient of thermal expansion: for all materials

$$\beta = 12 \times 10^{-6} \text{ per degree Celsius} \qquad \text{5/5.4.2.2}$$

Design resistance: table X.3.1 summarises the design resistances and relevant clause references in Parts 3, 4 and 5 of the Bridge Code.

X.3.4 Modular ratios

For short-term loading, $\alpha_{e1} = E_s/E_c = 6.6$

For long-term loading, $\alpha_{e2} = 2\alpha_{e1} = 13.2$ 5/4.2.3

For shrinkage, $\phi_c = 0.4$ and $\alpha_{e3} = E_s/\phi_c E_c = 16.5$, but for

convenience, take $\alpha_{e3} = \alpha_{e2}$ 5/5.4.3

X.3.5 Design crack widths and concrete cover to reinforcement

The relevant values obtained from Part 4 are given in table X.3.2. Crack widths need only be calculated for load combination 1 at SLS (4/4.2.2), and for this combination, where HB loading is considered, the number of units of HB may be reduced to 25.

Table X.3.2 Design crack widths and concrete cover

Location	Environment	Cover (mm)	Crack width (mm)
Surface of deck slabs protected by waterproofing	Moderate	30	0.25
Soffit of deck slab	Severe	35	0.25
Parapet plinths exposed to de-icing salts	Very severe	50	0.15

X.3.6 Method of construction

Unpropped construction, with the deck slab cast in five stages. The first four stages extend over the full width of the deck between the construction joints at the parapet plinth, as shown in fig. X.10.1. The sequence of casting is:

(1) A length of 18.5 m from the end supports of each 25 m span.

(2) A length of 12 m in the centre of the 37 m span.

(3) A length of 6 m at each end of second stage concrete in the 37 m span.

(4) A length of 13 m over each intermediate support.

(5) Both parapet plinths over the full length of the bridge.

The central section of the 37 m span is concreted in two stages to limit the compressive stress in the top flange at midspan before the steel girders act compositely with the deck slab. Once composite action has been developed between the slab and girder, its stiffness should be taken into account in the global analysis for subsequent stages of concreting. The final sections of deck slab over each support are cast last to reduce the longitudinal tensile strains and crack widths in the slab due to dead load, shrinkage and early thermal contraction. The parapet plinths would be cast in several stages in practice. Where, as in this example, they are cast onto the edge of a mature deck slab, adequate longitudinal reinforcement must be provided in the plinth to prevent excessive cracking due to restrained early thermal contraction. Further information on this can be found in reference 180 and Clause 5.8.9 of Part 4.

X.3.7 *Spacing of longitudinal girders and thickness of deck slab*

For highway live loadings in Britain, the economic spacing of longitudinal girders for a continuous deck slab with no intermediate cross girders is between 2.5 and 3.5 m. At spacings of 2.5 m or more, the critical loading case for punching shear failure of the slab occurs with two or more wheels of the HB vehicle bogie positioned on the slab panel between two adjacent girders. For 45 units of HB loading, the minimum thickness of slab for punching shear is about 190 mm (section 8.3.1), but this needs to be increased for transverse bending due to local wheel loads when the girder spacing exceeds 3 m. In practice, slab thicknesses of 200–250 mm are normally used with girder spacings of 2.5–3.5 m.

For the superstructure in this worked example, a spacing of **3.5 m** is suitable for the longitudinal girders, and the thickness of the deck slab is initially taken as **230 mm**.

X.4 Preliminary design moments and vertical shears – outer girder

For these preliminary calculations it is assumed that the composite section is uncracked and unreinforced throughout its length. Bending moments are calculated by moment distribution for a beam of uniform section. The whole of the wet concrete slab is assumed to be carried by the steel section alone. The weight of formwork and temporary construction loads has been neglected because its effect is small but, where this is not the case, the effect of applying it during construction and the effect of removing it after the deck has been completed should be calculated.

X.4.1 Dead load

(1) *Steelwork.* The weight of steel used in a design for a 30 m span, simply-supported bridge in 1970 was 2.2 kN/m^2, using rolled sections. With the use of more widely spaced plate girders, as in this example, assume initially that this will reduce to 1.5 kN/m^2 which is equivalent to 5.25 kN/m for each girder.

(2) *Concrete.* The detail of the deck slab at an outer girder is shown in fig. X.4.1.

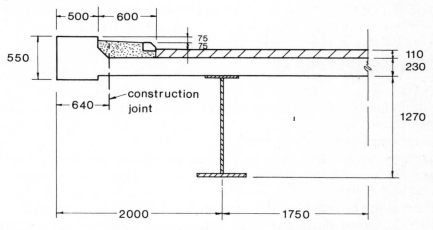

Fig. X.4.1   Cross section of outer girder

The nominal loading on the outer girder is:

Deck slab: $0.23 \times 3.11 \times 25 = 17.88 \text{ kN/m}$
Parapet plinth: $[(0.23 \times 0.64) + (0.32 \times 0.5)] \times 25 = 7.68 \text{ kN/m}$

Hence, total concrete dead load on outer girder $= 25.56 \text{ kN/m}$.

X.4.2 Superimposed dead load

Surfacing: 100 mm asphalt plus 10 mm waterproofing,
    $2.65 \times 0.11 \times 21 = 6.12 \text{ kN/m}$.
Other: infill concrete, $0.6 \times 0.185 \times 23 = 2.55 \text{ kN/m}$
    P1 type vehicle parapet $= 0.55 \text{ kN/m}$
                            Total 3.10 kN/m.

X.4.3 HA loading

NOTIONAL TRAFFIC LANES
For the purpose of calculating the intensity of HA live loading, BS 5400 Part 2 defines the carriageway as the width between raised kerbs, including all traffic lanes, hard shoulders, hard strips and marker strips. It then divides the carriageway into a number of notional traffic lanes. These do not correspond to the actual traffic lanes marked out on the carriageway.

In this example, the total carriageway width is 14.9 m, so from Clause 3.2.9 of Part 2, there are four notional traffic lanes, each 3.725 m wide.

LOADING INTENSITY
The intensity of HAU loading varies with loaded length. For maximum sagging moment in the 37 m span, the loaded length is 37 m and from the Department of Transport Interim Revised Loading Specification (IRLS), HAU loading is 27 kN per metre of lane.

For maximum hogging moment over the supports, the loaded length is $25 + 37 = 62$ m giving a uniformly distributed lane loading of 23.2 kN/m.

The knife-edge load (HAK) is 120 kN per notional lane.

APPLICATION OF TYPE HA LOADING
The HAU loading and the HAK loading are applied to any two notional lanes.

NOMINAL HA LOADING ON OUTER GIRDER
Assuming static distribution of loading (transversely) to each girder, the nominal loading on the outer girder is:

For 37 m loaded length: HAU = $\ \ 27 \times 2.65/3.725 = 19.2$ kN/m
 HAK = $120 \times 2.65/3.725 = 85.4$ kN

For 62 m loaded length: HAU = $23.2 \times 2.65/3.725 = 16.5$ kN/m
 HAK = $120 \times 2.65/3.725 = 85.4$ kN.

MAXIMUM HA MOMENT AT MIDSPAN AND MAXIMUM HA SHEAR AT INTERNAL SUPPORT
The maximum HA moment at midspan of the outer girder is 4076 kNm and is produced with the knife-edge load (HAK) positioned at midspan and a uniformly distributed load (HAU) of 19.2 kN/m over the 37 m span. The shear force at the support with this loading arrangement is 398 kN, but the maximum shear occurs with HAK at the support and is 441 kN.

MAXIMUM HA MOMENT AT INTERNAL SUPPORT

The maximum HA moment at the internal support is -2134 kNm. This occurs with HAU = 16.5 kN/m and the HAK at the centre of the 37 m span. The shear at the internal support in the 37 m span is then 367 kN, but the maximum shear occurs with HAK at the support and is 410 kN.

X.4.4 HB loading

APPLICATION OF TYPE HB LOADING

Type HB loading can occupy any transverse position within one notional lane, or may straddle two notional lanes. With the overall width of the HB vehicle at 3.5 m, the most severe loading on the outer girder occurs with the outermost wheel of the HB vehicle located 0.25 m from the kerb, as shown in fig. X.4.2.

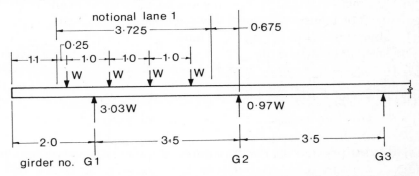

Fig. X.4.2 Transverse position of HB axle for maximum load on edge girder

The two adjacent notional lanes then carry full HA loading. At this preliminary stage the effect of HA loading in the other lanes is neglected in calculating the loading on the outer girder because static distribution of the wheel loads is likely to overestimate the actual loading carried by the outer girder.

Static distribution of each HB axle load gives the loading on the outer girder as 3.03 times the wheel load.

MAXIMUM HB MOMENT AT MIDSPAN

The maximum sagging moment at the centre of the 37 m span is produced with the shortest specified inner axle spacing (6 m), with the vehicle located as in fig. X.4.3. It should be noted that this position does not produce the maximum bending moment in the span. This occurs with the second axle 1.5 m nearer the support. However, since HB loading is always considered in combination with dead, superimposed dead and HA loading on other lanes, the maximum moment due to the combination of loadings generally occurs at midspan.

The maximum moment at midspan due to the HB vehicle, positioned as shown in fig. X.4.3 is $16.615W_a$, where $W_a = 3.03W$. The maximum HB nominal moment at midspan is then:

$$16.615 \times 3.03 \times 45 \times 10/4 = 5664 \text{ kNm}$$

MAXIMUM HB VERTICAL SHEAR AT INTERNAL SUPPORT
The maximum vertical shear of $3.58W_a = 1616$ kN occurs with the leading axle of the HB bogie almost at the support and the remaining axles in the 37 m span. If a grillage analysis is used to determine the maximum HB shear, the leading axle load should not be applied directly over a support, but a short distance away, otherwise the shear in the beam will be underestimated.

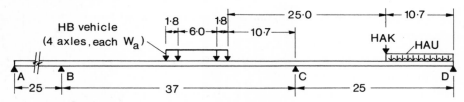

Fig. X.4.3   HB loading for maximum moments at midspan and at C

MAXIMUM HB MOMENT AT INTERNAL SUPPORT
Influence lines are required to determine the position of the HB vehicle which gives maximum moment at the internal support. Figure X.4.4 shows the influence line for bending moment at the internal support C. The position of the HB load in the 37 m span for maximum moment at C is very close to the position for maximum midspan moment and is here assumed to be the same. The moment at C in the outer girder due to HB loading is then:

$$12.81W_a = 12.81 \times 3.03 \times 45 \times 10/4 = 4367 \text{ kNm}$$

With the HB load in this position, it is possible for the HAU to occur in the same notional lane as the HB vehicle, over a length of 10.7 m in span CD, as shown in fig. X.4.3. The HAU loading intensity for this case is the same as for HAU occupying spans BC and CD (loaded length 62 m), so the HAU loading on the outer girder over part of span CD is 16.5 kN/m. The moment at C caused by HAU and HAK in span CD is 244 kNm. This is only 6% of the HB moment at C and is neglected at this preliminary stage because the HB moment will certainly have been overestimated by the simple static distribution.

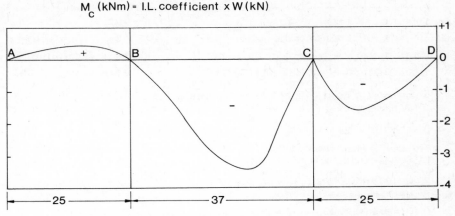

M_c (kNm) = I.L. coefficient x W (kN)

Fig. X.4.4   Influence line for bending moment at support C

X.4.5 Preliminary design moments and vertical shear carried by steel section alone

The design moments and shear at the centre of the 37 m span and at the internal support are summarised in table X.4.1.

At this stage, no account is taken of the sequence of casting of the deck slab because this will not affect the vertical shear carried by the steel section or the moment carried by the steel section at the internal support. The moment carried by the steel section at midspan will be overestimated.

Table X.4.1   Preliminary design bending moments and shear forces on steel section alone

	SLS		ULS	
Nominal loading	γ_{fL}	S-1	γ_{fL}	U-1
Steel: 1.5 × 3.5 = 5.25 kN/m	1.0	5.25	1.05	5.51
Concrete: 25.56 kN/m	1.0	25.56	1.15	29.39
Total load on steel section (kN/m)		30.81		34.90
BM at centre 37 m span = 0.0533 × 37² × W (kNm)		2248		2547
BM at support C = 0.0716 × 37² × W (kNm)		3020		3421
Shear at support C in 37 m span = WL/2 (kN)		570		646

X.4.6 Preliminary design moments and shears on composite section – outer girder

Preliminary design moments and vertical shears on the composite section are calculated in table X.4.2. HB loading produces higher

Table X.4.2 Preliminary HB with HA moments and shears in outer composite girder

	SLS		ULS	
Nominal loading	γ_{fL}	S-1	γ_{fL}	U-1
Midspan: BM for 37 m span (kNm)				
(1) Super DL (asphalt) $0.0533 \times 6.12 \times 37^2 = 447$	1.2	536	1.75	782
(other) $0.0533 \times 3.10 \times 37^2 = 226$	1.0	226	1.20	271
(2) HB with HA: $16.615 \times 3.03 \times 45 \times 10/4 = 5664$	1.1	6230	1.30	7363
or (3) HA alone: 4076 kNm	1.2	4891	1.50	6114
Design moment at midspan (1 + 2)		6992		8416
Max. shear at internal support (kN)				
(1) Super DLN: (asphalt) $6.12 \times 37/2 = 113$ kN	1.2	136	1.75	198
(other) $3.10 \times 37/2 = 57$ kN	1.0	57	1.20	68
(2) HB with HA: $3.58 \times 45 \times 10 = 1616$ kN	1.1	1778	1.30	2101
(3) HA alone: 441 kN	1.2	529	1.50	662
Design shear at support (1 + 2)		1971		2367
Support BM (kNm)				
(1) Super DL: (asphalt) $0.0716 \times 6.12 \times 37^2 = 600$	1.2	720	1.75	1050
(other) $0.0716 \times 3.10 \times 37^2 = 304$	1.0	304	1.20	365
(2) HB with HA: $12.81 \times 3.03 \times 45 \times 10/4 = 4366$	1.1	4803	1.30	5676
(3) HA alone: 2134 kNm	1.2	2561	1.50	3201
Design moment at support (1 + 2)		5827		7091

moments and vertical shears than HA, both at midspan and at the internal support. The design moments and shears due to superimposed dead loading plus HB with HA are therefore used to size the section.

X.5 Preliminary design of cross sections

X.5.1 Breadth of concrete flange acting with outer girder

To reduce the amount of computation required in calculating section properties, the cross-section of the deck slab shown in fig. X.4.1 is simplified in this example to a **slab of constant thickness 230 mm**, projecting **2 m** beyond the web of the outer girder, and **1.75 m** on the other side of the web.

X.5.2 First design of steel cross section at midspan

Assume initially a span:depth ratio of 25 for the composite section. Its overall depth is then 1.48 m, so round this up to 1.5 m. With a slab thickness of 230 mm, the depth of the steel section is then 1.27 m. If the bottom flange thickness is taken as 45 mm and the top flange thickness is 20 mm, the web is then 1205 mm deep. The web thickness will be determined by the vertical shear at the splice. For combination U-1, by static distribution, this is 1630 kN.

For a 10 mm web, the slenderness ratio $\lambda = 1205/10 = 120.5$ (Clause 9.9.9.2, Part 3). Assuming an aspect ratio, $\phi = 1.0$, and no tension field action, the limiting shear stress $\tau_\ell = 0.7\tau_y$ (fig. 11, Part 3), i.e.

$$\tau_\ell = 0.7 \times 355/\sqrt{3} = 143 \text{ N/mm}^2$$

The design shear resistance is:

$$V_D = 143 \times 1205 \times 10 \times 10^{-3}/(1.05 \times 1.1) = 1492 \text{ kN}$$

As no account has been taken of the reduction in HB shear that will occur due to the transverse distribution properties of the deck, or of tension field action in the web, a 10 mm web will probably be satisfactory. The moment that exists with the shear at the splice is likely to be less than $0.5M_R$, so a 10 mm web should also be adequate for combined shear and bending (Part 3, Clause 9.9.3.1, Note 2). Hence, assume a **10 mm web**.

At midspan, it is likely that the composite section will be compact (Clause 9.7.3, Part 3) but the steel section will almost certainly be slender before it acts compositely with the deck slab. Nevertheless, provided the top flange is adequately restrained against lateral torsional buckling, it should be able to develop its design yield strength in compression.

The cross-sectional area of the bottom flange at midspan (A_f) will be determined by the total moment on the section. The lever arm is 1.24 m for loading on the steel section alone and 1.365 m for loading on the composite section. The flange can develop its design yield strength in tension, so the area of steel required for the flange is:

$$A_f = [(2547/1.24) + (8416/1.365)]/(307 \times 10^{-3}) = 26774 \text{ mm}^2$$

Try a **bottom flange of area 600 $\times$ 45 mm** $= (27\ 000 \text{ mm}^2)$.

The area of the top flange will be determined principally by the load carried on the steel section before it acts compositely with the concrete slab. The limiting compressive stress in the steel flange at this stage will be controlled by lateral-torsional instability and will depend on the effectiveness of the lateral bracing provided to the compression flange. For economic design the lateral restraint provided should enable the flange to develop its design yield strength in compression, i.e.

$$\sigma_{yc} = 355/(1.20 \times 1.10) = 261 \text{ N/mm}^2$$

The area of flange required is then:

$$A_f = 2547/(1.24 \times 261 \times 10^{-3}) = 7870 \text{ mm}^2$$

Therefore try a **top flange of area 400 × 20 mm**.

The midspan cross section is now as shown in fig. X.5.1. The area of steel is 470.5 cm² and the weight is 3.623 kN/m. To allow for stiffeners and cross bracing, increase this to 4 kN/m. This is still less than the allowance made initially, so no corrections are required to the preliminary calculations.

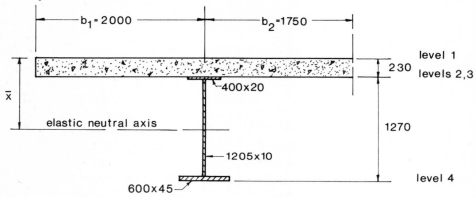

Fig. X.5.1   Preliminary cross section of outer girder at midspan

X.5.3 First design of steel cross section at internal supports

The overall depth of the section and the depth of the slab will be the same as at midspan. Try a bottom flange thickness of 60 mm and a top flange thickness of 30 mm. The web is then 1180 mm deep. The total design shear force on the web for load combination U-1 is $646 + 2367 = 3013$ kN. For an 18 mm web, the slenderness ratio $\lambda = 1180/18 = 65$ (Clause 9.9.9.2, Part 3). Taking the aspect ratio, ϕ, as 1.0 and $m_{fw} = 0$, the limiting shear stress (from fig. 11, Part 3), τ_ℓ equals $0.985\tau_y$, or $0.985 \times 355/\sqrt{3} = 202$ N/mm². The design shear resistance is:

$$V_D = 202 \times 1180 \times 18 \times 10^{-3}/(1.05 \times 1.1) = 3715 \text{ kN}$$

The moment that exists with the maximum design shear will be less than the maximum design moment at the support, but is unlikely to be less than $0.5M_R$, so some margin is required in the initial estimate of web thickness to allow for combined bending and shear. We assume that the margin between 3715 kN and 3013 kN will be sufficient, and so try an **18 mm thick web**.

Initially, the area of the bottom flange is determined assuming that the total moment is carried on the steel section alone and the lever arm is 1.225 m. If adequate lateral restraint is provided to prevent lateral torsional buckling, i.e. $\lambda_{LT} \leqslant 45$ in fig. 10, Part 3, then $\sigma_{\ell i} = \sigma_{yc} = 355\ N/mm^2$. Assuming the section is non-compact, the limiting compressive stress, $\sigma_{\ell c}$ is the lesser of $D\sigma_{\ell i}/2y_t$ or σ_{yc} (Part 3, Clause 9.8.3). If y_t is assumed to be 600 mm, then:

$$D\sigma_{\ell i}/2y_t = 444\ N/mm^2 > \sigma_{yc}, \qquad so\ \sigma_{\ell c} = \sigma_{yc} = 355\ N/mm^2$$

The area of flange required is given by:

$$A_{bf} = (7091 + 3421) \times 10^3/[1.225 \times 355/(1.2 \times 1.1)] = 31\,910\ mm^2$$

Assume, initially, a **600 × 60 mm bottom flange**.

The area of the top flange is determined principally by the loading carried on the steel section before it acts compositely. Assuming the flange is at its design yield strength in tension, the area required is then:

$$A_{tf} = 3421/(1.225 \times 307 \times 10^{-3}) = 9096\ mm^2$$

If a 400 × 30 mm flange is provided, the area of steel available in the top flange to resist moments on the composite section is $12\,000 - 9096 = 2904\ mm^2$ and the moment that can be carried is $3421 \times 2904/9096 = 1092\ kNm$. The total moment to be carried on the composite section is 7091 kNm (table X.4.2), so the moment to be resisted by reinforcement in the top flange is $7091-1092 = 5999\ kNm$. The lever arm for the cracked composite section is approximately 1.3 m, so the area of reinforcement required in the flange:

$$A_{rt} = 5999 \times 10^3/(1.3 \times 0.87 \times 460 \times 1.1) = 10\,482\ mm^2$$

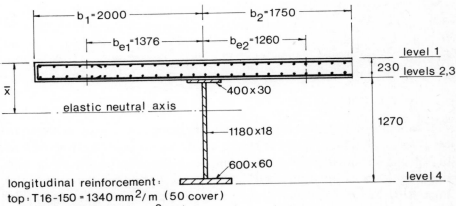

longitudinal reinforcement:
top: T16-150 = 1340 mm²/m (50 cover)
bottom: T20-150 = 2094 mm²/m (50 cover)

Fig. X.5.2 Preliminary cross section of outer girder at internal support

Assuming this is distributed in each face of the slab over a width of 3.5 m, the area required is $10\,482/3.5 = 2995$ mm^2/m. This can be provided by using T16–150 in the top face (1340 mm^2/m) and T20–150 in the bottom face (2094 mm^2/m), giving a total area of 3434 mm^2/m.

Hence, use a **top flange 400 × 30 mm** with **T16–150 longitudinally in the top** of the deck slab and **T20–150 longitudinally in the bottom** of the slab.

The support cross section is now as shown in fig. X.5.2. The area of steel is 692.4 cm^2 and the weight is 5.332 kN/m. To allow for stiffeners and bracing increase this to 5.5 kN/m. This is only 0.25 kN/m higher than that originally assumed, so the initial assumption for weight of steelwork is good enough.

X.6 Elastic properties of steel and composite cross sections

For global analysis, the effects of shear lag can be neglected at both ultimate and serviceability limit states, so the full breadth of the concrete flange is used in calculating section properties. For continuous beams, Part 5 first requires the designer to check whether the apparent tensile stress in the uncracked concrete flange over an internal support at SLS exceeds $0.1f_{cu}$ (section 8.2.1). The section properties for this are given in table X.6.1, and are the same as those required for calculating longitudinal shear at the ULS.

Table X.6.1 Elastic properties of outer girder, uncracked cross section at internal support

α_e	6.6	13.2
b_e (mm)	3750	3750
A (mm^2)	198 422	134 581
$\bar{x}$ (mm)	445	605
$10^{-6}I$ (mm^4)	59 050	48 227
$10^{-6}Z_1$ (mm^3)	875.8	1052.2
$10^{-6}Z_2$ (mm^3)	1812.7	1697.6
$10^{-6}Z_3$ (mm^3)	244.7	128.6
$10^{-6}Z_4$ (mm^3)	56.0	53.9
$A\bar{y}/I$ (m^{-1})	0.730	0.664

If the limiting stress is exceeded, either the concrete is assumed to be cracked, but reinforced, over a length of 15% of the span on either side of a support in subsequent calculations of stiffness and the analysis is repeated for a beam of non-uniform section, or the 'simple' method of increasing the midspan moment by $(40f_{tc}/f_{cu})$ per cent can be used. The latter becomes increasingly inaccurate where adjacent spans differ appreciably in length, and is not used in this worked example.

It will usually be found from preliminary calculations that the apparent tensile stress exceeds $0.1f_{cu}$.

For calculating the ultimate bending resistance of a composite section, the effects of shear lag can also be neglected, so the section properties at

Table X.6.2 Elastic properties of outer girder at midspan

	Steel section alone	Uncracked, unreinforced composite section for global analysis at SLS and ULS and for LS and stresses at ULS		Uncracked, unreinforced composite section for stresses at SLS	
Column	1	2	3	4	5
α_e	—	6.6	13.2	6.6	13.2
x_e (mm)	—	3750	3750	3580	3580
A (mm²)	47 050	177 732	112 391	171 808	109 429
$\bar{x}$ (mm)	1107*	378	530	387	541
$10^{-6} I_x$ (mm⁴)	11 963	46 502	38 816	46 061	38 665
$10^{-6} Z_1$ (mm³)	—	811.9	961.2	785.5	943.4
$10^{-6} Z_3$ (mm³)	13.64	314.2	128.1	293.4	124.3
$10^{-6} Z_4$ (mm³)	30.44	41.45	40.0	41.39	40.32
A_{cs} (mm²)	—	130 682	65 341	—	—
$A_{cs}\bar{y}/I$ (m⁻¹)	—	0.739	0.699	—	—

* $\bar{x}$ is measured from the top surface of the concrete slab

Table X.6.3 Elastic properties of outer girder at internal support

	Steel section	Cracked, reinforced composite section for global analysis at SLS and ULS and for stresses at ULS	Cracked, reinforced composite section for stresses at SLS
Column	1	2	3
b_e (mm)	—	3750	2636
A (mm²)	69 240	82 121	78 130
$\bar{x}$ (mm)	1068*	916	958
$10^{-6} I_x$ (mm⁴)	17 431	27 605	24 750
$10^{-6} Z_{tr}$ (mm³)	—	32.25	27.56
$10^{-6} Z_3$ (mm³)	20.80	40.24	34.00
$10^{-6} Z_4$ (mm³)	40.35	47.26	45.66
$10^{-6} I_y$ (mm⁴)	1240	16 312	Not required
r_y (mm)	133.8	446	

* $\bar{x}$ is measured from the top surface of the concrete slab

midspan and at the supports are the same as those used in the global analysis for the ULS.

For calculating stresses in the composite section at SLS, Part 5 requires the effects of shear lag to be taken into account, so a further set of section properties based on an effective breadth of concrete flange must be calculated.

Tables X.6.2 and X.6.3 give the elastic properties at midspan and at the support for the outer girder in the 37 m span.

In the tables, Z_1, Z_2, Z_3 and Z_4 refer to the levels 1, 2, 3 and 4 shown in figs X.5.1 and X.5.2. Z_{rt} refers to the top longitudinal reinforcement in the slab. The values of A, Z and I are given in 'steel' units, except for Z_1 and Z_2 in table X.6.1 which are in 'concrete' units. The small difference in values of Young's Modulus for structural steel and for reinforcement (section X.3.3) is neglected.

In table X.6.2 there is little difference between the section properties at midspan based on the full breadth of concrete flange and those based on the effective breadth of flange. In general where the effective breadth ratio, ψ, is greater than 0.95, it is sufficiently accurate in calculating stresses to neglect the effects of shear lag.

The effective breadth of flange used in column 3 of table X.6.3 is calculated from the effective breadth ratios given in tables 5 and 6, Part 3, with $\alpha = 0$.

For the portion of slab projecting beyond the web:

37 m span:
   $b_1/L = 2/37 = 0.054$, $\psi_1 = 0.56$ (table 5, support)
   $k = [1 - (0.15 \times 0.054)] = 0.99$, $k\psi_1 = 0.554$

25 m span:
   $b_1/L = 2/25 = 0.08$, $\psi_1 = 0.52$ (table 6, fixed end)
   $k = [1 - (0.15 \times 0.08)] = 0.99$, $k\psi_1 = 0.515$

The mean value of ψ_1 is 0.54 and of $k\psi_1$ is 0.535.
   For the portion between webs:

37 m span:
   $b_2/L = 1.75/37 = 0.047$, $\psi_2 = 0.610$ (table 5, support)
   $k = 1.0$, $k\psi_2 = 0.610$

25 m span:
   $b_2/L = 1.75/25 = 0.07$, $\psi_2 = 0.55$ (table 6, fixed end)
   $k = 1.0$, $k\psi_2 = 0.550$.

The mean value of ψ_2 is 0.58 and of $k\psi_2$ is 0.58.
   For cracked reinforced concrete flanges, Clause 5.2.3.2 of Part 5 allows

the effective breadth ratios calculated from Part 3 to be increased by an amount $(1 - \psi)/3$. The modified effective breadth ratios are therefore as follows.

Projecting portion of slab:
$(1 - \psi)/3 = (1 - 0.54)/3 = 0.153$, new $k\psi_1 = (0.153 + 0.535) = 0.688$

Portion between webs:
$(1 - \psi)/3 = (1 - 0.58)/3 = 0.14$, new $k\psi_2 = (0.14 + 0.58) = 0.72$

The **effective breadth of flange at the internal support** is then

$$(0.688 \times 2000) + (0.72 \times 1750) = \textbf{2636 mm}$$

Minor-axis properties I_y and r_y are also required at the internal support for the cracked composite section and for the steel section for calculating slenderness parameters in accordance with Clause 9.7.2 of Part 3. For the cracked composite section these are calculated assuming the equivalent thickness of each layer of reinforcement is equal to the area of the reinforcement divided by the width over which it acts. These minor-axis properties are:

steel section:
$10^{-6}I_y = 916.76$ mm^4, $r_y = 139.6$ mm

cracked composite section:
$10^{-6}I_y = 16\,312$ mm^4, $r_y = 446$ mm

X.7 Preliminary design of outer girder for bending, U-1

Design procedures at the ultimate limit state depend on whether the section is compact or non-compact (Clause 9.3.7, Part 3), but in either case the effects of shear lag can be neglected.

If the section is compact throughout the span considered and is not prone to lateral instability (Clause 9.2.1.3, Part 3), the effects of creep, shrinkage, temperature difference and settlement need not be considered at the ULS and the strength of the cross section is determined by simple 'plastic' theory.

If the section is non-compact, the bending resistance is determined by elastic theory.

The SLS is only considered in composite plate-girder construction when the effective breadth ratio ψ is less than 0.6 (Clause 9.2.3.1, Part 3) or when redistribution of stress is made from the tension flange to the web in longitudinally stiffened beams, or for the smaller flange in unsymmetrical beams which are compact.

X.7.1 Midspan section – compact or slender?

Steel section: from table X.6.2, $\bar{x} = 1107$ mm, so the depth of web in compression clear of welds is:

$$y_c = (1107 - 230 - 20 - 6) = 851 \text{ mm} = 85.1t_w$$

This is greater than the limit in Clause 9.3.7.2.1, Part 3 of $28t_w(355/\sigma_{yw})^{0.5}$ so the **steel section is not compact**.

Composite section: from table X.6.2, column 3, $\bar{x} = 530$ mm. The depth of web in compression is, therefore:

$$y_c = (530 - 230 - 20 - 6) = 274 \text{ mm} = 27.4t_w$$

This does not exceed the limit of $28t_w(355/\sigma_{yw})^{0.5}$, so the **composite section at midspan is compact**, provided the spacing of shear connectors perpendicular to the direction of compression does not exceed $24t_f(355/\sigma_{yf})^{0.5}$, or 480 mm, and, in the direction of compression, does not exceed half this value i.e., 240 mm (Clause 9.3.7.3.3, Part 3).

X.7.2 Section at internal support – compact or slender?

Steel section: from table X.6.3, $\bar{x} = 1068$ mm, so the compression web depth is:

$$y_c = (1500 - 1068 - 60 - 6) = 366 \text{ mm} = 20.3t_w < 28.8t_w(355/\sigma_{yw})^{0.5}$$

Therefore, **the web is compact**.

The compression flange outstand, $b_{fo} = 291$ mm $= 4.85t_{fo}$. This is less than the limit of $7t_{fo}(355/\sigma_{yf})^{0.5}$ given in Clause 9.3.7.3.1, Part 3, so the **steel section is compact**.

Composite section: the properties of the cracked reinforced section are used, neglecting shear lag effects. From table X.6.3, column 2, $\bar{x} = 916$ mm, so:

$$y_c = (1500 - 916 - 60 - 6) = 518 \text{ mm} = 28.78t_w > 28t_w(355/\sigma_{yw})^{0.5}$$

The **section is not compact**, but could be made compact by increasing the web thickness. For this example, it is assumed that at the internal support the section is slender both when the steel section alone carries loads and when the steel girder acts compositely with the deck slab. The effects of creep, shrinkage, temperature differences and differential settlement then have to be considered at ULS.

X.7.3 *Limiting compressive stress in steel flange*

In this example, calculations are not given for all the erection stages of the steel beams. In practice, this condition frequently determines the size of the top flange at midspan. Any restrictions on erection assumed in design must be stated clearly on the contract drawings. A typical erection sequence for the deck in this example would be to erect the crosshead steelwork above each pier first, clamping it to the piers. The girders in the end spans would then be assembled into 18.5 m lengths and erected individually with the bracings erected in sequence.

The 24 m long girders for the centre section of the 37 m span would then be erected individually with bracings erected in sequence.

Limiting compressive stresses for the steel flanges are now determined for the following conditions.

Midspan:
(i)   After erection of a 24 m long girder in the central section of the 37 m span.
(ii) After addition of cross-bracing (i.e. ready for sequential casting of deck slab).
(iii) Composite section, cross-bracing left in place.

Internal supports:
(iv) Steel beams erected with cross-bracing.
(v) Composite section, with cross-bracing left in place.

Full calculations are given only for case (i). The results for cases (ii) to (v) are summarised in table X.7.1 with a brief explanation of any factors leading to the choice of parameters that may not be obvious. Clause references are to BS 5400: Part 3, unless otherwise stated.

Table X.7.1   Limiting compressive stresses for outer girder (N/mm²)

Case	(i)	(ii)	(iii)	(iv)	(v)
ℓ_e (m)	24.0	8.0	0	6.5	6.5
r_y (mm)	139.6	139.6	—	133.8	446
k_4	1.0	1.0	—	1.0	1.0
M_A/M_M	0	—	—	—	—
M_B/M_A	—	—	—	—	—
η	0.94	1.0	—	1.0	1.0
λ_F	4.4	1.47	—	1.72	0.44
i	0.116	0.116	—	0.87	0.066
v	1.168	1.875	—	0.802	2.704
$\lambda_{LT}(\sigma_{yc}/355)^{0.5}$	189	108	—	39	39.4
σ_{ri}/σ_{yc}	0.135	0.39	1.0	1.0	1.0
$D/2y_t$	1.61	1.61	—	0.76	0.85
$\sigma_{\ell c}/\gamma_m\gamma_{f3}$ (N/mm²)	59	169	307	204	227

Note: A dash (—) indicates that the value need not be calculated

X.7.4 Midspan section

(i) *Girder erected in central span.* Immediately after erection, the girder is unrestrained over a 24 m length. The 6.5 m cantilevered girders to which it is connected are laterally restrained at the top and bottom flanges before the central portion of the girder is lifted into position, by bracing them to the adjacent cantilevered girders.

From 3/9.6.2, the effective length, $\ell_e = 24$ m.
From 3/9.7.2, the slenderness parameter, $\lambda_{LT} = (\ell_e/r_y)k_4\eta v$.
From section X.6, $r_y = 139.6$ mm.

The value of η may conservatively be taken as 1.0, or can be determined from fig. 9 of Part 3. The bending moment distribution in the girder will be as shown in fig. X.7.1, whence:

$$M_A = M_B \simeq 0, \qquad M_A/M_M = 0$$

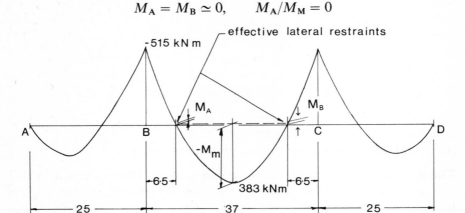

Fig. X.7.1　Bending-moment distribution due to self weight of girder

Hence, from fig. 9(b), Part 3, $\eta = 0.94$. Now:

$$\lambda_F = (\ell_e/r_y)(t_f/D) = (24\,000/139.6) \times (32.5/1270) = 4.4$$
$$k_4 = 1.0$$
$$I_c = 20 \times 400^3/12 = 106.67 \times 10^6 \text{ mm}^4$$
$$I_t = 45 \times 600^3/12 = 810 \times 10^6 \text{ mm}^4$$

Therefore, $i = I_c/(I_c + I_t) = 0.116$, and from table 9, Part 3, $v = 1.168$. Hence, $\lambda_{LT} = (24\,000/139.6) \times 1.0 \times 0.94 \times 1.168 = 189$. From fig. 10, Part 3, for $\lambda_{LT}(\sigma_{yc}/355)^{0.5} = 189$:

$$\sigma_{\ell i} = 0.135, \qquad \sigma_{yc} = 47.9 \text{ N/mm}^2$$

From 3/9.8.3, $\sigma_{\ell c} = D\sigma_{\ell i}/2y_t = 1270 \times 47.9/(2 \times 393) = 77.4$ N/mm².
The design strength, $\sigma_{\ell c}/\gamma_m\gamma_{f3} = 77.4/(1.2 \times 1.1) = 59$ N/mm².

(*ii*) *Steel beams erected with cross-bracing.* The calculations are similar to those for case (i), but it is assumed that the distance between effective lateral restraints at midspan is 8 m, so from 3/9.6.2, $\ell_e = 8.0$ m.

(*iii*) *Composite section.* The compression flange is continuously restrained by the deck at the level of the compression flange, so from 3/9.6.6.1, $\ell_e = 0$ and $\sigma_{\ell i} = \sigma_{yc} = 355$ N/mm^2.

The section is compact, so the design strength is $355/(1.05 \times 1.1) = 307$ N/mm^2.

A value of 1.05 can be used for γ_m for a steel flange in compression if it is connected to a concrete slab (table 2, Part 3).

X.7.5 Support section

(*iv*) *Steel beams erected with cross-bracing to compression and tension flanges.* The distance between the first effective lateral restraint and the crosshead is 6.5 m, so from 3/9.6.2, $\ell_e = 6.5$ m. Other results are in table X.7.1.

(*v*) *Composite section – permanent cross-bracing to bottom flange.* The distance between the first effective lateral restraint and the crosshead is 6.5 m, so from 3/9.6.2, $\ell_e = 6.5$ m.

X.7.6 Preliminary bending stresses in outer girder, U-1

The depth of web in compression at midspan for the steel section alone and for the composite section at the support both exceed the slenderness limit for a compact web given in Part 3, so the member is assumed to be slender for all stages of construction.

Shrinkage and temperature stresses then have to be considered in combinations U-1 and U-3 respectively, so stresses at SLS in the steel and concrete will not govern and need not be checked. However, stresses in the reinforcement at the internal support have to be calculated at SLS (combination 1) to ensure that crack widths are not excessive under the combined effects of global and local loading. If the spacing of longitudinal reinforcement in the bottom of the slab at a support is 150 mm or less, then in bridges designed for 45 units of HB loading, it will usually be found that the crack widths obtained with 25 units of HB loading in combination S-1 are within the 0.25 mm limit for waterproofed deck surfaces.

In this preliminary check, only the principal dead and live loads in combination U-1 are considered and this is usually sufficient for checking the initial member sizes. In the final analysis, combination U-3

with temperature effects included may be more onerous, but the lower
partial load factors in combination U-3 for HB and HA live loading
offset to some extent the stresses due to temperature effects. The dead
load of the concrete is initially assumed to be taken on the steel section
alone.

Bending stresses at midspan due to combination U-1 design loads are
summarised in table X.7.2.

Table X.7.2 Preliminary longitudinal bending stress (N/mm²) at midspan, combination U-1

Loading	Design BM (kNm)	σ_1	σ_2	σ_3	σ_4
(1) SW steel beam	402	0	0	−29.5 (−59.0)	13.2
(2) Dead load (concrete)	2145	0	0	−157.3	70.5
(3) Sub-total	2547	0	0	−186.8 (−169)	83.7 (307)
(4) Superimposed dead load	1053	−1.10	−0.62	−8.2	26.3
(5) HB with HA	7363	−9.07	−3.55	−23.4	177.6
(6) Total	10963	−10.17	−4.17	−218.4 (−307)	287.6 (307)

The design strengths for the steel flanges at each stage of construction
are shown in parentheses. The compressive stress in the steel flange is too
high after casting the deck slab, but this will reduce slightly when account
is taken of the sequence of casting the deck slab, and so is probably
satisfactory. The HB moment will also reduce when account is taken of
transverse distribution in the deck slab, so the preliminary section
sizes should be adequate.

Bending stresses at the internal support due to combination U-1
design loads are summarised in table X.7.3. The section properties used
are those given in columns 1 and 2 of table X.6.3.

At the internal support the tensile stress in the top and bottom flanges
of the steel girder is too high after the live loading is applied. This will not

Table X.7.3 Preliminary longitudinal bending stresses (N/mm²) at internal support, combination U-1

Loading	Design BM (kNm)	σ_{rt}	σ_3	σ_4
(1) SW steel beam	540	0	26.0	−13.4
(2) Dead load (concrete)	2881	0	138.5	−71.4
(3) Sub-total	3421	0	164.5 (307)	−84.8 (−204)
(4) Superimposed dead load	1415	43.9	35.2	−29.9
(5) HB with HA	5676	176.0	141.1	−120.1
(6) Total	10512	219.9	340.8 (307)	−234.8 (−227)

be reduced significantly by staged casting of the deck slab, but the moment due to HB loading will probably be reduced, as noted above. It is therefore concluded that the **preliminary section sizes should be adequate**.

X.8 Preliminary design of deck slab for local loads

X.8.1 Punching shear

The effects of both a single HA nominal wheel load of 100 kN and of a single or group of HB wheel loads have to be considered.

For combination U-1, the design wheel loads are:

$$\text{HA: } 100 \times 1.5 \times 1.1 = 165 \text{ kN} \qquad (2/6.2.5)$$

$$45 \text{ units HB: } (45 \times 10/4) \times (1.3 \times 1.1) = 161 \text{ kN} \qquad (2/6.3)$$

The contact pressure for the nominal wheel load in both cases is

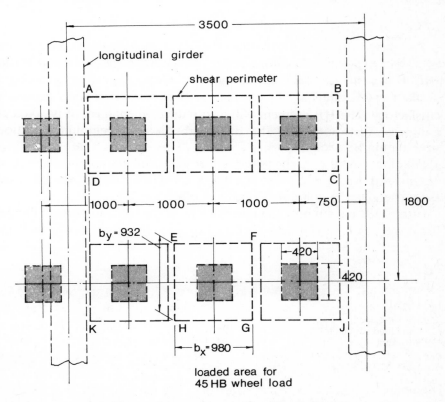

Fig. X.8.1   Shear perimeters for HB wheel loads

1.1 N/mm^2 at the carriageway surface, so this gives the contact area as a square of side 300 mm for HA and 320 mm for HB. Dispersal is permitted through the surfacing at a spread of 1 horizontal to 2 vertical, so at the surface of the deck slab, the loaded area with 100 mm asphalt is a square of side 400 mm for HA and 420 mm for HB.

From Clause 5.4.4.2, Part 4, the critical shear perimeter shown in fig. X.8.1 is found as follows. For a slab 230 mm thick, with 35 mm cover and assuming 16 mm diameter bars in each direction:

$$d_x = (230 - 35 - 8) = 187 \text{ mm}$$

$$d_y = (230 - 35 - 16 - 8) = 171 \text{ mm}$$

Then, for the HB wheel load:

$$b_x = (420 + 560) = 980 \text{ mm}$$

$$b_y = (420 + 512) = 932 \text{ mm}$$

and for the HA wheel load:

$$b_x = 960 \text{ mm}$$

$$b_y = 912 \text{ mm}$$

Thus, the single HA wheel load is more severe than the single HB wheel load. However, with longitudinal girders at 3.5 m centres it is possible for six wheels of an HB bogie to be located on the deck slab spanning between adjacent girders. The shear perimeter for each individual wheel load then almost overlaps with that of the adjacent wheel, so the critical shear perimeter for three wheels is ABCD, as shown in fig. X.8.1. The length of this shear perimeter is 7824 mm. The critical shear perimeter for six wheels is ABJK = 11 424 mm, which is clearly a worse case than for three wheels.

Thus, for the six wheel group:

$$b_x = 2980 \text{ mm}$$

$$b_y = 2732 \text{ mm}$$

Assuming that the tension reinforcement in each direction is not less than 0.15% of the gross concrete section, then from table 8 of Part 4, $v_c = 0.39$ N/mm^2 for $f_{cu} = 40$ N/mm^2 and from table 9, ε_s can conservatively be taken as 1.25 for $d \leqslant 200$ mm. Then:

$$V = 0.39 \times 1.25 \times (2980 + 2732) \times 2 \times 230 \times 10^{-3} = 1281 \text{ kN}$$

The resistance of 1281 kN is greater than the design loading of 966 kN for six HB wheel loads, so the **slab design is adequate for punching shear**.

Comment. Where the spacing of longitudinal girders exceeds about 2 m, groups of HB wheel loads will be more critical than a single HA wheel load for punching shear failure, but as shown in section 8.3.1, punching shear will rarely govern the design of the deck slab.

X.8.2 Transverse bending

Global transverse moments are not known until the grillage analysis has been completed, but these do not affect the cantilever edge, where local transverse bending determines the amount of top transverse reinforcement.

The amount of transverse reinforcement required in the bottom of the slab will usually be controlled by crack width limitations at the SLS under the combined effects of global and local bending but, for preliminary design at ULS, only local effects need be considered because these are usually greater than the global transverse moments, except in wide or heavily skewed decks. Global and local effects need not be combined at ULS (4/4.8.3).

Transverse moments due to dead and superimposed dead loading. We consider a transverse strip of deck slab of unit width continuous over the longitudinal girders, which are assumed at this stage to be undeflecting supports. From fig. X.4.1, the nominal transverse moment over the outer girder is:

$$\text{Concrete edge beam: } 0.55 \times 0.64 \times 25 \times 1.68 = 14.78$$
$$\text{Slab: } 0.23 \times 1.36 \times 25 \times 0.68 = 5.32$$
$$\text{Infill: } 0.175 \times 0.46 \times 23 \times 1.13 = 2.09$$

$$\overline{\qquad\qquad 22.19 \text{ kNm/m}}$$

$$\text{Surfacing + waterproofing: } 0.11 \times 0.9 \times 21 \times 0.45 = 0.94 \text{ kNm/m}$$
$$\text{Parapet } (0.55 \text{ kN/m}): \qquad\qquad 0.55 \times 1.68 = 0.92 \text{ kNm/m}$$

The distribution of transverse moments over the first four girders is as shown in fig. X.8.2.

Transverse moment due to collision of vehicle with parapet. This is considered in combination U-4 only. The nominal load applied to the member supporting the parapet is defined in Clause 6.8.1, Part 2 as the load which will bring about collapse of the parapet or the connection of the parapet to the element, whichever is the greater. For a P1 parapet designed to reference 190, the connection of the parapet must be capable of developing a moment of resistance at least 50% greater than the plastic

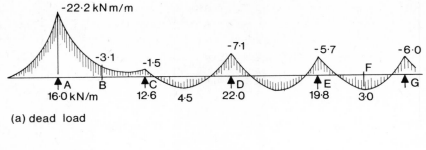

(a) dead load

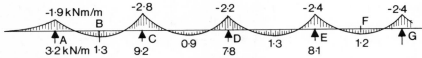

(b) superimposed dead load

Fig. X.8.2   Transverse moments due to dead and superimposed dead load (kNm/m)

moment of the post, so the connection will govern the design of the slab.

From reference 190, the force *P* in fig. X.8.3, acting at the level of each rail is (50/3) kN per post.

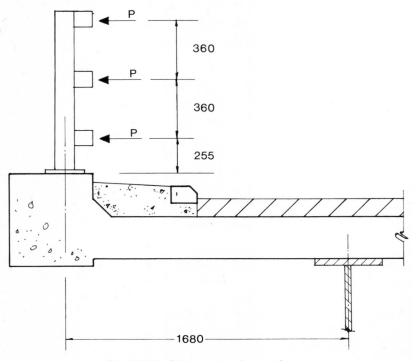

Fig. X.8.3   P1 parapet – impact forces

At the base of the post, the moment is:

$$M = (0.255 + 0.615 + 0.975) \times 50/3 = 30.75 \text{ kNm}$$

The moment to be developed by the post fixing[190] is:

$$M_p = 1.5 \times 30.75 = 46 \text{ kNm}$$

If it is assumed that this is dispersed in plan at 45°, then over the centreline of the outer girder the nominal transverse moment due to a vehicle collision is given by:

$$M_v = 46/(2 \times 1.68) = 13.6 \text{ kNm/m}$$

X.8.3 Bending due to local wheel loads

(*i*) *Transverse hogging moment over outer girder.* Outer verges are not required to be loaded with live load in the global analysis of the superstructure, but they must be capable of supporting any four wheels of 25 units of HB loading (Clause 6.4.3.3, Part 2).

For a 25 unit HB vehicle, the contact pressure for the nominal wheel load is 1.1 N/mm² at the carriageway surface, so the contact area is a square of side 240 mm.

For loading combination U-4, the four wheel loads are considered to act in combination with the loads due to a vehicle collision with the

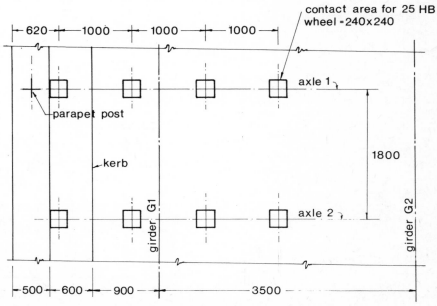

Fig. X.8.4   25-unit HB vehicle bogie on deck slab

parapet (Clause 6.8, Part 2). The four wheels are assumed here to be the two outer wheels on the two front axles, acting at the point of collision, as shown in fig. X.8.4. The wheel loads on the two rear axles are too far away to affect the local transverse moments caused by the two leading axles.

Dispersal through the surfacing is permitted by Clause 6.3.3, Part 2, but is neglected here because it has a small effect on local transverse bending moments. Using Pucher's influence charts[56] the local transverse moment over the outer girder due to the four wheel loads shown in fig. X.8.4 is 54 kNm/m.

The design transverse hogging moments in the deck slab over the outer girder are summarised in table X.8.1. The final line gives the design moment, M_D, where:

$$M_D = \text{nominal moment} \times \gamma_{fL} \times \gamma_{f3}$$

Table X.8.1 Design transverse hogging moments over outer girder (kNm/m)

Loading	Nominal moment M	U-1 γ_{fL}	U-1 $M\gamma_{fL}$	Combination U-4 γ_{fL}	Combination U-4 $M\gamma_{fL}$	S-1 γ_{fL}	S-1 $M\gamma_{fL}$
DL (concrete)	22.2	1.15	25.5	1.15	25.5	1.0	22.2
Surfacing	0.9	1.75	1.6	1.75	1.6	1.2	1.1
Parapet	0.9	1.20	1.1	1.20	1.1	1.0	0.9
25 units HB	54.0	1.30	70.2	1.50	81.0	1.1	59.4
Parapet impact	13.6	1.50	—	1.50	20.4	1.2	—
Total	91.6		98.5		129.7		83.6
Design moment, $M_D = M\gamma_{fL}\gamma_{f3}$			108.3		142.6		83.6

Table X.8.2 Design transverse sagging moments at B (kNm/m)

Loading	Nominal moment M	Combination S-1 γ_{fL}	Combination S-1 $M\gamma_{fL}$	Combination U-1 γ_{fL}	Combination U-1 $M\gamma_{fL}$
(1) DL (concrete)	−3.1	1.0	−3.1	1.0*	−3.1
(2) Superimposed DL	1.3	1.2	1.5	1.75	2.2
(3) HA wheel	27.9	1.2	33.4	1.5	41.8
(4) 25 units HB	27.6	1.1	30.4	N/A	—
(5) 45 units HB	49.7	N/A	—	1.3	64.6
			$\Sigma\,1, 2, 3 = 31.8$		$\Sigma\,1, 2, 5 = 63.7$
Design moment, $M_D = M\gamma_{fL}\gamma_{f3}$		—	31.8	—	70.1

* A reduced load factor is used for dead load where a more severe total effect is produced (Clause 6.1.2.2, Part 2)

The value of γ_{f3} is 1.1 for combinations U-1 and U-4 and is 1.0 for combination S-1.

(*ii*) *Longitudinal and transverse sagging moments at point B (fig. X.8.2)*. Transverse sagging moments at point B, midway between A and C in fig. X.8.2, due to dead and superimposed dead load are summarised in table X.8.2.

For HA loading, the effect of a single 100 kN wheel load is more severe than HAU loading. For HB loading, transverse and longitudinal bending due to groups of wheels on the 45 unit vehicle must be considered at ULS, but for crack control at SLS, a 25 unit HB vehicle is used.

The single HA wheel is located midway between longitudinal girders for maximum local sagging moments. The most adverse position of the HB wheel group is shown in fig. X.8.5(a).

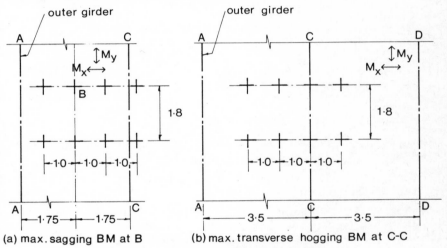

(a) max. sagging BM at B   (b) max. transverse hogging BM at C-C

Fig. X.8.5   Position of HB wheel loads

The portion of slab spanning between the outer girder and the first interior girder is assumed to be restrained over the interior girder and simply supported over the outer girder. Using Pucher's charts[56] the maximum transverse sagging moment M'_x due to a single 100 kN HA wheel load is 27.85 kNm/m and the longitudinal moment M'_y is 23.9 kNm/m. For 45 units HB, the eight-wheel group shown in fig. X.8.5(a) gives:

$$M'_x = 49.68 \text{ kNm/m}, \qquad M'_y = 31 \text{ kNm/m}$$

In table X.8.2, it is shown that for combination S-1, the single HA wheel load is more severe than the 25 unit HB bogie.

(iii) Transverse hogging bending at point C. From fig. X.8.2, transverse hogging moments at C due to dead and superimposed dead load are less than at A, but transverse hogging moments at C due to local wheel loads will be greater than at A, so both A and C must be considered.

The most adverse loading condition for local wheel loads is with the HB wheel group straddling the first interior girder as shown in fig. X.8.5(b). Using the method given in reference 56, the transverse moment for the 45 unit wheel group is -103 kNm/m.

Design moments for transverse hogging at point C are summarised in table X.8.3.

Table X.8.3 Design transverse hogging moments (kNm/m) at C

	Nominal moment M	Combination			
		S-1		U-1	
Loading		γ_{fL}	$M\gamma_{fL}$	γ_{fL}	$M\gamma_{fL}$
DL (concrete)	-1.5	1.0	-1.5	1.15	-1.8
Surfacing	-3.0	1.2	-3.6	1.75	-5.3
45 units HB	-103.0	1.1	N/A	1.3	-133.9
25 units HB	-57.0	1.1	-62.7	1.3	N/A
Total	—	—	-67.8	—	-141.0
Design moment, $M_D = M\gamma_{fL}\gamma_{f3}$	—	—	-67.8	—	-155.0

X.8.4 Top transverse reinforcement at A

Ultimate limit state. From table X.8.1, the design moment is 142.6 kNm/m (combination U-4). Neglecting compression reinforcement, M_u is the lesser of the values given by equations (1) and (2) in Clause 5.3.2.3 of Part 4.

Assuming T20 bars are provided in the top face, at a spacing of 150 mm and with a cover of 30 mm, then:

$$A_s = 2094 \text{ mm}^2/\text{m}$$

$$d = (230 - 30 - 10) = 190 \text{ mm}$$

$$z = \left[1 - \left(\frac{1.1 \times 460 \times 2094}{40 \times 1000 \times 190}\right)\right]190 = 163.5 \text{ mm}$$

If the reinforcement limits the resistance:

$$M_u = 0.87 \times 460 \times 2094 \times 163.5 \times 10^{-6} = 137 \text{ kNm/m} < 142.6$$

Therefore, reduce the spacing of bars to 125 mm. Then:

$$A_s = 2513 \text{ mm}^2/\text{m}$$

$$z = 158.2 \text{ mm}$$

If the reinforcement limits the resistance:

$$M_u = 0.87 \times 460 \times 2513 \times 158.2 \times 10^{-6} = 159 \text{ kNm/m}$$

If the concrete limits the resistance:

$$M_u = 0.15 \times 40 \times 1000 \times 190^2 \times 10^{-6} = 216 \text{ kNm/m}$$

Hence,

$$M_u = 159 \text{ kNm/m} > 142.6 \text{ kNm/m}$$

so, use **T20 bars at a spacing of 125 mm**.

Crack width. From table X.8.1, the design moment for combination S-1 is 83.6 kNm/m. Most of this moment is due to live load, so for convenience, the short-term modular ratio of 6.6 is used for the elastic analysis of the cross section, which gives a tensile stress in the top reinforcement of 197 N/mm². The design crack width calculated in accordance with Clause 5.8.8.2, Part 4 is then 0.18 mm which is less than the limiting crack width of 0.25 mm. A bar spacing of 125 mm is therefore satisfactory. The maximum compressive stress in the concrete is 15.4 N/mm², which is less than the permissible value of $0.5f_{cu}$ given in table 2 of Part 4.

X.8.5 Bottom transverse reinforcement at B

Ultimate limit state. From table X.8.2, the design moment for combination U-1 is 70 kNm/m. If the compression reinforcement is neglected, and the bottom reinforcement is initially assumed to be T16 bars at 150 mm spacing, then:

$$A_s = 1338 \text{ mm}^2/\text{m}$$

$$d = (230 - 35 - 8) = 187 \text{ mm}$$

$$z = \left[1 - \left(\frac{1.1 \times 460 \times 1338}{40 \times 1000 \times 187} \right) \right] 187 = 170 \text{ mm}$$

$$M_u = 0.87 \times 460 \times 1338 \times 170 \times 10^{-6} = 91 \text{ kNm/m}$$

This is greater than the design moment of 70 kNm/m, so use **T16 bars at 150 mm centres** and check for crack control at SLS.

Crack control. From table X.8.2, M_D is 32 kNm/m (combination S-1). Using a short-term modular ratio of 6.6, elastic analysis of the cross section gives a tensile stress in the bottom reinforcement of 120 N/mm² and a crack width of 0.12 mm.

X.8.6 Top transverse reinforcement at C

Ultimate limit state. From table X.8.3, the design moment for combination U-1 is 155 kNm/m. If the compression reinforcement in the bottom of the slab is neglected and **T20 bars at 125 mm spacing** are used in the top of the slab with 30 mm cover, the resistance from section X.8.4 is 159 kNm/m, which is satisfactory.

Crack width. From table X.8.3, the design moment is 67.8 kNm/m for combination S-1. This is less than the design moment at point A (table X.8.1), so the crack width will be less than 0.25 mm.

X.8.7 Longitudinal reinforcement in bottom of slab at B

Ultimate limit state. The deck slab is designed to span transversely between longitudinal girders, so longitudinal moments are due only to local wheel loads. For combination U-1, the effects of the 45 unit HB bogie are greater than the single 100 kN HA wheel load (section X.8.3(ii)), so the design moment is:

$$M_D = 31 \times 1.3 \times 1.1 = 44.3 \text{ kNm/m}$$

Neglecting compression reinforcement, M_u is the lesser of the values from equations (1) and (2) in Clause 5.3.2.3 of Part 4. Near the internal support, the longitudinal reinforcement has initially been assumed to be T20 bars at 150 mm centres, but at midspan, where the deck slab is in global longitudinal compression, the bottom longitudinal reinforcement can probably be reduced to **T16 bars at 150 mm spacing**, so this is used for checking the resistance to local wheel loads at ULS. Then:

$$A_s = 1338 \text{ mm}^2/\text{m}$$

$$d = (230 - 30 - 16) = 179 \text{ mm}$$

$$z = \left[1 - \left(\frac{1.1 \times 460 \times 1338}{40 \times 1000 \times 179} \right) \right] 179 = 162 \text{ mm}$$

If the reinforcement limits the resistance:

$$M_u = 0.87 \times 460 \times 1338 \times 162 \times 10^{-6} = 87 \text{ kNm/m}$$

This is greater than the design moment of 44.3 kNm/m, so check the crack width at SLS.

Crack width. From section X.8.3(ii) for combination S-1, the design longitudinal moment due to a single 100 kN wheel load is:

$$M'_y = 23.9 \times 1.2 = 28.7 \text{ kNm/m}$$

This is greater than the moment due to the 25 unit HB wheel group. Elastic analysis of the cross section gives the tensile stress in the reinforcement as 141 N/mm^2 and a crack width of 0.19 mm. This will be satisfactory where global longitudinal bending produces compression in the deck slab, but in negative moment regions, where the deck slab is in longitudinal global tension, the crack width will have to be checked later (section X.13).

X.9 Global analysis

X.9.1 Grillage analysis

The layout of the grillage members is shown in fig. X.9.1. Each longitudinal girder in the bridge deck is represented by a string of eighteen connected members.

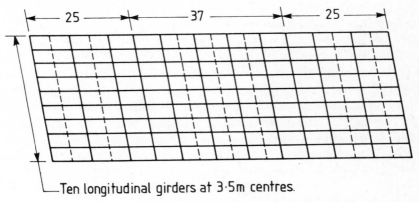

Ten longitudinal girders at 3·5m centres.

Fig. X.9.1 Layout of members in grillage analysis

The crosshead steelwork and permanent bracings are represented in the grillage by the transverse members shown by full lines. Additional intermediate transverse members, shown dashed, are provided in the grillage where the ratio between the length of a longitudinal member and the length of a transverse member framing into it exceeds two. This gives a total of 200 joints and 361 members. The grillage analysis used takes account of the shear flexibility of the transverse bracings.

The required section properties are the second moment of area, I, and torsional inertia, J, of each grillage member. The effects of superimposed dead load are small compared with those of HA and HB loading, so the short-term modular ratio (6.6) was used for all loads. The second moments of area of the longitudinal grillage members are those given in table X.6.2 for the uncracked unreinforced section, except for those members within a distance of 6.5 m on either side of an internal support where the concrete flange was assumed to be cracked. For these members the second moment of area was that given in table X.6.3, neglecting concrete in tension, but including longitudinal reinforcement in the deck slab. The length of 6.5 m is slightly greater than 15% of the span given in Clauses 5.1.1.2 and 6.1.1.4.1 of Part 5. The effects of shear lag are neglected throughout.

The use of cracked section properties in the negative moment region for global analysis at SLS assumes that the maximum tensile stress in the uncracked composite section at the internal support, f_{tc}, exceeds $0.1f_{cu}$ (5/5.1.2.2). This assumption is now checked, using the properties of the uncracked section given in table X.6.1 and the preliminary design moments for combination S-1 given in table X.4.2.

$$f_{tc} = \frac{(720 + 304)}{1052.2} + \frac{4803}{875.8} = 6.46 \text{ N/mm}^2$$

This is $0.16f_{cu}$, so the assumption is correct.

The torsional inertia J is essentially $\Sigma bt^3/3$, but the contribution of the slab is halved[68] and that of the top flange of the steel beam is neglected; so, for an internal girder:

$$J_x = \tfrac{1}{3}\left[\left(\frac{3500 \times 230^3}{2 \times 6.6}\right) + (1180 \times 18^3) + (600 \times 60^3)\right]$$

$$= 1120.8 \times 10^6 \text{ mm}^4$$

The concrete flange is assumed to be uncracked in calculating J because the torsional stiffness is not reduced in the same way as the flexural stiffness, due to aggregate interlock.

The section properties of the transverse members in the grillage representing cross-girders and permanent bracing should include the portion of the deck slab associated with that member.

The additional transverse members representing only portions of deck slab have the following section properties:

$$I_y = bh_c^3/12\alpha_e(1 - v_c^2)$$

$$J_y = 0.5bh_c^3/3\alpha_e$$

The torsional stiffness of the steel I-sections is negligible in comparison with that of the deck slab, so torsional stresses in the steel members need not be considered.

The end transverse members are reinforced concrete diaphragms. The section properties of these members should be based on the uncracked unreinforced concrete section of the diaphragm and the portion of deck slab that acts with it.

X.9.2 Loads and load combinations

A full description of all live load combinations which would need to be considered for the analysis of this three-span bridge deck is outside the scope of this book, but they should be chosen with care to reduce the cost of computation and must include cases which give the following values.

(1) Maximum longitudinal moment at midspan. (Different load cases will need to be considered for the edge girder and for the internal girders.)
(2) Maximum longitudinal moment at internal supports.
(3) Maximum transverse hogging moments at midspan.
(4) Maximum transverse sagging moments at midspan.
(5) Maximum vertical shear at the supports.
(6) Maximum/minimum vertical shear at or near quarter span.
(7) Maximum/minimum vertical shear at midspan.
(8) Maximum vertical reactions at the bearings.

The grillage analysis should have a facility to produce an envelope of the maximum and minimum bending moments and shear forces at each nodal point.

For the design of the edge girder in this example, HB loading combinations cause larger bending moments and shear forces than HA loading combinations, but both have to be considered because HA may be more severe than 25 units of HB loading in calculations for control of cracking at the internal supports.

Loadings for moments in edge girder. Superimposed dead loading should be included as a loading case in the grillage analysis because the edge girder carries a higher proportion of the superimposed dead load than the internal girders.

For maximum longitudinal sagging moment in the edge girder the HB vehicle is positioned in the nearside lane with the second axle at midspan (fig. X.9.2(a)). From influence lines it is evident that the maximum sagging moment is produced with the minimum spacing of 6 m between the inner axles. The second and third lanes are loaded with full HA loading with the knife-edge load at midspan. All other lanes are loaded with 0.6 HAU loading, with 0.6 HAK at midspan. No live loading is applied in the 25 m end spans. Under HA loading alone, the two outer lanes in

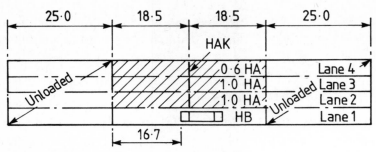

(a) Maximum sagging BM in outer girder

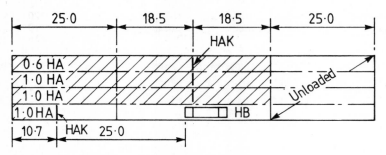

(b) Maximum hogging moment in outer girder

Fig. X.9.2   HB load combinations for outer girder

the 37 m span are loaded with HA and all others with 0.6 HA.

For maximum longitudinal hogging moment, it is not immediately evident, for HB loading, whether the maximum axle spacing of 26 m with one bogie in each span will produce a higher negative moment at an internal support than with the minimum axle spacing of 6 m and the HB vehicle in the 37 m span or in the 25 m span. Influence lines for bending moments at the internal support and for moments at other points in the span must be constructed first, therefore, to determine the most adverse position for the HB vehicle. In this example, this was with the HB vehicle in the centre span with full HA loading over a length of 10.7 m in the adjacent end span. The second and third lanes are loaded with full HA loading and all other lanes with 0.6 HA loading. The remote end span carries no live load. For HA loading, the two outer lanes are loaded with full HA, with 0.6 HA in all other lanes.

X.9.3 *Longitudinal moments – edge girder*

The nominal bending moments obtained from the grillage analysis for superimposed dead loading and for live loading are summarised in table

X.9.1. From these, the design moments at midspan, and at the internal support, for combinations S-1, U-1 and U-3 are deduced. Combination S-1 is required for crack control at the internal support only with 25 units HB combined with the appropriate HA loading.

Table X.9.1 Design moments – edge girder

Loading	Nominal BM (kNm)	Design BM (kNm)					
		γ_{fL}	S-1	γ_{fL}	U-1	γ_{fL}	U-3
Midspan BM							
(1) Super DL							
(asphalt)	588	1.2	706	1.75	1029	1.75	1029
(other)	157	1.0	157	1.20	188	1.20	188
(2) 45 HB with HA	6405	1.1	7046	1.30	8326	1.10	7046
(3) HA alone	3328	1.2	3994	1.50	4992	1.25	4160
Σ 1, 2			7909		9543		8263
Support BM							
(1) Super DL							
(asphalt)	−600	1.2	−720	1.75	−1050	1.75	−1050
(other)	−304	1.0	−304	1.20	−365	1.20	−365
(2) 45 HB with HA	−3144	1.1	−3458	1.30	−4087	1.10	−3458
(3) HA alone	−2363	1.2	−2836	1.50	−3545	1.25	−2954
(4) 25 HB with HA	−2630	1.1	−2893	1.30	−3419	1.0	−2893
Σ 1, 2			−4482		−5502		−4873
Σ 1, 4			−3860		—		—

X.10 Longitudinal stresses due to sequential casting of the deck slab

In continuous composite bridge decks, it is common practice to cast the deck slab in stages. This reduces the amount of cross-bracing that is required to prevent lateral-torsional buckling at midspan before the top flange acts compositely with the deck slab.

In this example, the deck slab is assumed to be cast in five stages, in the sequence shown in fig. X.10.1, which has been discussed in section X.3.

From fig. X.10.1 and section X.4, the nominal dead load of the deck slab carried by the outer girder is 17.88 kN/m for stages 1 to 4. It is assumed here that the dead load of the parapet strip (7.68 kN/m) is carried entirely by the outer girder. This will reduce if proper account is taken of the transverse distribution properties of the deck slab.

The stiffness of the composite section used in this analysis for checking the steelwork during erection is assumed to be that of the uncracked section at midspan, with $\alpha_e = 6.6$ (column 2, table X.6.2). Clause 12.1 of Part 5 permits this assumption to be made if the cube strength of the concrete at the time considered is not less than $0.75f_{cu}$, otherwise the

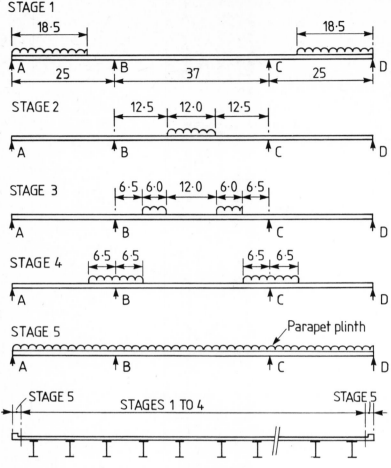

Fig. X.10.1   Sequence of casting deck slab

stiffness should be appropriately reduced.

The stiffness of the steel section is assumed to be that at midspan. This is not correct but is sufficiently accurate for girders of constant depth.

The bending moment and shear force distribution at each stage of construction is easily calculated by hand using Mohr's moment area method, or can be calculated by computer using an elastic stiffness method of analysis. The results of the hand analysis are summarised in fig. X.10.2.

X.10.1 Stresses at each stage of casting the deck slab

Nominal stresses at each stage of casting the slab are now calculated at

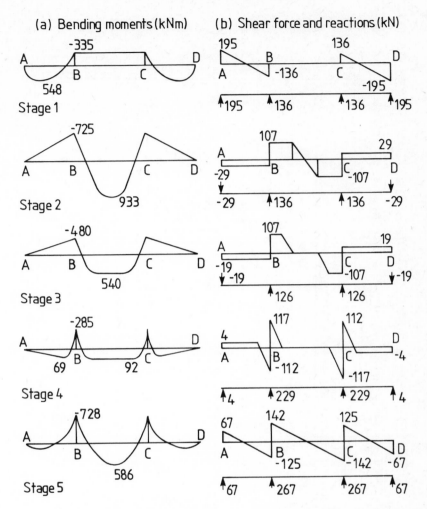

(a) Bending moments (kNm)   (b) Shear force and reactions (kN)

Fig. X.10.2   Bending moment and shear force at each stage of casting deck slab

the internal support and at midspan, and are summarised in table X.10.1, with the design stresses for combination U-1. All these stresses exclude those already present in the steel due to its own weight.

The section properties given in columns 1 and 3 of table X.6.2 for midspan and those given in columns 1 and 2 of table X.6.3 for the internal support were used. These properties are not strictly the correct ones because:

(1) They include part of the concrete parapet edge strip which is cast in stage 5.

Table X.10.1 Stresses (N/mm²) due to dead load of concrete at each stage of deck slab construction

Casting stage	Midspan					Internal support				
	Nominal BM (kNm)	σ_1	σ_2	σ_3	σ_4	Nominal SF (kN)	Nominal BM (kNm)	σ_{rt}	σ_3	σ_4
1	-335	—	—	24.6	-11.0	0	-335	—	16.1	-8.3
2	933	—	—	-68.4	30.7	-107	-725	—	34.8	-18.0
3	540	-0.56	-0.32	-4.2	13.5	-107	-480	—	23.1	-11.9
4	92	-0.09	-0.05	-0.7	2.3	-116	-285	—	13.7	-7.1
5	586	-0.61	-0.35	-4.6	14.7	-142	-728	22.6	18.1	-15.4
Total nominal stresses		-1.26	-0.72	-53.3	50.2	-472	-2553	22.6	105.8	-60.7
Design stresses combination U-1		-1.45	-0.83	-61.3	57.7	-543	—	26.0	121.7	-69.8

(2) They take no account of the effect of shear lag in the partially-completed span.

The effect of shear lag need only be considered at the serviceability limit state in Part 5 (Clause 5.2.5.4) and, as it has only a small effect in this example, is neglected. The error resulting from the first approximation is also small.

If the design stresses in table X.10.1 are compared with the preliminary stresses given in tables X.7.2 and X.7.3, it can be seen that at the internal support the compressive stress in the bottom flange is effectively the same in both analyses, but the tensile stress in the reinforcement is higher with the slab cast in stages because the parapet plinth dead load is carried on the composite section. The total moment acting on the section is 22% greater than in the preliminary analysis. This is due to the variation in stiffness at the different stages of construction.

At midspan, the maximum compressive stress in the top flange of the steel section due to loading carried by the steel section alone reduces from -157.3 N/mm^2 to $(+24.6 - 68.4) \times 1.15 = -50.4$ N/mm^2 and only increases to -61.3 N/mm^2 in the completed deck. The tensile stress in the bottom flange reduces from 70.5 N/mm^2 to 57.7 N/mm^2.

X.11 The effects of temperature

The various effects of temperature that have to be considered in the design of a composite beam-and-slab deck are summarised in section 7.3.4.

In this worked example, the steel and concrete have the same coefficient of thermal expansion, so uniform changes of temperature cause no stresses in the deck, other than those caused by restraint to movement at the bearings. This is neglected here but, in practice, would have to be considered in combination 5 if the restraint were due to friction at roller or sliding bearings, or in combination 3 if the restraint to movement were due to the shear stiffness of elastomeric bearings or to the flexural resistance of a pier (see Clause 5.4, Part 2).

The only other temperature effects that cause stresses in the composite section are temperature differences through the depth of the section. For beam-and-slab decks, two distributions are given in Part 2 of the Bridge Code. For the composite deck in this example, the resulting temperature differences are shown in fig. X.11.1. For the analysis of stresses due to reverse temperature difference, it is convenient to take the datum temperature as that of the bottom flange and lower part of the web, as shown in fig. X.11.1(d).

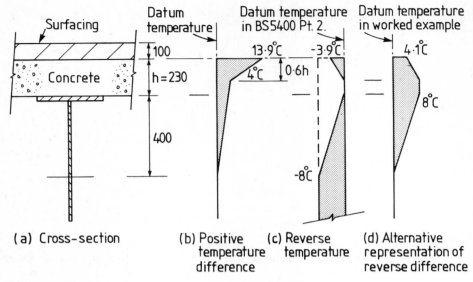

(a) Cross-section   (b) Positive temperature difference   (c) Reverse temperature   (d) Alternative representation of reverse difference

Fig. X.11.1   Temperature differences

X.11.1 *Primary stresses due to positive temperature difference – midspan*

For the midspan cross section of the outer girder shown in fig. X.5.1 the primary stresses due to positive temperature difference are now calculated using the method described in Appendix A. The section is divided into six slices as shown in fig. X.11.2. The calculation of $\bar{F}$ and $\Sigma F_r \bar{y}_r$ is set out in table X.11.1. Columns 1 to 3 give the depth, temperature and free strain at the boundaries of the slices. The mean strains $\bar{\varepsilon}_r$ are found from equation (A.3), and the depths of the centroids of the strain blocks, $\bar{y}_r$, from equation (A.4). It should be noted that where the temperature varies through a slice, $\bar{\varepsilon}_r$ is not free strain at depth $\bar{y}_r$. The breadth, b_r, for slices 1 and 2 is the full slab breadth (3750 mm) divided by the modular ratio for temperature effects, $\alpha_e = 6.6$.

The force F_r, from equation (A.7), is the product of E_s and the entries in the three preceding columns. From the table, $\bar{F} = 2279$ kN. Stresses are calculated neglecting the effects of shear lag in accordance with Clause 5.4.1 of Part 5, and using the following properties from column 2 of table X.6.2:

$$y_g = 378 \text{ mm}$$

$$A = 177\,700 \text{ mm}^2$$

$$I = 46\,502 \times 10^6 \text{ mm}^4$$

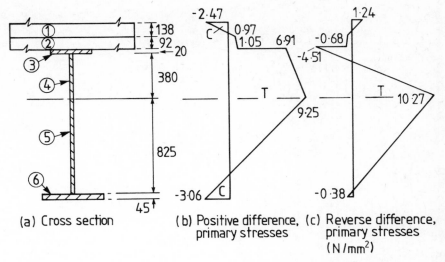

(a) Cross section

(b) Positive difference, primary stresses

(c) Reverse difference, primary stresses (N/mm^2)

Fig. X.11.2 Primary stresses at midspan due to temperature differences

Then, from equation (A.8):

$$\bar{M} = \sum_1^6 F_r\bar{y}_r - \bar{F}y_g = 203 - (2279 \times 0.378) = -658 \text{ kNm}$$

The free hogging curvature is:

$$\bar{M}/EI = -658/(205 \times 46\ 502) = -6.902 \times 10^{-5} \text{ m}^{-1}$$

and $\bar{F}/A = 12.8 \text{ N/mm}^2$, $\bar{M}/I = -0.01415 \text{ N/mm}^3$.

From equation (A.14), the stress at level 4 in the steel section ($y = 1500$ mm) is:

$$\sigma_4 = [+12.8 - 0.01415(1500 - 378)] = -3.1 \text{ N/mm}^2$$

From equation (A.15), the stress at the top of the deck slab (level 1) is:

$$f_{c1} = \frac{1}{6.6}[(-168 \times 207 \times 10^{-3}) + 12.8 + (0.01415 \times 378)]$$

$$= -2.47 \text{ N/mm}^2$$

Stresses at other levels are found in the same way and are plotted in fig. X.11.2(b).

The total longitudinal force in the concrete slab (Q) is required only for the design of the shear connection in the end spans. From table X.11.1, the longitudinal force in each slice is the product of the mean stress in the slice and the area of the slice, so the total force Q, in elements 1 and 2, is $(-388 + 347) = -41$ kN. This is distributed over a distance ℓ_s from each

Table X.11.1 Effects of positive temperature difference – midspan

Depth y_r (mm)	T (°C)	$10^6 \, \varepsilon_r$ (A.1)	Element No.	$10^6 \, \bar{\varepsilon}_r$ (A.3)	b_r (A.5) (mm)	h_r (A.6) (mm)	F_r (A.7) (kN)	$\bar{y}_r$ (A.4) (mm)	$F_r \bar{y}_r$ (kNm)	f_s (A.14) (N/mm²)	f_c (A.15) (N/mm²)	Q (kN)
0	14	168	1	108	568.2	138	1736	56.2	97.6	—	−2.47	−387.7
138	4	48	2	43.5	568.2	92	466	182.4	85.0	—	+0.97	+347.3
230	3.25	39	3	38.0	400	20	62.4	239.9	15.0	+6.91	+1.05	+55.3
250	3.09	37.08	4	18.5	10	380	14.4	376.7	5.4	+7.03		+30.9
630	0	0	5	0	10	825	0	0	0	+9.25		+28.2
1455	0	0	6	0	600	45	0	0	0	−2.42		−74.0
1500	0	0								−3.06		
							$\bar{F} = 2279$		$\Sigma F_r \bar{y}_r = 203$			$\Sigma = 0$

end support. Using equation (8.63):

$$\ell_s = 2(0.003 \times 41 \times 10^9/200)^{\frac{1}{2}} = 1568 \text{ mm}$$

The maximum shear at the beam ends is $41 \times 2/1.568 = 52.3 \text{ kN/m}$.

X.11.2 Primary stresses due to reverse temperature difference – midspan

The calculation for reverse temperature difference is similar to that described above and gives the stresses shown in fig. X.11.2(c) and a nominal force Q in the slab of -19 kN. $\bar{M} = -570$ kNm.

X.11.3 Primary stresses due to positive temperature difference – internal supports

The primary stresses due to positive temperature difference acting on the uncracked unreinforced composite section at the internal support are shown in fig. X.11.3. It can be seen that the difference between these and

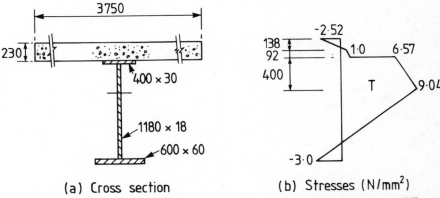

(a) Cross section
(b) Stresses (N/mm²)

Fig. X.11.3   Primary stresses at internal support due to positive temperature difference

the stresses calculated for the midspan is about 2%, and for all practical purposes the change in cross section of the steel member can be neglected. The moment $\bar{M} = -817$ kNm, which is 32% higher than at midspan, but the free hogging curvature $\phi = -6.752 \times 10^{-5} \text{ m}^{-1}$ is within 2% of that at midspan. It will be shown in the following paragraphs that changes in flange and web thickness in this constant-depth composite girder can also be neglected in calculating moments and reactions due to the secondary effects of temperature differences. As a general rule, variations in the steel cross section can be ignored in calculating primary and secondary temperature and shrinkage effects except where the depth of the steel section varies.

X.11.4 Secondary effects of temperature – positive temperature difference

The secondary moments and reactions caused by the positive temperature difference are initially calculated assuming the steel girder is of
uniform section throughout the three spans. The section properties are
those from column 2 of table X.6.2 for the uncracked section at midspan,
with $\alpha_e = 6.6$. Shear lag is neglected (Clause 5.4.2, Part 5). The secondary
moments are here calculated using Mohr's moment-area method but,
alternatively, may be determined by computer using a stiffness method of
analysis.

The moment diagram due to the 'free' primary hogging moment
$\bar{M} = -658$ kNm applied at each end of the three-span girder, with the
internal supports B and C removed, is as shown in fig. X.11.4(b). In figs
X.11.4(c) and (d), the deflection at A relative to midspan, Δ_A, is equal to
the first moment of area about A of the $\bar{M}/EI$ diagram between A and
midspan. Thus:

$$\Delta_A = 658 \times 43.5^2/2EI = 622\ 500/EI$$

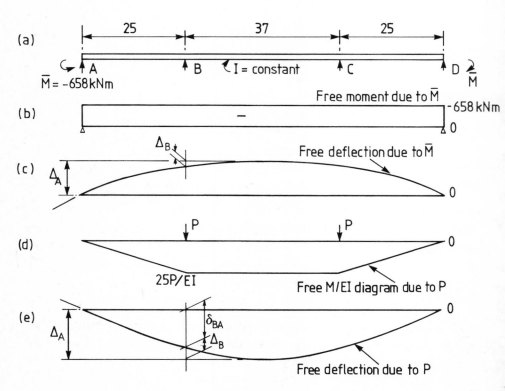

Fig. X.11.4   Moment-area diagrams for girder of uniform stiffness

Similarly:

$$\Delta_B = 112\ 600/EI$$

The upward deflection at B relative to A, $\delta_{BA} = \Delta_A - \Delta_B$. For $E_s = 205\ kN/mm^2$ and $I = 46\ 502 \times 10^6\ mm^4$:

$$\delta_{BA} = -509\ 900 \times 10^3/(205 \times 46\ 502) = -53.4\ mm \quad (upwards)$$

Let the downward reactions at B and C which are required to produce a deflection at B and C on the 'free' girder of $+53.4$ mm be P kN.

The M/EI diagram due to those reactions is shown in fig. X.11.4(d). The deflection at A relative to midspan is:

$$\Delta_A = \frac{1}{EI} [(\tfrac{1}{2} \times 25P \times 25^2 \times \tfrac{2}{3}) + (25P \times 18.5 \times 34.25)]$$

$$= 21\ 049P/EI$$

Similarly:

$$\Delta_B = \frac{1}{EI} \left[25P \times \frac{18.5^2}{2} \right] = 4278P/EI$$

Hence:

$$\delta_{BA} = \Delta_A - \Delta_B = 16\ 771P/EI$$

Therefore:

$$P = \frac{53.4 \times 205 \times 46\ 502 \times 10^6}{16\ 771 \times 10^9} = 30.4\ kN$$

and the secondary moment over the internal support is:

$$M_B = 30.4 \times 25 = 760\ kNm$$

Using the cracked section properties from column 2 of table X.6.3, the nominal stresses at the internal support due to the secondary effects of positive temperature difference are as follows:

$$\sigma_{rt} = -760/32.25 = -23.6\ N/mm^2\ (C)$$
$$\sigma_3 = -760/40.24 = -18.9\ N/mm^2\ (C)$$
$$\sigma_4 = 760/47.26 \quad = 16.1\ N/mm^2\ (T)$$

The stresses in the reinforcement and steelwork are of opposite sign to those caused by live loading, so the secondary effects of positive temperature difference need not be calculated at internal supports. The same is true of reverse differences, because these also cause sagging curvature at the internal supports.

At midspan, using the uncracked properties from column 2 of table X.6.2, the nominal stresses due to secondary effects of positive temperature difference are:

$$f_1 = -760/811.9 = -0.94 \text{ N/mm}^2$$
$$f_2 = -760/314.2 \times 6.6 = -0.37 \text{ N/mm}^2$$
$$\sigma_3 = -760/314.2 = -2.4 \text{ N/mm}^2$$
$$\sigma_4 = 760/41.45 = +18.3 \text{ N/mm}^2$$

In calculating the secondary moments and reactions due to positive temperature difference, it was assumed that the stiffness of the steel cross section was constant throughout the three spans. If account is taken of the variation in web and flange thickness of the steel section on the stiffness of the composite girder, the secondary moment at internal support B increases by 9%. This difference is sufficiently small to be neglected and, as a general rule, where only the flange and web thickness vary throughout the length of a girder, no account need be taken of the variation in stiffness of the girder in calculating secondary effects of temperature differences.

X.11.5 Secondary effects due to reverse temperature difference

The primary effects of reverse temperature difference cause a 'free' hogging moment, $\bar{M}$ of -570 kNm on the uncracked composite section at midspan. The secondary effects can be determined directly by proportion from the secondary moments due to positive temperature differences. Assuming the girder is of uniform section throughout:

$$M_B = (570/658) \times 760 = +658 \text{ kNm}$$

Table X.11.2 Nominal stresses due to temperature differences

	Midspan				Internal support		
	f_1	f_2	σ_3	σ_4	σ_{rt}	σ_3	σ_4
Positive							
Primary	−2.47	1.05	6.9	−3.1	—	6.6	−3.0
Secondary	−0.94	−0.37	−2.4	+18.3	−23.6	−18.9	+16.1
Total	−3.41	+0.68	+4.5	+15.2	−23.6	−12.3	+13.1
Reverse							
Primary*	1.24	−0.68	−4.5	−0.4	—	−4.5	−0.4
Secondary	−0.81	−0.32	−2.1	+15.8	−20.4	−16.4	13.9
Total	−0.43	−1.00	−6.4	+15.4	−20.4	−20.9	13.5

* Primary stresses at internal support due to reverse difference are assumed to be the same as at midspan

and the secondary stresses at midspan and at the internal support are 0.866 times those calculated for the positive temperature difference.

X.11.6 Nominal stresses due to temperature differences

The nominal stresses due to positive and reverse temperature differences are summarised in table X.11.2.

At midspan, it can be seen that primary effects of positive temperature difference are adverse in the concrete but are unlikely to govern design in plate girder bridges. The secondary effects of positive temperature difference are adverse at midspan in both concrete and steel, but can usually be neglected at ULS because the section at midspan will be compact.

Primary and secondary stresses at midspan due to reverse temperature differences are less than for positive temperature difference everywhere except at level 3. But, as midspan sections are usually compact, reverse temperature differences will never govern design at midspan.

At the internal support, primary stresses due to positive temperature differences are adverse in the steel section at level 4, but are unlikely to govern because secondary effects are always favourable and usually cancel out primary effects. Similarly, secondary effects of reverse temperature difference are favourable at internal supports and will almost always cancel out any adverse primary effect in the concrete and steel, so it will usually be found that the effects of temperature differences do not govern design at an internal support. A general summary of the effects of temperature differences on the design of composite sections to Part 3 of the Bridge Code is given in section 8.7.3.

X.12 The effects of shrinkage

For the reasons explained in section 8.7.1 no account is taken of the sequence of casting the deck slab in calculations for longitudinal bending stresses and interface shears due to the primary effects of shrinkage, but in calculating secondary effects it will be shown that it may be beneficial to take account of the sequence of casting.

For constant depth girders, changes in web and flange thickness at internal supports of continuous girders can be neglected as shown for temperature differences in section X.11.

X.12.1 Stresses due to primary effects of shrinkage

Stresses due to the primary effects of shrinkage are now calculated using equations (A.18) to (A.20). The nominal free shrinkage strain is -200×10^{-6}. The relevant section properties for the outer girder are those given in column 5, table X.6.2 for midspan, with the concrete uncracked and unreinforced, including the effects of shear lag.

The modular ratio α_e is taken as $2E_s/E_c$ (i.e. 13.2) as permitted by Clause 5.4.3, Part 5. Other parameters are:

$E_s = 205 \text{ kN/mm}^2$, $h_c = 230 \text{ mm}$
$b_e = 3580 \text{ mm}$, $A = 109\,429 \text{ mm}^2$, $y_g = 541 \text{ mm}$
$I = 38\,665 \times 10^6 \text{ mm}^4$, $h = 1500 \text{ mm}$

From equation (A.18):

$$\bar{F} = -200 \times 10^{-6} \times 3580 \times 230 \times 205/13.2 = -2679 \text{ kN (C)}$$

From equation (A.19):

$$\bar{M} = -2679 \,(0.115 - 0.541) = 1141 \text{ kNm (sagging)}$$

Hence $\bar{F}/A = 24.5 \text{ N/mm}^2$, $\bar{M}/I = 0.0295 \text{ N/mm}^3$ and $E_s \varepsilon_{sc} = -41.0 \text{ N/mm}^2$.

From equations (A.14) and (A.15), the nominal shrinkage stresses at levels 1 to 4 in the cross section as shown in fig. X.5.1 are:

top of slab

$$f_{c1} = [41.0 - 24.5 - (0.0295 \times 541)]/13.2 = +0.04 \text{ N/mm}^2$$

bottom of slab

$$f_{c2} = [41.0 - 24.5 - (0.0295 \times 311)]/13.2 = +0.55 \text{ N/mm}^2$$

top of steel flange

$$\sigma_3 = [(-24.5 - 0.0295 \times 311)] = -33.7 \text{ N/mm}^2$$

bottom of steel flange

$$\sigma_4 = [-24.5 + 0.0295 \times 959] = +3.8 \text{ N/mm}^2$$

These stresses are shown in Fig. X.12.1.

The free sagging curvature, $\phi_{sc} = \bar{M}/E_s I = +1.439 \times 10^{-4} \text{m}^{-1}$.

X.12.2 Longitudinal shear forces due to shrinkage

The longitudinal shear force, Q, at each end of the girder is obtained from equation (A.20):

$$Q = \frac{3580 \times 230}{13.2}\,[41.0 - 24.5 - 0.0295\,(541 - 115)]10^{-3} = 257 \text{ kN}$$

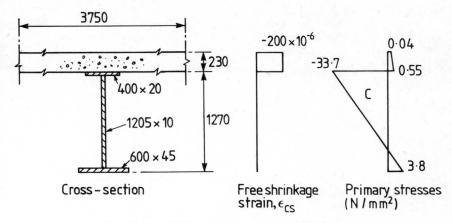

Fig. X.12.1   Primary stresses due to shrinkage

This reduces the longitudinal shear due to superimposed dead and live loading and it need not be considered in the design of the shear connection.

X.12.3 Secondary effects of shrinkage

The stiffness of the composite section is initially assumed to be uniform throughout the length of the girder, as in the calculation of temperature difference effects, with no account taken of the sequence of casting the deck slab.

The secondary moment due to shrinkage is therefore directly proportional to the previous result for temperature difference (section X.11.4). For positive temperature difference, $\bar{M} = -658$ kNm and $M_B = +760$ kNm. From section X.12.1, for shrinkage, $\bar{M} = +1141$ kNm. Hence, for shrinkage:

$$M_B = (760 \times 1141)/(-658) = -1318 \text{ kNm (hogging)}$$

The stresses due to primary and secondary effects of shrinkage are summarised in table X.12.1 for midspan and for the internal support. At the internal support, the section properties are those for the cracked section in column 2, table X.6.3, neglecting the effects of shear lag (5/5.4.2.1). The secondary moment at the internal support due to shrinkage is approximately half that due to HA loading, so it is worth taking account of the sequence of casting the deck slab in calculating the secondary effects of shrinkage. The distribution of secondary moments due to shrinkage has the shape shown in fig. X.11.4(d), but with a value along BC of -1318 kNm.

Table X.12.1   Nominal stresses (N/mm²) due to shrinkage

Section	Effect	Nominal stresses (N/mm²)				
		f_1	f_2	σ_{rt}	σ_3	σ_4
Internal support	Primary	0.04	0.55	—	−33.7	3.8
	Secondary	—	—	40.9	32.7	−58.7
	Total	0.04	0.55	40.9	−1.0	−54.9
Midspan	Primary	0.04	0.55	—	−33.7	3.8
	Secondary	1.36	0.76	—	10.2	−33.1
	Total	1.40	1.31	—	−23.5	−29.3

X.12.4   Effect of staged casting on secondary effects of shrinkage

The sequence of casting the deck slab is shown in fig. X.10.1. To simplify the calculations for shrinkage effects in this worked example, it is assumed that stages 2 and 3 are cast together and the effect of stage 5 is neglected.

Shrinkage on stage 1. After stage 1 of the deck slab is cast, the stiffness of the girder is as shown in fig. X.12.2(a) with:

$I_1 = 38\,816 \times 10^6$ mm⁴ (column 3, table X.6.2)
$I_2 = 17\,431 \times 10^6$ mm⁴ $= 0.449 I_1$, (column 1, table X.6.3)
$I_3 = 11\,963 \times 10^6$ mm⁴ $= 0.308 I_1$, (column 1, table X.6.2)

The moment caused by primary shrinkage effects, $\bar{M} = 1141$ kNm, is distributed as shown in fig. X.12.2(b).

Using the moment-area method as before, the reactions P at B and C are found to be 4.56 kN (upwards) and the secondary moments at this stage are as shown in fig. X.12.2(c).

Shrinkage on stages 2 and 3. The sagging moments due to primary shrinkage on the second and third stages of the slab cast are shown in fig. X.12.2(d). The stiffness of the girder is as shown in fig. X.12.2(e).

Using the moment-area method as before, the reactions P at B and C moments at B and C are as shown in fig. X.12.2(f).

Shrinkage on stage 4 of deck slab. The secondary moments and reactions due to shrinkage on stage 4 of the deck slab are shown in fig. X.12.3.

Stresses due to secondary effects of shrinkage – staged casting. At the *internal support*, the secondary moments due to shrinkage on stages 1, 2 and 3 of the deck slab act on the steel section alone, so the relevant section properties are those from column 1 of table X.6.3. The shrinkage on stage 4 of the deck slab acts on the cracked composite section, so the

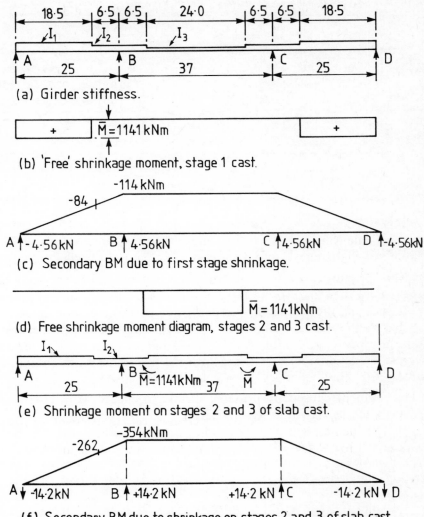

(a) Girder stiffness.

(b) 'Free' shrinkage moment, stage 1 cast.

(c) Secondary BM due to first stage shrinkage.

(d) Free shrinkage moment diagram, stages 2 and 3 cast.

(e) Shrinkage moment on stages 2 and 3 of slab cast.

(f) Secondary BM due to shrinkage on stages 2 and 3 of slab cast.

Fig. X.12.2 Secondary bending moments and reactions due to shrinkage

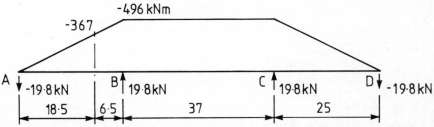

Fig. X.12.3 Secondary bending moments and reactions due to shrinkage of concrete cast in stage 4

Table X.12.2 Stresses due to secondary effects of shrinkage – internal support

Casting stage	Nominal moment (kNm)	Nominal stresses (N/mm²)				
		f_1	f_2	σ_{rt}	σ_3	σ_4
1 to 3	−468	—	—	—	22.5	−11.6
4	−496	—	—	15.4	12.3	−10.5
Total	−964	—	—	15.4	34.8	−22.1

Table X.12.3 Stresses due to secondary effects of shrinkage – midspan

Casting stage	Nominal moment (kNm)	Nominal stresses (N/mm²)			
		f_1	f_2	σ_3	σ_4
1	−114	—	—	8.4	−3.7
2 to 4	−850	0.88	0.49	6.6	−21.3
Total		0.88	0.49	15.0	−25.0

relevant properties are those in column 3 of table X.6.3. The secondary stresses are then as given in table X.12.2.

At *midspan*, the secondary moments due to shrinkage on stage 1 are carried by the steel section alone, so the relevant section properties are those from column 1, table X.6.2. The shrinkage on stages 2, 3 and 4 is carried by the composite section, with the properties from column 3 of table X.6.2. The secondary stresses are then as given in table X.12.3.

X.12.5 *Nominal stresses due to shrinkage, taking account of casting sequence*

Nominal stresses due to shrinkage, taking account of the sequence of casting the deck slab are summarised in table X.12.4. These are

Table X.12.4 Nominal stresses (N/mm²) due to shrinkage (taking account of casting sequence)

Section	Effect	Nominal stresses (N/mm²)				
		f_1	f_2	σ_{rt}	σ_3	σ_4
Internal support	Primary	0.04	0.55	—	−33.7	3.8
	Secondary	—	—	15.4	34.8	−22.1
	Total	0.04	0.55	15.4	1.1	−18.3
Midspan	Primary	0.04	0.55	—	−33.7	3.8
	Secondary	0.88	0.49	—	15.0	−25.0
	Total	0.92	1.04	—	−18.7	−21.2

consistent with the general comments on shrinkage effects given in table 8.5.

It can be seen that at the internal supports, the stresses in the reinforcement and the bottom flange due to the secondary effects, given in table X.12.4, are almost one third of the stresses when no account is taken of the sequence of casting (table X.12.1). It must be remembered, however, that in the calculations which allow for the sequence of casting, it is assumed that all the shrinkage on one section takes place before the next stage is cast. This assumption will not be correct when sections of slab are cast at intervals of a few days. Therefore, if advantage is taken of this in the design, the designer must ensure that his assumptions are compatible with the sequence of casting that is permitted on site.

X.13 Design of deck slab for global and local bending

In the preliminary design of the deck slab (section X.8), the local effects of wheel loads were considered in combination U-1. Once the results of the grillage analysis are available, the effects of global transverse bending moments in the deck slab must also be considered at ULS. If each of these effects, considered separately, is satisfied, then no further checks are required in Part 4 at ULS, but the combined effect of global and local bending has to be considered in combination S-1 (Clause 4.8.3, Part 4).

X.13.1 *Transverse bending, S-1 and U-1*

The maximum global transverse sagging moment occurs at midspan with the HB vehicle in lane 4 and HA loading in lanes 3 and 5, as shown in fig. X.13.1. This loading configuration also gives the worst combination of global and local transverse bending moments.

From the grillage analysis, the maximum transverse moment in the deck slab, due to HA and 45 units of HB live load combined, is 42 kNm/m. For crack width calculations in combination S-1 only, HB loading of 25 units is considered. The global transverse moment then reduces to 28.7 kNm/m. The coexistent transverse moment due to wheels of the 25 unit HB bogie can be found using Pucher's charts.[56] Here it is taken to be the same as at point B (fig. X.8.2). This is a conservative assumption because in calculating the local moment at point B, it was assumed that the slab was fully restrained over one longitudinal girder and simply supported over the edge girder, whereas at point F (fig. X.8.2) it is continuous over both longitudinal girders.

The design moments for combinations U-1 and S-1, including γ_{f3}, are

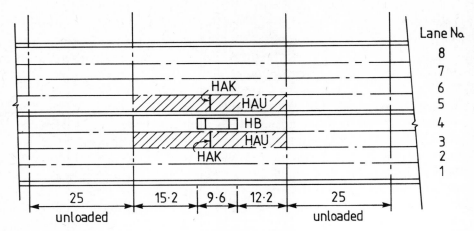

Fig. X.13.1   Live loading for maximum global transverse sagging moment

summarised in table X.13.1. The moments due to DL and SDL are for point F in fig. X.8.2.

The design moment for combination U-1 due to global transverse sagging bending is 66 kNm/m. This is less than the design moment in table X.8.2 for point B, so no check is required if the same transverse reinforcement is provided in the bottom of the slab.

The design moment for combination S-1 due to combined global and local effects is 66.3 kNm/m. This is larger than the moment at B in table X.8.2, so crack widths must be calculated in accordance with Clause 5.8.8.2 of Part 4. Taking the top reinforcement as T20-125 and the bottom transverse reinforcement as T16-150, elastic analysis of the cross section gives a tensile stress in the bottom reinforcement of 293 N/mm^2 and a crack width of 0.28 mm, which exceeds the limit of 0.25 mm in table X.3.2. The spacing of transverse bars in the bottom of the slab is

Table X.13.1   Design transverse moments for point F (fig. X.8.2)

Loading	Nominal BM (kNm)	S-1 γ_{fL}	S-1 BM	U-1 γ_{fL}	U-1 BM
Dead (concrete)	2.97	1.0	2.97	1.15	3.4
Superimposed dead	1.16	1.2	1.39	1.75	2.0
Global HA + 45 HB	42.0	1.1	—	1.3	54.6
Global HA + 25 HB	28.7	1.1	31.57	1.3	—
Local 25 HB	27.6	1.1	30.36	1.3	—
Total			66.3		60.0
Design moments (including γ_{f3})			66.3		66.0

therefore reduced to **T16 at 125 mm**. The tensile stress is then 246 N/mm^2 and the crack width is 0.235 mm.

X.13.2 *Longitudinal bending, U-1*

Near an internal support, the deck slab is subjected to almost uniform tensile strain due to global longitudinal bending. The strain on the soffit of the slab is increased further by local longitudinal bending due to wheel loads as the HB vehicle approaches the internal support. The combined effect of global and local loadings must then be considered at SLS. The worst loading configuration is shown in fig. X.13.2. The leading axle of the HB vehicle should be close to the internal support, but sufficiently far away to ensure local longitudinal bending is not reduced by the crosshead girder. The inner axle spacing is taken as 11 m to produce maximum global longitudinal bending from the two trailing axles at midspan. The two adjacent lanes in the 25 m and 37 m span are loaded with full HA to produce maximum global longitudinal bending over the internal support. From section X.8.3(ii) the local longitudinal sagging moment due to 25 units of HB is:

$$M'_y = 31 \times 25/45 = 17.2 \text{ kNm/m}$$

This is less than the local longitudinal moment of 23.9 kNm/m due to the single 100 kN HA wheel load, but this can only coexist with global longitudinal bending if it replaces one lane of HA. It is unlikely then that it would produce a more severe loading condition than the one considered.

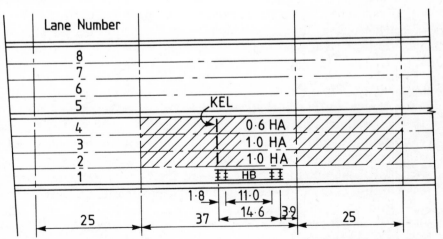

Fig. X.13.2 Loading for maximum combined global and local longitudinal tensile strain in deck slab

From the grillage analysis the global longitudinal moment near the internal support due to the live loading configuration in fig. X.13.2 is -2457 kNm. The only other loadings causing global tension in the deck slab at this section are due to staged casting of the deck slab and secondary moments due to shrinkage. The longitudinal moment on the composite section due to staged casting of the deck slab is -728 kNm from table X.10.1. The secondary moment due to the effects of shrinkage, -496 kNm, is from table X.12.2. These design moments for combination S-1 are summarised in table X.13.2.

Table X.13.2 Design longitudinal moments (kNm) near internal support

Loading	Nominal BM	γ_{fL}	S-1 BM
Global effects			
Dead (concrete)	-728	1.0	-728
Superimposed dead (asphalt)	-600	1.2	-720
(other)	-304	1.0	-304
25 HB with HA	-2457	1.1	-2703
Shrinkage (secondary effect)	-496	1.0	-496
Total	—	—	-4951
Local effects			
25 HB	-17.2	1.1	-19

Using the section properties in table X.6.3, the longitudinal stress in the steel at the steel-concrete interface due to global effects is:

$$\sigma_3 = 4951/34 = 145.6 \text{ N/mm}^2$$

The global tensile strain is:

$$\varepsilon_1{}^G = 145.6/(205 \times 10^3) = 710 \times 10^{-6}$$

With this large global tensile strain, it might be thought that the sagging moment of 19 kNm/m due to local wheel loads must be resisted entirely by the longitudinal reinforcement. Such an approach would lead to a very substantial increase in the amount of reinforcement that is normally provided, and is inappropriate, because the compressive strains at the top surface of the slab due to local wheel load effects cause the cracks from global tensile strains to close. For this reason, tensile strains due to bending from local wheel loads are normally calculated at SLS by elastic theory assuming the concrete in compression to be effective.

Assuming the longitudinal reinforcement in the bottom of the slab is T20-150, elastic analysis of the cross section due to the local moment, M, of 19 kNm/m with $h = 230$ mm and $d_1 = (230 - 35 - 16 - 10) = 169$ mm gives a tensile stress in the reinforcement of 60 N/mm^2 and a neutral-axis depth of 56 mm. The tensile strain at the soffit of the deck slab:

$$\varepsilon_1{}^L = \left[\frac{(230 - 56)}{(169 - 56)}\right] \times 60 \times 10^{-3}/205 = 450 \times 10^{-6}$$

The reduction in strain due to tension stiffening is now calculated in accordance with Clause 5.8.8.2 of Part 4. Where the ratio of the live load moment to the permanent moment M_q/M_g exceeds 1.0, the effect of tension stiffening is taken as zero, and then $\varepsilon_m = \varepsilon_1$.

For global effects, the ratio of live-load moment to the moment due to permanent loads is:

$$M_q/M_g = 2703/(728 + 720 + 304 + 496) = 120 \quad \text{(table X.13.2)}$$

Hence, from equation (25), Part 4, $\varepsilon_m = \varepsilon_1{}^G = 710 \times 10^{-6}$.

For local bending, the ratio of live load moment to dead load moment, M_q/M_g is 19/0, i.e. it approaches infinity, so for local effects, $\varepsilon_m = \varepsilon_1{}^L = 450 \times 10^{-6}$.

The total longitudinal strain at the slab soffit due to global and local effects is then the algebraic sum of the strains calculated separately (5.8.8.2(c), Part 4). Thus:

$$\varepsilon_m = (710 + 450)10^{-6} = 1160 \times 10^{-6}$$

The design crack width from Clause 5.8.8.2(b) of Part 4, is $w = 3a_{cr}\varepsilon_m$.

For T20 bars at 150 mm spacing with a concrete cover of 51 mm:

$$a_{cr} = [75^2 + (51 + 10^2)]^{\frac{1}{2}} - 10 = 87 \text{ mm}$$

and:

$$w = 3 \times 87 \times 1160 \times 10^{-6} = 0.30 \text{ mm}$$

This exceeds the limiting crack width of 0.25 mm, so reduce the spacing of longitudinal bars in the bottom of the slab to **T20 at 125 mm** in regions of global longitudinal hogging moments. No account is taken in subsequent calculations of the additional bending resistance that results from this reduced spacing.

X.14 Longitudinal stresses in edge girder, U-1

Stresses at midspan are calculated using section moduli from columns 2 and 3 of table X.6.2 as appropriate. Stresses at the internal support are calculated using the section moduli in column 2 of table X.6.3. The bending moments in the edge girder due to superimposed dead loading and live loading are those obtained from the grillage analysis and given in table X.9.1. The dead load stresses for concrete are from table X.10.1 and take account of the staged casting of the deck slab. Stresses due to shrinkage are from table X.12.4 multiplied by $\gamma_{fL} = 1.2$ for combination U-1.

The stresses in the steel section are everywhere lower than the design strengths calculated in section X.7 and table X.3.1, but interaction between vertical shear and longitudinal bending will have to be examined before it can be concluded that the section is adequate.

The stresses in the concrete and reinforcement in tables X.14.1 and X.14.2 must be multiplied by γ_{f3} ($= 1.1$) before they are compared with the limiting stresses in Part 4.

The maximum tensile stress in the reinforcement

$$\sigma_{rt} = 215.1 \times 1.1 = 237 \text{ N/mm}^2.$$

Table X.14.1 Longitudinal stresses (N/mm²) at midspan, outer girder, U-1

Loading	Design BM(kNm)	f_1	f_2	σ_3	σ_4
(1) DL steel beam	402	—	—	−29.5	−13.2
(2) DL concrete	—	−1.45	−0.83	−61.3	57.7
(3) Sub-total	—	−1.45	−0.83	−90.8	70.9
BS 5400 limit				−169	307
(4) Superimposed DL	1217	−1.27	−0.72	−9.5	30.4
(5) HB with HA	8326	−10.26	−4.01	−26.5	200.8
(6) Shrinkage*	—	(1.10)	(1.25)	−22.4	(−25.4)
Total		−12.98	−5.56	−149.2	302.1
BS 5400 limit				−203	307

* Figures in parentheses are neglected where they reduce the total effect considered

Table X.14.2 Longitudinal stresses (N/mm²) at internal support, outer girder, U-1

Loading	Design BM (kNm)	σ_{rt}	σ_3	σ_4
(1) DL steel beam	−540	—	26.0	−13.4
(2) DL concrete	—	26.0	121.7	−69.7
(3) Sub-total	—	26.0	147.7	−83.1
BS 5400 limit			307	−204
(4) Superimposed DL	−1415	43.9	35.2	−29.9
(5) HB with HA	−4087	126.7	101.6	−86.5
(6) Shrinkage	—	18.5	1.3	−22.0
Total		215.1	285.8	−221.5
BS 5400 limit			307	−227

This is less than $0.87f_{ry}$ or 400 N/mm^2 and, so, is satisfactory.

The maximum compressive stress in the concrete is $12.98 \times 1.1 = 14.3 \text{ N/mm}^2$, which is less than $0.45f_{cu}$, or 18 N/mm^2.

X.15 Design of web for vertical shear and bending, U-1 and U-3

In the 37 m centre span, the critical sections for vertical shear in the web are at the internal support and at the change in steel section, 6.5 m from the support. Calculations are given here only for the section at the internal support, for two cases of loading:

(1) maximum vertical shear with coexisting bending moment.
(2) maximum bending moment with coexisting shear.

It will usually be found that case (2) governs the design of the web at internal supports, case (1) governs the design of the web at end supports, and both must be considered at changes in section within a span.

X.15.1 *Design moments and vertical shears – outer girder, internal support*

In Part 3 of the Bridge Code, the design of webs is considered only at ULS. Combination U-3 need not be considered because the secondary effects of the positive and reverse temperature differences specified in Part 2 of the Code both produce sagging bending moments at the internal support and, in this example, no vertical shears in the internal span. Wind (combination 2) is not considered in this worked example.

Design moments and vertical shear forces for combination U-1 are summarised in table X.15.1 for the two combinations of moment and shear identified above, using the results obtained from the grillage analysis and from sections X.10 and X.12. The effects of HB loading are more severe than HA loading in both cases.

X.15.2 *Design of web at internal support for combined bending and shear*

For a non-compact section where the beam is built in stages and the loading is applied in stages, the design of the web for combined bending and shear should be in accordance with Clause 9.9.5 of Part 3. The shear and bending resistance at each stage of construction are considered separately, and then combined in accordance with Clause 9.9.5.3. Initially the resistance of the web in pure shear is determined from 3/9.9.2.2, as follows, assuming that the spacing of stiffeners, a, is equal to the effective depth of the web, d_{we}, so that $\phi = 1.0$.

Table X.15.1 Design moments (kNm) and shear forces (kN) at internal support, combination U-1

| | Case 1 | | | | | Case 2 | | | |
| | Nominal | | γ_{fL} | U-1 | | Nominal | | U-1 | |
Loading	BM	VS		BM	VS	BM	VS	BM	VS
(1) DL steel girder	−515	−97	1.05	−541	−102	−515	−97	−541	−102
(2) DL concrete	−2553	−472	1.15	−2936	−544	−2553	−472	−2936	−544
(3) SDL									
(a) asphalt	−600	−113	1.75	−1050	−198	−600	−113	−1050	−198
(b) other	−304	−57	1.20	−365	−68	−304	−57	−365	−68
(4) HA live load	−2010	−555	1.5	−3015	−833	−2363	−512	−3545	−768
(5) HB with HA	−2324	−930	1.3	−3021	−1209	−3144	−654	−4087	−850
(6) Shrinkage	−964	0	1.2	−1157	0	−964	0	−1157	0
Σ 1, 2, 3, 5, 6				−9070	−2121			−10136	−1762

Shear resistance of web under pure shear. Using Clause 9.9.2.2 of Part 3 or equation (8.11), with $d_{we} = 1180$ mm, $t_w = 18$ mm and $\sigma_{yw} = 355$ N/mm^2:

$$\lambda = (d_{we}/t_w)(\sigma_{yw}/355)^{\frac{1}{2}} = 65.6$$

When there is no tension-field action from the flanges, $m_{fw} = 0$, $V_D = V_R$ and from fig. 11, Part 3, with $\phi = 1.0$.

$$\tau_\ell/\tau_y = 0.985.$$

Hence:

$$\tau_\ell = 0.985 \times 355/\sqrt{3} = 202 \text{ N/mm}^2$$

and:

$$V_R = 18 \times 1180 \times 202 \times 10^{-3}/(1.05 \times 1.1) = 3713 \text{ kN}$$

The ability of the flanges to anchor tension-field forces depends on the resistance of the smaller flange on the web boundary and in Part 3 is measured by the value m_{fw}. No guidance is given in Part 3 on the evaluation of m_{fw} for a cracked composite flange (section 8.4.3), but it is clearly safe at an internal support to use the value for the steel top flange alone.

The value b_{fe} is the smaller of the following two values:

(1) $10t_f(355/\sigma_{yw})^{\frac{1}{2}}$
(2) $b_f/2$

In this example, the value of (1) is $10 \times 30 = 300$ mm and of (2) is 200 mm.

From Part 3, $m_{fw} = \sigma_{yf}b_{fe}t_f^2/2\sigma_{yw}d_{we}^2t_w$. With $t_f = 30$ mm:

$$m_{fw} = 355 \times 200 \times 30^2/(2 \times 355 \times 1180^2 \times 18) = 0.0036$$

Interpolating between figs. 11 and 12 in Part 3 gives $\tau_\ell/\tau_y = 1.0$. Hence, from equation (8.10):

$$V_D = 18 \times 1180 \times 205 \times 10^{-3}/(1.05 \times 1.1) = 3770 \text{ kN}.$$

Bending resistance of steel section alone (Clauses 9.9.1 and 9.9.3.1, Part 3). The bending resistance, M_D, is the lesser of the values from equations (8.7) and (8.8). From section X.3 and table X.6.3:

$\sigma_{\ell c} = 269$ N/mm^2, $Z_{xc} = 40.35 \times 10^6$ mm^3 and $\gamma_m = 1.2$
$\sigma_{yt} = 355$ N/mm^2, $Z_{xt} = 20.80 \times 10^6$ mm^3 and $\gamma_m = 1.05$

Equation (8.7) then gives:

$$M_D = 40.35 \times 10^6 \times 269 \times 10^{-6}/(1.2 \times 1.1) = 8223 \text{ kN}$$

Equation (8.8) gives:

$$M_D = 20.8 \times 10^6 \times 355 \times 10^{-6}/(1.05 \times 1.1) = 6393 \text{ kNm}$$

Tension in the top flange therefore governs the ultimate moment of resistance and $M_D = 6393$ kNm.

The limiting force in the flange, F_f, is then:

$$F_f = 355 \times 400 \times 30 \times 10^{-3} = 4260 \text{ kN}$$

and from equation (8.13), with $d_f = 1180 + 15 + 30 = 1225$ mm:

$$M_R = 4260 \times 1.225/(1.05 \times 1.1) = 4518 \text{ kNm}$$

These resistances are written as M_{DS} and M_{RS}. Together with the values V_D and V_R, they define the interaction diagram for the steel section, which is plotted in fig. X.15.1.

Bending resistance of cracked composite section. If the compressive stress in the bottom flange governs, then from equation (8.7), with $\sigma_{\ell c} = 300 \text{ N/mm}^2$ (section X.7) and $Z_{xc} = 47.26 \times 10^6 \text{ mm}^3$ (table X.6.3):

$$M_D = 47.26 \times 10^6 \times 300 \times 10^{-6}/(1.2 \times 1.1) = 10\,741 \text{ kNm}.$$

If tension governs, then from equation (8.8), with $\sigma_{yc} = 355 \text{ N/mm}^2$ and $Z_{xt} = 40.24 \times 10^6 \text{ mm}^3$ (table X.6.3):

$$M_D = 40.24 \times 10^6 \times 355 \times 10^{-6}/(1.05 \times 1.1) = 12\,368 \text{ kNm}$$

Hence, compression governs and $M_D = 10\,741$ kNm.

The limiting force in the compression flange is:

$$F_f = 300 \times 600 \times 60 \times 10^{-3} = 10\,800 \text{ kN}.$$

The limiting force in the tension flange, F_{ft} and the lever arm, d_f, are calculated by the methods given in section 8.4.3. From table X.15.1, $M_s = 3477$ kNm and $M_c = 5593$ kNm for loading U-1. As compression governs M_D, we use section moduli Z_4 from table X.6.3 in calculating the design bending moment:

$$M_x = 5593 + (3477 \times 47.26/40.35) = 9665 \text{ kNm}$$

Putting $A_r = 7854 + 5027 = 12\,881 \text{ mm}^2$, $A_f = 12\,000 \text{ mm}^2$ and $\sigma_{yt} = 355 \text{ N/mm}^2$:

$$F_{ft} = 0.355[12\,000 + 12\,881(5593/9665)] = 6906 \text{ kN}$$

For the accurate calculation of the lever arm, $d_{fs} = 1225$ mm and:

$$d_{fr} = [(7854 \times 1410) + (5027 \times 1300)]/12\,881 = 1367 \text{ mm}$$

Whence:

$$d_f = \frac{(1225 \times 12) + (1367 \times 12.88 \times 5593/9665)}{12 + (12.88 \times 5593/9665)} = 1279 \text{ mm}$$

The approximation $d_f \simeq d_{fs}$ gives a result only 4% lower.

For this cross section, F_{ft} is more than 15% below F_f, and so governs. Hence:

$$M_R = 6906 \times 1.279/(1.05 \times 1.1) = 7647 \text{ kNm}$$

For the composite section, V_D and V_R are as for the steel section, so the interaction diagram can now be plotted (fig. X.15.1).

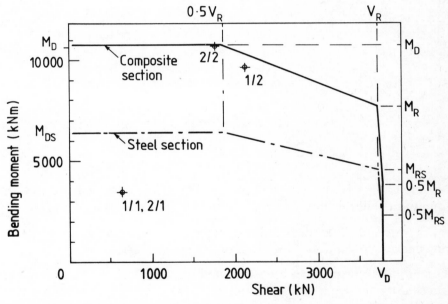

Fig. X.15.1 Moment-shear interaction diagrams for cross section at internal support

X.15.3 *Resistance to combined bending and shear, case 1*

Stage 1 – loads carried on the steel section alone. From table X.15.1, the design moments and shears for the steel section alone are:

$$M = 541 + 2936 = 3477 \text{ kNm}$$
$$V = 102 + 544 = 646 \text{ kN.}$$

These are plotted on fig. X.15.1 as point 1/1, which lies well below the resistance of the steel section.

Stage 2 – loads carried on the composite section. The design moment for the composite section, calculated above, is $M_x = 9665$ kNm. The coexisting vertical shear (table X.15.1) is 2121 kN. This combination is plotted on fig. X.15.1 as point 1/2.

X.15.4 *Resistance to combined bending and shear, case 2*

The stage 1 loads are the same as for case 1, point 2/1.

For stage 2, from table X.15.1, $M_s = 3477$ kNm and $M_c = 6659$ kNm. As before, the design moment that allows for sequential construction is:

$$M_x = 6659 + (3477 \times 47.26/40.35) = 10\,731 \text{ kNm}$$

and the coexisting shear is 1762 kN.

These are plotted on fig. X.15.1 as point 2/2. This is the most critical of the three points, and shows that **the section is just adequate**.

X.16 Longitudinal shear

The distribution of longitudinal shear force at the steel-concrete interface is required for:

(1) The design of the shear connection for static loads at the SLS.

(2) The design of transverse reinforcement in the deck slab to prevent longitudinal shear failure of the slab under static loads at the ULS.

(3) The design of the shear connection for repeated loading at the SLS to prevent fatigue failure of one or more shear connectors.

Longitudinal shear forces are calculated directly from the vertical shear forces obtained from the global analysis for longitudinal bending. For both limit states, the value of $A\bar{y}/I$ is calculated assuming the concrete to be uncracked and unreinforced in both hogging and sagging moment regions. At the ULS, the full breadth of the concrete flange is used in calculating $A\bar{y}/I$, so the values are as given in tables X.6.1 and X.6.2. At the SLS, Part 5 requires the effect of shear lag to be taken into account, but allows a constant effective breadth of flange (that at quarter span) to be used in any span. This requires two further sets of section properties to be calculated, one for short-term loading and one for long-term loading. In beam and slab decks the difference in $A\bar{y}/I$ for the full flange breadth and for the quarter span effective breadth is usually small enough to be neglected. In the calculations that follow, $A\bar{y}/I$ is calculated using the full breadth of the flange for both limit states. The longitudinal shears at the SLS can then be obtained simply by factoring those at the ULS.

X.16.1 *Loadings to be considered*

Longitudinal shear forces at the steel-concrete interface arise only from loadings applied to the composite section. For the outer girder in this worked example these loads are:

(1) Staged casting of the deck slab.
(2) Superimposed dead loading.
(3) HB combined with HA.
(4) Shrinkage.
(5) Temperature differences.

Staged casting of the deck slab. Nominal vertical shear forces due to staged casting of the deck slab are summarised in table X.10.1 and fig. X.10.2. In the 37 m span, only stage 5 of casting applies vertical shear to the composite section, the value at the internal supports being 142 kN, as shown in table X.16.1. Loads from other stages are carried on the steel

Table X.16.1 Design vertical shear on composite section in 37 m span

Loading	Nominal VS (kN)	S-1		U-1		Section
		γ_{fL}	VS (kN)	γ_{fL}	VS (kN)	
DL concrete (stages 1 to 5)	—	1.0	—	1.15	—	
SDL (asphalt)	—	1.2	—	1.75	—	Midspan
(other)	—	1.0	—	1.2	—	
HB with HA	− 332	1.1	− 365	1.3	− 432	
Total	− 332		− 365		− 432	
DL Concrete (stages 1 to 5)	− 92	1.0	− 92	1.15	− 106	
Super DL (asphalt)	− 73	1.2	− 88	1.75	− 128	Point X 6.5 m from
(other)	− 37	1.0	− 37	1.2	− 44	support C
HB with HA	− 690	1.1	− 759	1.3	− 897	
Total	− 892		− 976		− 1175	
DL concrete (stages 1 to 5)	− 142	1.0	− 142	1.15	− 163	
SDL (asphalt)	− 113	1.2	− 136	1.75	− 198	Internal support C
(other)	− 57	1.0	− 57	1.2	− 68	
HB with HA	− 930	1.1	− 1023	1.3	− 1209	
Total	− 1242		− 1358		− 1638	

section before composite action is developed. In general, this is not the case, but, here it is, due to the symmetry of the structure and the symmetry of the chosen casting sequence.

Superimposed dead loading. This is uniformly distributed over the span and the nominal vertical shear forces at the three sections referred to in table X.16.1 can be determined from the shear given in table X.15.1 at the internal support.

HB live loading with HA. For the outer girder, the maximum vertical shear occurs with the HB vehicle in the lane nearest the parapet with HA in the two adjacent lanes and 0.6HA in other lanes. The longitudinal disposition of the loads for maximum shear at the three sections considered in table X.16.1 can be found using the influence line in fig. X.16.1.

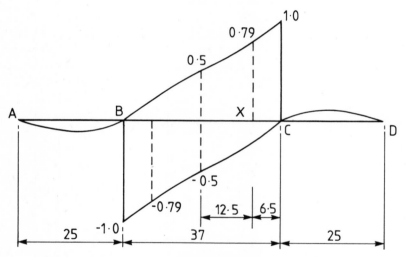

Fig. X.16.1 Influence line for vertical shear in 37 m span

(1) Maximum shear in the 37 m span at internal support C occurs with the leading axle of the HB vehicle at C and the other axles in the 37 m span. The shortest axle spacing of 6 m is obviously the most severe. HAU loading in other lanes is applied over the 37 m span only with the HAK just to the left of C. An alternative arrangement should also be considered with HB loading and the HAK in the same positions, but with HAU occupying spans BC and CD.

(2) Maximum shear in the 37 m span at point X, 6.5 m from the internal support, occurs with the leading axle of the HB vehicle at point X and the

remaining axles at 8.3 m, 14.3 m and 16.1 m from C towards B. HAU loading is applied over 30.5 m between B and X in the two adjacent lanes, with HAK applied at X.

The alternative arrangement with HAU occupying a 30.5 m length BX and the 25 m span CD should also be considered.

(3) Maximum shear at the centre of the 37 m span occurs with the leading HB axle at midspan and the shortest spacing of 6 m between inner axles. HAU occupies the two adjacent lanes in the same half of the 37 m span as the HB load. The intensity of lane loading is based on a loaded length of 18.5 m. The HAK is positioned at midspan.

The envelope of maximum vertical shear forces for the 37 m span due to HB loading combined with HA, as obtained from the grillage analysis, is shown in fig. X.16.2 for the outer girder. Vertical shear due to HA loading alone is everywhere less than for HB loading.

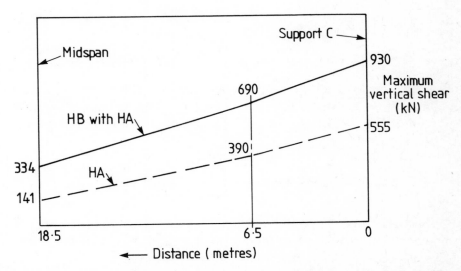

Fig. X.16.2   Envelope of nominal live-load vertical shear force for outer girder in 37 m span

Longitudinal shears due to shrinkage and temperature differences. The primary effects of TSC cause longitudinal shear forces at the free ends of the 25 m spans only (see sections X.11 and X.12) and do not affect the design of the outer girder in the centre span.

Secondary effects of shrinkage and temperature difference give zero vertical shear in this symmetric central span. Hence, for girders in the 37 m span, combination 3 will never be critical for longitudinal shear.

X.16.2 *Longitudinal shear force, S-1 and U-1*

Vertical shear forces acting on the composite section are summarised in table X.16.1 for combinations S-1 and U-1 for the three positions considered in the 37 m span. From these, longitudinal shear forces at the steel-concrete interface are calculated in table X.16.2 using the values of

Table X.16.2 Longitudinal shear force for combinations S-1, U-1

Loading	$A\bar{y}/I$ (m^{-1})	VS (kN) U-1	LS (kN/m) U-1	Section	VS (kN) S-1	LS (kN/m) S-1
DL + SDL	0.699	—	—	Midspan	—	—
HB with HA	0.739	− 432	− 320		− 365	− 270
Total		− 432	− 320		− 365	− 270
DL + SDL	0.699	− 268	− 179	6.5 m from	− 217	− 145
HB with HA	0.739	− 897	− 664	support C	− 759	− 564
Total		− 1165	− 843	.X	− 976	− 709
DL + SDL	0.664	− 429	− 285	Internal	− 335	− 222
HB with HA	0.73	− 1209	− 883	support C	− 1023	− 747
Total		− 1638	− 1168		− 1358	− 969

$A\bar{y}/I$ from table X.6.1 for the internal support and the values from columns 2 and 3, Table X.6.2, for midspan. At point X, there is a change in steel section, so the slightly higher values of $A\bar{y}/I$ for the 10 mm web section are used.

X.17 Design of shear connection for static loading, S-1

For girders in the 37 m span, only combination S-1 need be considered, for the reasons explained in the previous section. Part 5 of the Bridge Code assumes that the design loads include for the effects of γ_{f3}, but for combination S-1, $\gamma_{f3} = 1.0$, so the design values are as given in column 7 of table X.16.2. The design envelope for combination S-1 is plotted for a half span in fig. X.17.1.

From Part 5, the design resistance of a 100×22 mm diameter stud connector in Grade 40 concrete is:

$$P_u/\gamma_m = 139/1.85 = 75 \text{ kN}$$

At the internal support:

$$Q = 969 \text{ kN/m}$$

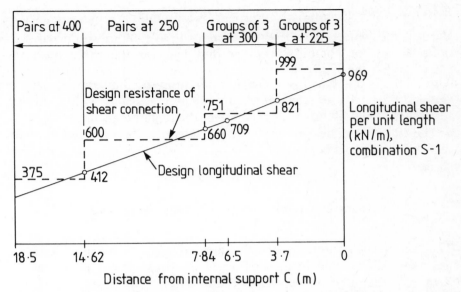

Fig. X.17.1 Envelope of design longitudinal shear force for outer girder in 37 m span

The number of connectors required per metre is:

$$N_c = 969/75 = 12.9$$

So, for connectors in groups of three, the required spacing is:

$$s = 3/12.9 = 0.232 \text{ m}.$$

Connectors should therefore be provided in **groups of three at 225 mm centres** at the internal support. This spacing should be continued for 10% of the span, or 3.7 m (5/5.3.3.6).

At 3.7 m from the internal support, the longitudinal shear in fig. X.17.1 reduces to 821 kN/m, so from 5/5.3.3.6, the connector resistance required at this point must be at least $821 \times 100/110 = 746$ kN/m.

Taking $Q = 746$ kN/m, the required spacing of studs in groups of three is:

$$s = 3 \times 75/746 = 0.302 \text{ m}$$

Hence, use **studs at 300 mm centres in groups of three,** giving a design resistance of $746 \times 302/300 = 751$ kNm.

From 5/5.3.3.1, the longitudinal spacing of 100×22 mm diameter connectors should not exceed 400 mm. The design resistance of **pairs of connectors at 400 mm spacing** is

$$Q = 2 \times 75/0.4 = 375 \text{ kN/m}$$

From fig. X.17.1, this arrangement is suitable at midspan where $Q = 270 \text{ kN/m}$ and to a point on the envelope where, from 5/5.3.3.6:

$$Q = 375 \times 110/100 = 412 \text{ kN}$$

This point is 14.62 m from the internal support. Similarly, it can be shown that **pairs of connectors at 250 mm spacing**, with a design resistance of 600 kN/m, can be used over the remaining length of 6.78 m.

The envelope of design resistance provided by the proposed arrangement of connectors is plotted in Fig. X.17.1 with the design longitudinal shear force.

X.18 Transverse reinforcement for longitudinal shear, U-1

The transverse reinforcement calculated in sections X.8 and X.13 for transverse global and local bending of the deck slab is shown in fig. X.18.1 for the outer girder. With 35 mm cover to the soffit and 16 mm diameter transverse reinforcement in the bottom of the slab, the clear distance between the underside of the stud head and the bottom transverse reinforcement is 34 mm for a 100 mm stud with a 15 mm deep head. This is less than the 40 mm minimum recommended in 5/5.3.3.3 to prevent separation, so the minimum stud height that can be used is 110 mm. The nominal static strength of the connector is unchanged. If the studs are located as shown, the length of the shear surface, L_s, for the failure plane EFGH (type 2 in Part 5) is 550 mm.

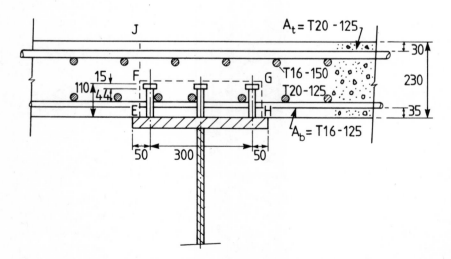

Fig. X.18.1   Detail at shear connection near an internal support

X.18.1 Design longitudinal shear

The equations given in Clauses 6.3.3.2 and 6.3.3.3 of Part 5 (equations (8.58) to (8.61) in section 8.6) assume that γ_{f3} is included in the design loadings, so the design longitudinal shears for combination U-1 are the values given in column 7 of table X.16.2 multiplied by $\gamma_{f3} = 1.1$.

X.18.2 *Shear plane type 1-1, internal support (EFJ, fig. X.18.1)*

Design longitudinal shear force per unit length on plane 1-1 is:

$$q_p = 1168 \times 1.1/2 = 642 \text{ kN/m}$$

and $L_s = 230 \text{ mm}$, $A_e = (A_b + A_t) = 3853 \text{ mm}^2/\text{m}$, $f_{ry} = 460 \text{ N/mm}^2$, $f_{cu} = 40 \text{ N/mm}^2$.

From equations (8.58) and (8.60):

$$q_p \not> (0.9 \times 230) + (0.7 \times 3853 \times 460 \times 10^{-3}) = 1448 \text{ kN/m}$$

and:

$$q_p \not> 0.15 \times 230 \times 40 = 1380 \text{ kN/m}$$

so **reinforcement provided for transverse bending is sufficient** to prevent longitudinal shear failure on plane EFJ.

X.18.3 *Shear plane type 2-2, internal support (EFGH, fig. X.18.1)*

$$q_p = 1168 \times 1.1 = 1284 \text{ kN/m}$$

and

$$L_s = 550 \text{ mm}, \quad A_e = 2A_b = 2680 \text{ mm}^2/\text{m}.$$

From 5/6.3.3.2:

(1) $q_p \not> (0.9 \times 550) + (0.7 \times 2680 \times 460 \times 10^{-3}) = 1358 \text{ kN/m}$
(2) $q_p \not> 0.15 \times 550 \times 40 = 3300 \text{ kN/m}$.

Therefore, the **reinforcement provided is sufficient**.

X.18.4 *Interaction between longitudinal shear and transverse bending*

Transverse bending moments in the deck slab above the outer girder are almost always hogging, i.e. they cause compression on the faces EF and GH of the shear plane around the connectors (fig. X.18.1). When this is the case, no account need be taken of the effect of transverse bending on the longitudinal shear resistance of the slab (5/6.3.3.3).

However, in midspan regions over internal girders, live loading can cause transverse global sagging moments in the deck slab which exceed the transverse hogging moments due to dead and superimposed dead loading. The shear resistance of a failure plane passing around the connectors (EFGH in fig. X.18.1) is then reduced. As an example, we assume a transverse bending moment of 50 kNm/m (including γ_{f3}) exists in the deck slab above the outer girder at a point 6 m from midspan where the design longitudinal shear force per unit length at the interface, q, is $1.1 \times 580 = 638$ kN/m.

Assuming the lever arm for transverse bending in the slab is 0.8 times the effective depth, or $0.8 \times 187 = 150$ mm, the area of transverse reinforcement required for transverse sagging bending is:

$$A_{bb} = 50 \times 10^6/(150 \times 0.87 \times 460) = 833 \text{ mm}^2/\text{m}$$

The area available to resist longitudinal shear is:

$$A_{bv} = 1340 - 833 = 507 \text{ mm}^2/\text{m}$$

From 5/6.3.3.3(b), for plane EFGH:

$$q_p \not> (0.9 \times 550) + (1.4 \times 507 \times 460 \times 10^{-3}) = 822 \text{ kN/m}$$

This is greater than the design longitudinal shear of 638 kN/m, so **the reinforcement is sufficient**.

X.19 Design of shear connection for repeated loading

The shear connector spacing determined in section X.17 and summarised in fig. X.17.1 is now checked for fatigue failure of the connector weld using the simplified procedure for highway bridges given in Clause 8 of BS 5400 Part 10. This procedure is applicable to stud connector welds which are classified as Class S details in table 17(c) of Part 10. In the method, the range of stress on the weld throat produced by the passage of a single standard fatigue vehicle of 320 kN is compared with a limiting stress range σ_H given in Part 10 for different loaded lengths and different types of road.

The first stage is to determine the maximum and minimum vertical shears at predefined positions on the girder due to the passage of the standard fatigue vehicle (SFV) along each slow lane and each adjacent lane in turn. For this purpose the traffic lanes are those actually marked out on the carriageway. The slow, adjacent and fast lanes are as defined in fig. 5 of Part 10. For the dual three-lane motorway bridge used in this example, the designation of lanes for fatigue assessment is as shown in fig. X.19.1.

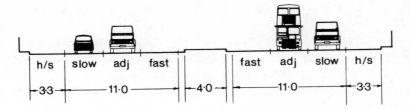

Fig. X.19.1 Designation of traffic lanes for fatigue assessment

The SFV has the same axle spacing as the shortest HB vehicle (10/7.2.2.2), but its width, load, and wheel arrangement on each axle is different (fig. 10.1). With appropriate factoring, it is sometimes possible to use the results of previous global analysis for the HB vehicle to determine the maximum and minimum vertical shears due to the SFV. However, where the positions of the HB vehicle and the SFV on the carriageway differ appreciably, as in this example, further analyses may be required for the SFV.

X.19.1 *Transverse position of standard fatigue vehicle on carriageway*

The centreline of the SFV is assumed to travel along a path parallel to, and within 300 mm of, the centreline of the slow or adjacent lanes shown in fig. X.19.1. The hard shoulder is considered to be unloaded for fatigue assessment, so longitudinal shear forces on the connectors on the outer girder due to the SFV in the slow lane will be small. For this reason, the fatigue assessment in this example will now be made on the connectors on the first interior girder, assuming the section properties and shear connector arrangement are the same as for the outer girder. The most severe loading on the first interior girder is produced with the centre of the SFV positioned 300 mm from the centre line of the slow lane, as shown in fig. X.19.2. The range of vertical shear on the girder is not changed significantly if the SFV takes a path along the adjacent lane in the same carriageway or in the slow or adjacent lanes of the second carriageway, so it is here only considered in one lane.

X.19.2 *Longitudinal position of standard fatigue vehicle for vertical shear range*

The most likely location on a girder for fatigue failure of connectors is near midspan, where the number of connectors required for static loading is small, and the range of live load shear, which determines the fatigue life of a connector weld, is high. Therefore, sections which should be checked are:

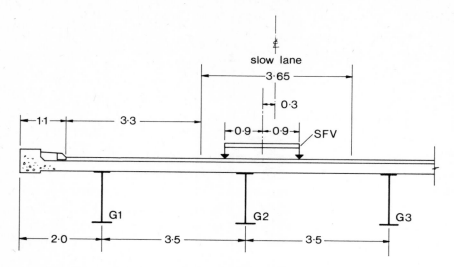

Fig. X.19.2   Lateral location of vehicle for fatigue check on girder G2

(1) within the span, where the connector density changes, and
(2) at the points of maximum shear.

In this example, three sections are examined: at the internal support C, and at 7.84 m and 14.62 m from C (fig. X.17.1).

Range of vertical shear at C. Using the influence line for vertical shear in the 37 m span, fig. X.16.1, the position of the SFV for maximum shear at C is with the vehicle positioned in span BC with the leading axle at C. The minimum vertical shear at C occurs with the vehicle in span AB, with the axles positioned at 6.0, 7.8, 13.8 and 15.6 m from B.

7.84 m from C in span BC. Maximum vertical shear occurs with the leading axle 7.84 m from C and the remaining axles nearer B than C. Minimum vertical shear occurs with the leading axle at the section and the remaining axles nearer to C.

14.62 m from C in span BC. Maximum vertical shear occurs with the leading axle 14.62 m from C and the remaining axles nearer B than C. Minimum vertical shear occurs with the leading axle at the section and the remaining axles nearer to C. These positions are shown in fig. X.19.3.

Impact allowance. As there is no discontinuity in the road surface within the 37 m span, no further allowance for impact is required (10/7.2.4).

X.19.3 *Nominal vertical shear forces due to standard fatigue vehicle*

The maximum and minimum nominal vertical shear forces determined

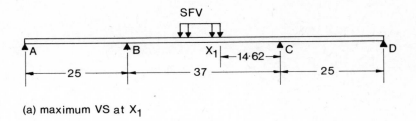

(a) maximum VS at X_1

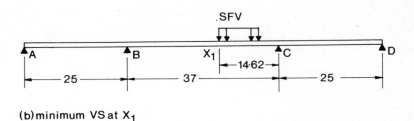

(b) minimum VS at X_1

Fig. X.19.3   Positions of fatigue vehicle for maximum range of shear at point X_1

from the grillage analysis for the passage of the SFV along the slow lane
are summarised in table X.19.1 for the three sections considered.

Table X.19.1   Nominal vertical shears due to standard fatigue vehicle – first interior girder

Section	Nominal vertical shear (kN)		
	Max.	Min.	Range
Support C	170	−19	189
7.84 m from C in span BC	118	−19	137
14.62 m from C in span BC	90	−55	145

X.19.4 *Nominal stress range in connector weld*

The range of longitudinal shear on each stud, Q_c, is calculated in table
X.19.2, using the appropriate values of $A\bar{y}/I$ from tables X.6.1 and X.6.2,
for $\alpha_e = 6.6$ and the vertical shear range from table X.19.1, as follows:

$$Q_c = (VA\bar{y}/I)\left[\frac{\text{longitudinal spacing of connectors(m)}}{\text{number of connectors across flange width}}\right] kN$$

Table X.19.2 Nominal stress range in connector weld (N/mm²)

Section	Range of VS (kN)	$A\bar{y}/I$ (m^{-1})	Range of LS (kN/m)	Stud spacing (m)	No. of studs	Range of LS per stud (kN)	Range of stress on weld (N/mm²)
Support C	189	0.730	138	0.225	3	10.35	31.6
7.84 m from C	137	0.739	101	0.25	2	12.66	38.7
14.62 m from C	145	0.739	107	0.4	2	21.4	65.4
				0.25	2	13.375	40.9

Then, from 10/6.4.2, the range of stress in the connector weld is:

$$\sigma_r = 425 \left[\frac{\text{longitudinal shear per stud, } Q_c}{\text{nominal static strength of connector, } P_u} \right] \text{N/mm}^2$$

where $P_u = 139$ kN (table X.3.1).

X.19.5 *Limiting stress range in connector weld*

The limiting stress on the connector weld is determined from fig. 8(a) of Part 10, using the appropriate loaded length and the curve given for a class S detail on a dual three-lane motorway.

At support C. For a loaded length of 37 m, the limiting stress range is:

$$\sigma_H = 36 \text{ N/mm}^2 > 31.6 - \text{satisfactory}$$

At 7.84 m from C in span BC. For a loaded length of 29.16 m, the limiting stress range is:

$$\sigma_H = 39 \text{ N/mm}^2 > 38.7 - \text{satisfactory}$$

At 14.62 m from C in span BC. For a loaded length of 22.38 m, the limiting stress range is:

$$\sigma_H = 42 \text{ N/mm}^2$$

If this result is compared with the range of stress on the connector weld given in table X.19.2, it will be seen that the proposed arrangement of pairs of connectors at 400 mm centres from midspan to the section considered is not satisfactory. To satisfy the fatigue criteria, the spacing of pairs of connectors over this 7.76 m length of girder at midspan should not exceed $42 \times 400/65.4 = 257$ mm. Therefore, **reduce spacing of pairs of connectors to 250 mm** over this section to prevent fatigue failure in connector welds.

The results obtained above are typical for longitudinal composite girders, where it will usually be found that the number and spacing of connectors required for static loading at the SLS is sufficient to prevent fatigue failure, except near midspan.

Appendix Y: Worked Example

Box Girder Footbridge with Precast Deck Slab using High-strength Friction-grip Bolts as Shear Connectors

Y.1 Description of the footbridge and introduction

Design calculations are given for a simply supported composite box girder footbridge spanning 36 m over a river. The bridge superstructure comprises an open-type steel box girder acting compositely (for superimposed dead and live loading only) with precast concrete deck slab units, nominally 2 m in length. The shear connection is achieved by 16 mm diameter HSFG bolts passing through the precast slab and the top flange of the steel beam. The principal dimensions of the midspan cross section shown in fig. Y.1.1 were obtained from preliminary calculations (not given), and were used to calculate the elastic properties of the steel and composite cross sections given in tables Y.3.1 and Y.3.2.

No detailed calculations are given for the design of the steel box and its stiffeners, or for the deck slab (other than for longitudinal shear), and it is assumed that prior to composite action being developed, the compression flanges of the steel box girder are adequately restrained by cross-frames or diagonal bracing against lateral-torsional buckling. In practice, the bottom flange and webs of the steel box would be fabricated as stiffened plates, and their thickness may vary over the end regions of the beam, but in this example, only the midspan cross section shown in fig. Y.1.1 is considered.

The design of transverse joints between adjacent precast concrete deck slab units is discussed in section 6.2.1.

Throughout this example, as in BS 5400, the term 'characteristic

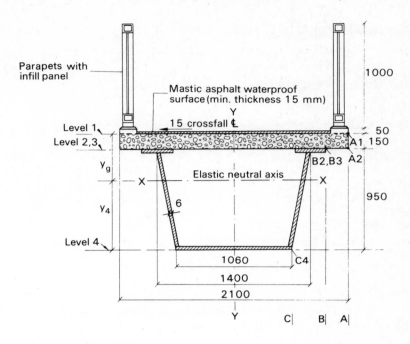

Fig. Y.1.1   Cross section of footbridge superstructure

Steel box section:

> Top flanges (each): 290x30 (gross), 250x30 (net)
> Webs: 6 mm
> Bottom flange, midspan :   25 mm

Deck slab:

> Precast concrete slab units, nominal length 2 m, f_{cu} = 30 N/mm^2

strength' is used for concrete, reinforcement and shear connectors, but 'nominal' is used to describe the strength of structural steel and for loadings. The reasons for this are explained in section 7.1. References to Clauses in BS 5400: Part 5, 1979 are given thus, 5/6.2.1, where '6.2.1' is the clause number. A similar notation is used for clauses in other parts of the Bridge Code.

Y.2 Loading, material properties and limit-state requirements

Nominal loads
Dead loads:
  – unit weight of structural steel = 77 kN/m^3
  – unit weight of reinforced concrete = 25 kN/m^3

Superimposed dead loads:
- mastic asphalt waterproofing = $0.45 \, \text{kN/m}^2$
  (average 22.5 mm thickness)
- parapets = 0.4 kN/m (each)

Live loads:
- footway loading as specified in Part 2 of the Bridge Code.

The uniformly distributed loading (udl) is $k \times 5.0 \, \text{kN/m}^2$, where k is 1.0 for loaded lengths not exceeding 30 m. For loaded lengths exceeding 30 m:

$$k = \frac{\text{HAU lane loading (kN/m) for loaded length, } l, \text{ metres}}{30[1 + \{(L - 30)/300\}]}$$

Thus, for a loaded length of 36 m:

$$w = 5 \times 27.5/[30(1 + 0.02)] = 4.5 \, \text{kN/m}^2$$

wind and temperature loadings as specified in Part 2 of the Bridge Code for an inland site in the West Midlands of Britain at 100 m above sea level, for a design life of 120 years.

A horizontal load on each parapet of 1.4 kN/m acting one metre above footway level should also be applied. This affects only the design of the parapet and the transverse bending reinforcement in the deck slab, and is not considered further in this example.

Material properties
Structural steel:
- Grade 43C (mild steel) to BS 4360.
From Part 4, short-term modulus $E_c = 28 \, \text{kN/mm}^2$, $v_c = 0.2$.

Concrete:
- Grade 30, with $f_{cu} = 30 \, \text{N/mm}^2$ at 28 days.
From Part 4, short-term modulus $E_c = 28 \, \text{kN/mm}^2$, $v_c = 0.2$.

Free shrinkage:
$\varepsilon_{cs} = 200 \times 10^{-6}$ (5/5.4.3).

Reinforcement:
hot-rolled deformed bars with $f_{ry} = 460 \, \text{N/mm}^2$.

Shear connectors:
high-strength friction-grip bolts to BS 4395:
Part 1, and BS 4604.

Coefficients of thermal expansion:
   taken as 12×10^{-6} per degC for concrete, reinforcement and structural steel.

Modular ratios
   for short-term loading, $\alpha_{e1} = 205/28 = 7.3$.
   for long-term loading, $\alpha_{e2} = 2\alpha_{e1} = 14.6$ (5/4.2.3).
   for shrinkage, from 5/5.4.3, $\phi_c = 0.4$, so $\alpha_{e3} = 7.3/0.4 = 18.3$.

   For simplicity, α_{e3} will be taken as 14.6 in calculating the effects of shrinkage, as permitted by 5/5.4.3.

Concrete cover to reinforcement
   top of slab: 35 mm.
   soffit of slab: 35 mm.

Design loads. In this example, the term 'design loads' is always used to mean the nominal loads multiplied by γ_{fL}. For steel elements designed to Part 3 of the Bridge Code, these design loads must be less than or equal to the design resistance given in Part 3.

   For concrete elements designed to Part 4, and for shear connectors and transverse reinforcement designed to Part 5 of the Bridge Code, the design loads are multiplied by γ_{f3} to obtain the design load effects, which are then compared with the design resistances given in Part 4 or Part 5.

   The reasons for this are explained in section 7.1.2.

Design strengths. The strength limitations recommended in Parts 3, 4 and 5 of the Bridge Code are summarised in table Y.2.1. In most circumstances, it will not be necessary to check that all of these are satisfied.

Table Y.2.1   Design strengths

	Design strengths (N/mm²)	
Material	Serviceability	Ultimate
Steel (tension): $f_y/\gamma_m\gamma_{f3}$	$245/1.0 \times 1.0 = 245$	$245/1.05 \times 1.1 = 212$
Steel (compression):	As above when the steel flange acts compositely with the concrete deck (3/4.3.3)	
Steel (shear)	—	3/9.9.2
Concrete* (compression)	$0.5f_{cu} = 15$	$0.67f_{cu}/\gamma_m = 13.4$, or $0.4f_{cu} = 12$
Reinforcement*	$0.75f_{ry} = 345$	$f_{ry}/\gamma_m = 400$

   * The limiting stresses for concrete and reinforcement are taken directly from Part 4 of the Bridge Code and do not include γ_{f3}.

Y.3 Elastic properties of the steel and composite cross sections

Shear lag. In design to Part 3, the effects of shear lag need only be considered where the additional requirements of Clause 9.2.3.1 have to be satisfied at the SLS. None of these apply to this example, so the full flange breadth is used in calculating section properties.

Elastic properties of the steel and composite cross sections. Section properties for bending about the x–x axis (fig. Y.1.1) are given in table Y.3.1 for the steel and composite cross sections, using the net area of steel top flange (to allow for bolt holes) and the full breadth of the concrete

Table Y.3.1 Elastic properties for bending about the x–x axis

Property	Units	Steel box alone	Composite section $\alpha_e = 14.6$	Composite section $\alpha_e = 7.3$
Area, A	mm²	52 240	73 815	95 390
Height of centroid above soffit, y_4	mm	372	563	667
$I_{xx} \times 10^{-6}$	mm⁴	9005	15 573	19 136
$Z_1 \times 10^{-6}$	mm³	—	423.4	322.6
$Z_2 \times 10^{-6}$	mm³	—	587.5	493.6
$Z_3 \times 10^{-6}$	mm³	15.58	40.24	67.62
$Z_4 \times 10^{-6}$	mm³	24.21	27.66	28.69
$A\bar{y}/I$ (interface)	m⁻¹	—	0.640	0.807

flange. For the composite sections the concrete is assumed to be uncracked and unreinforced and the presence of the parapet plinths is neglected.

The section moduli Z_1, Z_2, Z_3, Z_4 refer to levels 1 to 4 in fig. Y.1.1, and the values of A, Z, and I are in terms of the equivalent steel section, except that Z_1 and Z_2 are in 'concrete' units.

Table Y.3.2 Elastic properties for bending about y–y axis

Property	Units	Composite section $\alpha_e = 7.3$
$I_{yy} \times 10^{-6}$	mm⁴	29 830
$Z_A \times 10^{-6}$	mm³	207.4
$Z_B \times 10^{-6}$	mm³	35.43
$Z_C \times 10^{-6}$	mm³	56.28

Table Y.3.2 gives the section properties of the composite box for bending about the y–y axis (for transverse wind load effects). The modular ratio is taken as 7.3, appropriate to short-term loading. The effects of wind loading on the steel box alone (a condition which may arise during erection) have not been considered. The values of I and Z are given in terms of 'steel' units, except that Z_A is in 'concrete' units. The section moduli Z_A, Z_B, Z_C refer to reference lines A, B, C respectively in fig. Y.1.1.

Y.4 Nominal loads, midspan moments and vertical shear forces at the ends of the beam

Values of the nominal load per metre of superstructure, the nominal midspan bending moments and the nominal vertical shears due to the principal dead and live loadings (assumed to be uniformly distributed over the span) are calculated in table Y.4.1.

Table Y.4.1 Nominal loads, midspan moments and vertical end shears

Loading type	Nominal load (kN/m)	Nominal midspan BM (kNm)	Nominal VS at support (kN)
Dead:			
Steel box	$52\,240 \times 10^{-6} \times 77 = 4.02$	652	72.3
Cross-frames and stiffeners	Allow 10% = 0.40	65	7.2
Concrete deck	$2100 \times 150 \times 25 \times 10^{-6} = 7.88$	1277	141.8
Superimposed dead:			
Mastic asphalt	$0.45 \times 1.77 = 0.80$	130	14.4
Parapets	$2 \times 0.4 = 0.80$	130	14.4
Live load:			
Pedestrians	$4.5 \times 1.77 = 7.97$	1337	148.6

Y.5 Bending moments at midspan due to principal loads

Design midspan bending moments due to the principal dead and live

Table Y.5.1 Design midspan bending moments due to principal loads

Loading	Nominal midspan BM (kNm)	Limit state and load combination					
		U-1		U-2, U-3		S-1, S-2, S-3	
		γ_{fL}	BM	γ_{fL}	BM	γ_{fL}	BM
Dead:							
Steel box	717	1.05	753	1.05	753	1.0	717
Concrete deck	1277	1.15	1469	1.15	1469	1.0	1277
Superimposed dead:							
asphalt	130	1.75	228	1.75	228	1.2	156
parapets	130	1.20	156	1.20	156	1.0	130
Live:	1337	1.50	2066	1.25	1671	1.0	1337
Carried by steel box alone	1994		2222		2222		1994
Carried by composite section	1597		2450		2055		1623

loads are calculated in table Y.5.1 from the nominal moments in table Y.4.1 and the partial load factors γ_{fL} for the limit state and load combination considered. Values of γ_{fL} are given in Part 2 of the Bridge Code and are summarised in section 7.3.

Y.6 Bending stresses at midspan due to principal loads

The distance from the elastic neutral axis of the composite section to the compressive edge of the web, measured parallel to the web, is:

$$(950\text{-}30\text{-}563)/\cos 10.47° = 363 \text{ mm} = 60.5t_w$$

This exceeds the limit of $28t_w(355/\sigma_{yw})^{\frac{1}{2}}$ for compact sections given in Clause 9.3.7 of Part 3, so the section is slender. Elastic analysis is therefore used for analysis of sections at the ULS and for determining the bending resistance of the steel and composite sections. Longitudinal stiffeners will be required on the webs and bottom flange to enable the full thickness of web and flange plates to be considered as effective.

The stresses at levels 1 to 4 due to design ultimate loadings are

Table Y.6.1 Longitudinal stresses at midspan (N/mm²) at ULS due to principal design loads

Loading	U-1				U-2 and U-3			
Level	1	2	3	4	1	2	3	4
Dead	—	—	−143	92	—	—	−143	92
Superimposed dead ($\alpha_e = 14.6$)	−0.91	−0.65	−10	14	−0.91	−0.65	−10	14
Live ($\alpha_e = 7.3$)	−6.40	−4.19	−31	72	−5.18	−3.39	−25	58
	−7.31	−4.84	−184	178	−6.09	−4.04	−178	164

calculated in table Y.6.1 for combinations U-1, U-2, U-3 using the design moments given in table Y.5.1 and section moduli given in table Y.3.1.

The compressive stresses at level 3 of the composite box under dead plus superimposed dead plus live loading are acceptable because buckling of the steel flanges in the composite box is prevented by the shear connection. But, prior to composite action being developed, the compression flanges of the steel box would have to be braced sufficiently to enable them to develop a stress of 143 N/mm² resulting from the self weight of the steel box and the precast concrete deck slabs. The additional stresses due to wind loading, temperature, and shrinkage modified by creep are now considered.

Y.7 Wind loading effects

In Part 2 of the Bridge Code, the effects of wind loading acting in combination with other sources of loading are considered only in combination 2. Combination 2 consists essentially of all compatible permanent and primary live loads, including the effects of wind, but excluding the effects of temperature (which are considered only in combination 3). The wind gust speed to be used in deriving the wind loading on the structure depends on the other loadings with which it is combined. The effects of wind loading during erection have not been considered.

Wind gust speed
(a) *For bridges without live load.* The equation for the gust speed given in Part 2 of the Bridge Code is:

$$v_c = K_1 v S_1 S_2$$

For the site location and altitude specified in section Y.S, assuming the superstructure to be 5 m above ground level: $v = 28$ m/s, $S_1 = 1.0$, $S_2 = 1.43$, and $K_1 = 1.0$. Thus:

$$v_c = 40 \text{ m/s}$$

(b) *For bridges with live load.* In Part 2 of the Bridge Code it is assumed that the live loading on a bridge decreases at gust speeds in excess of 35 m/s. For footbridges with live load, the maximum gust speed is the lesser of 35 m/s and $0.7 K_1 v S_1 S_2$ as calculated in (a) above. Thus, v_c is taken as 28 m/s.

Nominal transverse wind load, P_t. The equation for P_t given in section 7.3.3 (and in Part 2 of the Bridge Code) is:

$$P_t = q C_D A_1 \text{ (newtons) acting horizontally}$$

where $q = 0.613 v_c^2$ (N/m^2)

When live load is not present, the depth of superstructure exposed to the wind (assuming parapets 1 m high with mesh infill) is equal to the overall depth of the composite box plus parapet (i.e. $1.15 + 1.0 = 2.15$ m).

When live load is present, the resistance afforded by it is assumed (for footbridges) to be equivalent to an additional depth of 1.25 m of superstructure, measured above the surface of the footway. Thus, the net exposed area, A_1 is:

$$(36 \times 2.15) = 77.4 \text{ m}^2 \text{ when live load is not present,}$$

and:

$(36 \times 2.4) = 86.4 \text{ m}^2$ when live load is present.

The drag factor, C_D, for a footbridge of the cross section used in this example ($b/d > 1.0$) is given in Part 2 of the Bridge Code as 2.0 for both the loaded and the unloaded bridge, so nominal transverse wind loads are:

(1) with live load present,

$$P_t = 0.613 \times 28^2 \times 2.0 \times 86.4 \times 10^{-3} = 83 \text{ kN}$$

(2) with live load not present,

$$P_t = 0.613 \times 40^2 \times 2.0 \times 77.4 \times 10^{-3} = 152 \text{ kN}.$$

Nominal vertical wind load. From Part 2, the upwards or downwards nominal vertical wind load, P_v, is given by:

$$P_v = qC_LA_3 \text{ (newtons)}$$

where C_L is a lift coefficient, equal to 0.4 for a superstructure of the proportions used here, and A_3 is the net plan area of the deck ($2.1 \times 36 = 75.6 \text{ m}^2$).

The nominal vertical wind loading is thus:

(1) with live load present,

$$P_v = 0.613 \times 28^2 \times 0.4 \times 75.6 \times 10^{-3} = 15 \text{ kN}$$

(2) with no live load present,

$$P_v = (40/28)^2 \times 15 = 31 \text{ kN}.$$

Nominal longitudinal wind forces. These have no effect on the longitudinal stresses in the bridge superstructure, and are therefore not considered further.

Application of wind loads. From Part 2, the only combinations of P_t and P_v that need be considered in this simple type of structure are:

(1) P_t alone
(2) P_t with P_v acting downwards.

P_v is not considered to act alone.

Bending stresses at midspan due to a unit transverse wind load of 10 kN uniformly distributed over the span. The midspan bending moment due to this unit loading is:

$$M_t = 10 \times 36/8 = 45 \text{ kNm}$$

Using the section moduli in table Y.3.2, the corresponding bending stresses at reference lines A, B and C in the cross section (as defined in fig. Y.1.1) are:

$$\sigma_A = 45/207.4 = \pm 0.22 \text{ N/mm}^2$$
$$\sigma_B = 45/35.43 = \pm 1.27 \text{ N/mm}^2$$
$$\sigma_C = 45/56.28 = \pm 0.80 \text{ N/mm}^2$$

Bending stresses at midspan due to a unit vertical wind load of 10 kN (*acting downwards*), *uniformly distributed over the span*. The midspan moment, $M_v = 45$ kNm.

Using the section properties in table Y.3.1 for $\alpha_e = 7.3$, the corresponding bending stresses at levels 1 to 4 (fig. Y.1.1) are:

$$\sigma_1 = 45/322.6 = -0.14 \text{ N/mm}^2$$
$$\sigma_2 = 45/493.6 = -0.09 \text{ N/mm}^2$$
$$\sigma_3 = 45/67.62 = -0.67 \text{ N/mm}^2$$
$$\sigma_4 = 45/28.69 = 1.57 \text{ N/mm}^2.$$

Bending stresses at midspan due to design wind loads. These stresses are computed in table Y.7.1 for various points on the cross section using the stresses due to unit wind loads previously calculated and the appropriate load factors for the ultimate limit state (combination U-2).

Table Y.7.1 Stresses at midspan due to design wind loading (combination U-2)

		Ratio of nominal load to unit load	γ_{fL}	Stress (N/mm²)			
				A1	A2	B3	C4
Stresses due to unit loads	P_t	1.0	—	±0.22	±0.22	±1.27	±0.80
	P_v	1.0	—	−0.14	−0.09	−0.67	+1.57
Max. wind, no live load	P_t	15.2	1.4	±4.68	±4.68	±27.0	±17.0
	P_v	3.1	1.4	−0.61	−0.39	−2.9	+6.8
	Σ			−5.29	−5.07	−29.9	+23.8
Wind with live load	P_t	8.3	1.1	±2.01	±2.01	±11.6	±7.3
	P_v	1.5	1.1	−0.23	−0.15	−1.1	+2.6
	Σ			−2.24	−2.16	−12.7	+9.9

Y.8 Temperature effects

In this example, longitudinal stresses due to temperature effects need only be considered at the ultimate limit state (5/4.2.2 and 3/9.2.1.2), but longitudinal shear forces arising from temperature effects must be considered at the serviceability limit state for the design of the shear

Table Y.8.1 Stresses due to nominal temperature difference, type 1

Depth y_r (mm)	T (°C)	$10^6 \varepsilon_r$ (A.1)	Slice No.	$10^6 \bar{\varepsilon}_r$ (A.3)	b_r (mm)	h_r (A.6) (mm)	F_r (A.7) (kN)	$\bar{y}_r$ (A.4) (mm)	$F_r \bar{y}_r$ (kNm)	f_s (A.14) (N/mm²)	f_c (A.15) (N/mm²)	Q (kN)
0	13									−16.37	−2.24	−156
		48	(1)	102	288	90	541	37	20.0			
90	4									4.36	0.60	77
		42	(2)	45	288	60	159	119	18.9			
150	3.5									4.65	0.64	72
		38.4	(3)	40.2	500	30	124	165	20.4			
180	3.2									4.92	—	27
		0	(4)	19.2	12.2	370	18	304	5.4			
550	0									7.00	—	19
		0	(5)	0	12.2	525	0	—	0			
1075	0									−1.22	—	−39
		0	(6)	0	1060	25	0	—	0			
1100	0									−1.61	—	
Total							$\bar{F} = 842$		Σ 64.7			0 (check)

connectors, and at the ultimate limit state for the design of the deck slab against longitudinal shear failure (5/5.4.1 and 5/6.1.5).

It is assumed that the superstructure is free to translate in the longitudinal direction and that the steel and concrete have the same coefficients of thermal expansion (12×10^{-6} per degree Celsius), so that uniform changes of temperature cause no stress in the superstructure. Thus, only the effects of temperature differences through the depth of the cross section need be considered. The nominal temperature differences are those shown in figs Y.8.1(b) and (e).

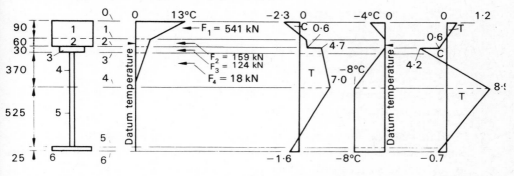

(a) Cross-section (b) Temperature difference (type 1) (c) Forces that would produce free strains due to temperature (d) Longitudinal stresses, N/mm² (type 1) (e) Temperature difference (type 2) (f) Longitudinal stresses, N/mm² (type 2)

Fig. Y.8.1 Nominal temperature differences and temperature stresses

Table Y.8.2 Stresses due to nominal temperature difference, type 2

Slice No.	f_s (N/mm²)	f_c (N/mm²)	Q (kN)
	8.67	1.19	
(1)			77
	−2.74	−0.38	
(2)			−57
	−3.79	−0.52	
(3)			−50
	−2.84	—	
(4)			14
	8.89	—	
(5)			28
	−0.28	—	
(6)			
	−0.72	—	−12
Total			0

$\bar{F} = -834$ kN, $\bar{M} = -334$ kNm.

Detailed calculations are given for type 1 in table Y.8.1, and the results of similar calculations for type 2 are given in table Y.8.2. The method, notation, and sequence of calculations are exactly as explained in Appendix A, and the numbers of the equations used are given in parentheses in the column headings.

Notes on the calculation are now given.

(1) The six slices into which the cross section is divided are shown in fig. Y.8.1(a).

(2) The section properties A, I, and y_g are as given in table Y.3.1 for $\alpha_e = 7.3$.

(3) The sum of the values of $F_r \bar{y}_r$ is 64.7 kNm; putting $y_g = 433$ mm in equation (A.8) then gives $\bar{M} = -299.8$ kNm.

(4) The distribution of the stresses f_s and f_c due to the nominal temperature differences are shown in fig. Y.8.1(d).

(5) The last column in the table gives the longitudinal force Q for each element, calculated from its area and the stresses at its boundaries. These forces should sum to zero, which is a useful check. The longitudinal force required for the design of the shear connection is $-156 + 77 = -79$ kN (elements 1 and 2).

Stresses due to temperature differences. These are only considered in load combination U-3 for which γ_{fL} is 1.0. The design stresses are thus the nominal stresses shown in fig. Y.8.1.

Both temperature cases must be considered, because type 1 increases the maximum resultant compressive stresses in the deck slab at level 1; but at level 2, type 2 gives the worst effect.

Longitudinal shear forces due to temperature differences. Again these are only considered in load combination 3, but the design of the shear connectors must be considered at SLS.

Type 2 causes a net tensile force in the slab of $77 - 57 = 20$ kN which reduces the maximum resultant compressive force due to superimposed dead and live load, so need not be considered further.

Type 1 causes a compressive longitudinal force of 79 kN in the slab. This is distributed uniformly over a distance of one fifth of the span at each end of the beam, as permitted in 5/5.4.2.3 for stud shear connectors. The distribution length ℓ_s is thus $36/5 = 7.2$ m, giving a nominal longitudinal shear force per unit length of 11 kN/m. For the SLS, γ_{fL} is 0.8, so the design longitudinal shear force per unit length for combination S-3 is 8.8 kN/m, and that for combination U-3 including γ_{f3} is 12 kN/m.

The assumption that the force can be distributed uniformly over the end regions of the beam is satisfactory when friction-grip bolts are used as connectors, because their deformation capacity is of the same order as that of stud connectors of the same shank diameter.

Y.9 Shrinkage modified by creep

The effects of shrinkage modified by creep are calculated using the equations in Appendix A.

Approximately 50% of the total free shrinkage strain will occur within two months of the deck slabs being cast. It is assumed that the precast deck slab units do not act compositely within this period, so that for the following calculations, the free shrinkage strain of 200×10^{-6} given in section Y.2 is halved. The modular ratio for shrinkage modified by creep, α_e, is calculated as in Part 5, using a creep reduction factor $\phi_c = 0.4$. Thus:

$$\alpha_e = E_s/\phi_c E_c = 205/(0.4 \times 28) = 18.3$$

This value is used in equations (A.17), (A.18) and (A.20) for calculating the longitudinal force in the concrete and the stresses in the composite section, but it is more convenient (and sufficiently accurate) to use the section properties of the transformed composite section for $\alpha_e = 14.6$, so avoiding the calculation of another set of section properties.

From table Y.3.2: $A = 73\,815 \text{ mm}^2$, $I = 15\,573 \times 10^6 \text{ mm}^4$, $y_g = 537$ mm, and $\varepsilon_{sc} = -0.5 \times 200 \times 10^{-6} = -100 \times 10^{-6}$. From equations (A.18) and (A.19):

$$\bar{F} = -205 \times 100 \times 10^{-6} \times 2100 \times 150/18.3 = -353 \text{ kN}$$

and:

$$\bar{M} = 353(0.537 - 0.075) = 163 \text{ kNm}$$

The nominal stresses in the concrete and steel at levels 1 to 4 are calculated from equations (A.14) and (A.15) as follows:

$f_1 = (20.5 - 4.78 - 5.47)/18.3 = 0.56 \text{ N/mm}^2$
$f_2 = (20.5 - 4.78 - 3.97)/18.3 = 0.64 \text{ N/mm}^2$
$f_3 = -4.16 - 3.97 = -8.13 \text{ N/mm}^2$
$f_4 = -4.16 + 5.49 = 1.33 \text{ N/mm}^2$

In the steel beam, the stresses at levels 3 and 4 increase those due to the principal dead and live loads, so cannot be neglected, but in the deck slab the resulting stresses are tensile so the effects of shrinkage can there be neglected. Longitudinal shear forces due to shrinkage can also be neglected. As with temperature difference, in slender sections the effects of shrinkage modified by creep need only be considered at the ultimate limit state (5/6.1.5), but they have to be included in all load combinations.

The partial load factor for the ULS, γ_{fL}, is 1.2, so the design stresses for combinations U-1, U-2 and U-3 are:

$$\left. \begin{array}{l} f_1 = 0.67 \text{ N/mm}^2 \\ f_2 = 0.77 \text{ N/mm}^2 \end{array} \right\} \begin{array}{l} \text{not required} \\ \text{in design} \end{array}$$
$$f_3 = -9.76 \text{ N/mm}^2$$
$$f_4 = 1.60 \text{ N/mm}^2$$

The distribution over the cross section is as shown in fig. Y.9.1(c).

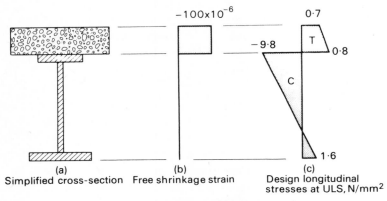

(a)	(b)	(c)
Simplified cross-section	Free shrinkage strain	Design longitudinal stresses at ULS, N/mm²

Fig. Y.9.1   Shrinkage modified by creep

Y.10 Summary of bending stresses at midspan

Table Y.10.1 summarises the longitudinal stresses in the cross section at levels 1 to 4 (fig. Y.1.1) given in tables Y.6.1 and Y.7.1 and figs Y.8.1 and Y.9.1 for load combinations U-1, U-2, and U-3. Load combinations U-4 and U-5 are obviously less severe in this example, so are not considered.

The maximum stresses in steel are those given in line 10 for combination U-2: (dead + superimposed dead + live + reduced wind + shrinkage), and are less than the limiting value of 212 N/mm² for tension and compression given in table Y.2.1. Prior to composite action being developed, the maximum compressive stress in the top flange of the steel box is -143 N/mm² (line 1 of table Y.10.1).

The maximum compressive stress in the concrete is 8.33 N/mm² (line

Table Y.10.1 Summary of design longitudinal stresses (N/mm²), ultimate limit state

Loading	U-1				U-2				U-3			
Level in cross section	1	2	3	4	1	2	3	4	1	2	3	4
(1) Dead (steel box plus conc. slab)	—	—	−143	92	—	—	−143	92	—	—	−143	92
(2) Superimposed dead	−0.91	−0.65	−10	14	−0.91	−0.65	−10	14	−0.91	−0.65	−10	14
(3) Shrinkage	—	—	−10	2	—	—	−10	2	—	—	−10	2
(4) Live	−6.40	−4.19	−31	72	−5.18	−3.39	−25	58	−5.18	−3.39	−25	58
(5) Temp. difference	—	—	—	—	—	—	—	—	−2.24	−0.52	−4	—
(6) Max. wind	—	—	—	—	−5.29	−5.07	−30	24	—	—	—	—
(7) Reduced wind	—	—	—	—	−2.24	−2.16	−13	10	—	—	—	—
(8) Σ 1, 2, 3, 4	−7.31	−4.84	−194	180	−6.09	−4.04	−188	166	−6.09	−4.04	−188	166
(9) Σ 1, 2, 3, 6					−6.20	−5.72	−193	132				
(10) Σ 1, 2, 3, 4, 7					−8.33	−6.20	−201	176				
(11) Σ 1, 2, 3, 4, 5									−8.33	−4.56	−192	166

11, table Y.10.1) which must be multiplied by $\gamma_{f3} = 1.1$ (giving 9.16 N/mm^2) before it is compared with the limiting stress of $0.67f_{cu}/\gamma_m$ ($= 13.5$ N/mm^2) given in Part 4 of the Code. In this case, the design stress is less than the limit in Part 4, but where this is not the case, the bending resistance of the composite section should be calculated by an elastic-plastic analysis as described in section 8.4.3.

Y.11 Vertical shear

Design vertical shear forces at the end of the beam are calculated in table Y.11.1 from the nominal vertical shears given in table Y.4.1.

Table Y.11.1 Design vertical shear at end of beam

| | Nominal VS at end of beam (kN) | Design VS at end of beam (kN) | | | | | |
| | | S-1 | S-2 | U-1 | | U-2 | U-3 |
Loading		γ_{fL}	VS	γ_{fL}	VS	γ_{fL}	VS
Dead:							
Steelwork	79.5	1.0	79.5	1.05	83.5	1.05	83.5
Concrete slab	141.8	1.0	141.8	1.15	163.1	1.15	163.1
Superimposed dead							
Asphalt	14.4	1.2	17.3	1.75	25.2	1.75	25.2
Parapets	14.4	1.0	14.4	1.20	17.3	1.20	17.3
Live	148.6	1.0	148.6	1.50	222.9	1.25	185.8
Totals			401.6		512.0		474.9

Table Y.12.1 Design longitudinal shears per unit length due to live and superimposed dead load

| | $\dfrac{A\bar{y}}{I}$ (m^{-1}) | midspan | | | quarter span | | | support | | |
Loading		S-1	U-1	U-2 U-3	S-1	U-1	U-2 U-3	S-1	U-1	U-2 U-3
Superimposed dead	0.640	0	0	0	11	14	14	21	27	27
Live	0.807	32	48	40	72	109	90	120	180	150
Design LS kN/m		32	48	40	83	123	104	141	207	177
Design LS $\times \gamma_{f3}$		32	53	44	83	135	114	141	228	195

Y.12 Longitudinal shear

The equations given in Part 5 for the resistance of shear connectors and for transverse reinforcement for longitudinal shear assume that the effect of γ_{f3} is included in the design loadings, so the design longitudinal shears

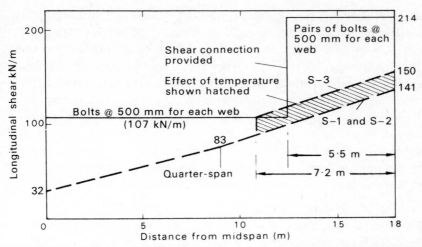

Fig. Y.12.1   Envelope of design longitudinal shear force per unit length at serviceability limit state

in table Y.12.1 and in figs Y.12.1 and Y.12.2 allow for this. Envelopes of design longitudinal shear force per unit length of beam at the interface between the concrete slab and steel beam are required

(1) at the serviceability limit state – for the design of the shear connectors (5/5.4.1 and fig. Y.12.1) and

(2) at the ultimate limit state – for the design of the transverse reinforcement (5/6.1.5 and fig. Y.12.2).

The longitudinal shears due to wind loading are found to be negligible,

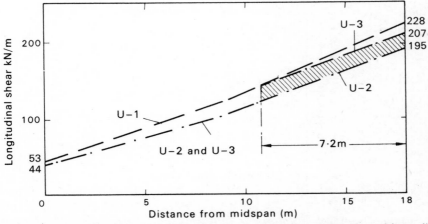

Fig. Y.12.2   Envelope of design longitudinal shear force per unit length at ultimate limit state

so that load combination 2 need not be considered. The longitudinal shear due to temperature difference, including γ_{f3}, is found in section Y.8 to be 8.8 kN/m and 12 kN/m over lengths of 7.2 m adjacent to each support for loadings S-3 and U-3 respectively. Apart from this, the shears for S-1 and S-3 are the same, so only the former are calculated at SLS. The effects of shrinkage are beneficial, and so have been omitted.

The maximum longitudinal shear forces due to live and superimposed dead load are calculated in table Y.12.1 at midspan, quarter span, and a support, using values of $A\bar{y}/I$ from table Y.3.1, and the nominal vertical shear at a support given in table Y.11.1. Account has been taken of the application of live load to part of the span only. The results, with temperature effects added, are plotted in figs Y.12.1 and Y.12.2.

Shear connectors. The method given in Part 5 of the Bridge Code for the design of friction-grip bolts as shear connectors in composite beams assumes that under service loading, composite action is developed by friction between the slab and the steel beam, and that at the ultimate limit state the longitudinal shear force is carried by the bolts in shear alone. In practice the ability to carry further load after first slip derives in part from the bolts bearing directly onto the concrete and partly from the shear carried by the bolt. These assumptions are not valid if the tolerance between the bolt shank and the hole is large and permits excessive slip before the shear strength of the bolt can be mobilised, as may be necessary when precast slabs are used. In these circumstances, the bolt should be grouted after tightening. Alternatively, it should be assumed that the ultimate shear strength is developed only by the interface friction.

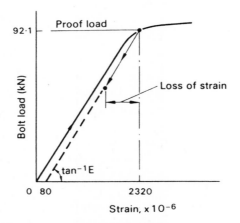

Fig. Y.12.3   Loss of prestrain in friction-grip bolts

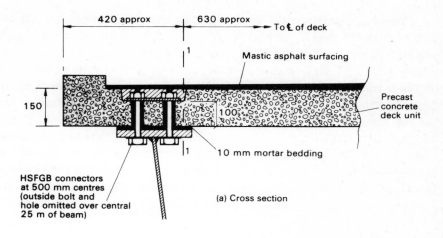

(a) Cross section

HSFGB connectors at 500 mm centres (outside bolt and hole omitted over central 25 m of beam)

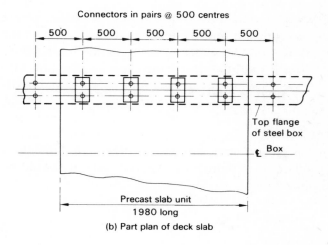

Connectors in pairs @ 500 centres

(b) Part plan of deck slab

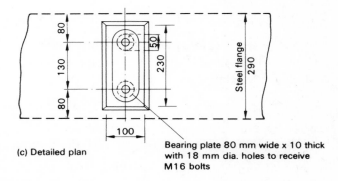

(c) Detailed plan

Bearing plate 80 mm wide x 10 thick with 18 mm dia. holes to receive M16 bolts

Fig. Y.12.4 Detail of pockets in precast slabs to receive HSFG bolts

In this example, the former method, as given in Part 5, is used. No check on the static strength at the ultimate limit state or on the fatigue strength of the connection is then necessary.

From 5/10.2.1 the initial tensile force in the friction-grip bolt is taken as the proof load obtained from BS 4604: Part 1. For 16 mm diameter bolts, the proof load is 92.1 kN. It is assumed that the bolts pass through the top flange of the steel box (30 mm), the deck slab units (100 mm), a 10 mm thick steel bearing plate and a mortar bed that is specified as 10 mm thick (fig. Y.12.4), but is taken as 15 mm thick in this calculation, to allow for possible variation. The grip length is therefore 155 mm, and the initial prestrain in the bolts over this length is:

$$(0.0125/155) + \{92.1/(64\pi \times 205)\} = 0.00232.$$

The elongation over 155 mm is 0.36 mm.

Loss of prestrain. The initial prestrain in the bolt is reduced by shrinkage and creep of the concrete and by relaxation of the bolt itself. Elastic deformation of the steel and concrete occur as the bolt is tightened and so do not constitute a strain loss.

(1) Shrinkage: assuming that 50% of the total free shrinkage has occurred in the precast slab by the time the bolts are tightened, then the free shrinkage is given by:

$$\varepsilon_{sc} = 0.5 \times 200 \times 10^{-6} = 100 \times 10^{-6}$$

The reduction in thickness of the slab and bedding is:

$$\delta_s = 100 \times 10^{-6} \times 115 = 0.012 \text{ mm}$$

(2) Creep of concrete: assuming the creep strain, ε_c, to be as given for prestressed concrete in Part 4 of the Bridge Code, then for $f_{cu} = 30 \text{ N/mm}^2$:

$$\varepsilon_c = 36 \times 10^{-6} \times 40/30 = 48 \times 10^{-6} \text{ per N/mm}^2$$

If the compressive stress in the concrete directly under the bolt bearing plate is limited to 10 N/mm², then the reduction in thickness of the slab and bedding due to creep, δ_c is:

$$\delta_c = 48 \times 10^{-6} \times 10 \times 115 = 0.055 \text{ mm}$$

(3) Relaxation of bolt material: assuming the loss is 3% of the initial strain in the bolt, the relaxation (extension) of the bolt, δ_b, is then:

$$\delta_b = 0.03 \times 0.00232 \times 155 = 0.011 \text{ mm}.$$

Then, the total strain loss in the bolt is

$$(0.012 + 0.055 + 0.011)/155 = 5.0 \times 10^{-4}$$

and, the loss of force in the bolt is:

$$5.0 \times 10^{-4} \times 205 \times 64\pi = 20.6 \text{ kN}.$$

The remaining tension in the bolt is $92.1 - 20.6 = 71.5$ kN, so the loss of prestress is $(20.6/92.1) \times 100 = 22\%$, of which over two-thirds is due to creep of concrete.

To limit the direct compressive stress in the concrete to 10 N/mm^2, the area of the bearing plate or washer must be at least $71.5 \times 10^3/10.0 = 7150$ mm^2 (e.g. plate 85 mm square).

Spacing of connectors. At the serviceability limit state, the design frictional resistance per bolt is:

$$\mu \times \text{bolt tension}/1.2 = 0.45 \times 71.5/1.2 = 26.8 \text{ kN}$$

Shear forces in this structure are low, so it is convenient next to calculate the maximum spacing of the connectors. Clause 5.3.3.1 of Part 5 recommends that this should not exceed the least of 600 mm, three times the slab thickness (450 mm), and four times the height of the connector (600 mm). In this structure, the risk of loss of interaction and of vertical separation between the slabs and the steel beam caused by using a wider spacing of connectors is very low, as there are no local effects from wheel loads.

A maximum spacing of 500 mm is therefore used. Single bolts at this spacing above each steel flange transfer a shear of $4 \times 26.8 = 107$ kN/m. Figure Y.12.1 shows that this is sufficient for the central 25 m of the span. In the end regions it is convenient to use pairs of bolts at the same spacing, even though this provides more shear connection than necessary.

Details of the shear connection are shown in fig. Y.12.4. Holes 50 mm in diameter are left in the concrete slab, and are grouted after the bolts have been finally tightened. The bearing area of the 10 mm steel plates is 7610 mm^2 per bolt.

Transverse reinforcement. The possibility of a shear failure occurring at the interface between the steel beam and the deck slab is prevented by the shear connectors, but longitudinal shear failure is still possible by the formation of planes within the slab itself. In Part 5, this is considered at the ultimate limit state and for this example, the only possible plane of shear failure is through the full depth of the slab (plane 1-1, fig. Y.12.4a).

For the end 3.6 m of the beam (10% of the span), the design longitudinal shear force per unit length of each web (from fig. Y.12.2) is $228/2 = 114$ kN/m. Hence, the design shear for the plane 1-1 of length (L_s) 100 mm is:

$$q_p = 114 \times 630/1050 = 68 \text{ kN/m (approx.)} \qquad (Y.12.1)$$

assuming this plane to be 630 mm from the longitudinal centreline of the deck. This shear is a low value, so it is convenient to next calculate the shear resistance of the minimum quantity of transverse reinforcement allowed in Part 5 of the Bridge Code. This is given in 5/6.3.3.4 as:

$$A_e \not< 0.8 \, sh_c/f_{ry} = 0.8 \times 100/460 = 0.17 \text{ mm}^2/\text{mm}$$

From 5/6.3.3.2, for plane 1-1 with this reinforcement:

$$q_q \not> k_1 s L_s + 0.7 \, A_e f_{ry}$$

that is:

$$q_p \not> [(0.9 \times 100) + (0.7 \times 0.17 \times 460)] = 145 \text{ kN/m} \quad (Y.12.2)$$

From equation (Y.12.1), this is satisfied everywhere. The area, A_e, of steel is only 0.17% of the cross-sectional area of the slab, so the bars are likely to be at the maximum spacing allowed.

From 5/6.3.3.7, the maximum spacing is then 600 mm but 500 mm would be more convenient here. At least half of the steel has to be placed near the bottom of the slab (5/6.3.3.4), so 8 mm high yield bars top and bottom at 500 mm pitch are proposed. This gives $A_e = 0.201$ mm^2/mm, which is sufficient.

In practice, the design of the precast deck slab for handling stresses and the footway loading would lead to the provision of more top reinforcement than this, and bottom reinforcement would probably be provided at a closer spacing.

Y.13 Vibration serviceability

The response of the bridge deck to vibrations induced by pedestrians using the bridge is calculated using the method given in Appendix C of Part 2 of the Bridge Code. The relevant transformed section properties of the composite box are those given in table Y.3.1 for a modular ratio of 7.3 (i.e. derived using the short-term modulus of elasticity and assuming the concrete to be uncracked and unreinforced). The stiffening effect of the parapets is here neglected.

The equation given in section 8.8.2 for the fundamental natural frequency of a simply-supported span of length L is:

$$f_0 = (\pi/2)(E_s I g/W L^3)^{\frac{1}{2}} \qquad (8.67)\text{bis}$$

where EI is the flexural rigidity of the full width of the deck and W is the nominal dead plus superimposed dead load. From table Y.4.1, $W = 13.9 \times 36 = 500$ kN.   Putting $L = 36$ m, $E = 205$ kN/mm^2, $I = 19.14 \times 10^9$ mm^4, and $g = 9.81$ m/s^2 in equation (8.67) gives:

$$f_0 = 2.02 \text{ Hz}.$$

This is less than the minimum value of 5 Hz given in Part 2, so the maximum vertical acceleration, a, must be checked. This is given in Part 2 as:

$$a = 4\pi^2 f_0^2 y_s K \psi$$

where K (configuration factor) is 1.0 and ψ (dynamic response factor) is 11.3 for a simply-supported composite bridge deck of 36 m span.

The static deflection at midspan, y_s, due to a standard central point load of 0.7 kN is easily calculated as 0.173 mm from the usual expression

$$y_s = WL^3/48EI$$

Hence:

$$a = 4\pi^2 \times 2.02^2 \times 0.173 \times 10^{-3} \times 1.0 \times 11.3 = \pm 0.315 \text{ m/s}^2$$

From equation (8.70), this acceleration should not exceed $\pm\frac{1}{2}(f_0)^{\frac{1}{2}}$, or 0.71 m/s^2 for this bridge. The dynamic behaviour is therefore satisfactory.

References

1. **BS 5400**, *Steel, concrete and composite bridges:*
   Part 1: General statement, 1978
   Part 2: Loads, 1978
   Part 3: Design of steel bridges, 1982
   Part 4: Design of concrete bridges, 1984
   Part 5: Design of composite bridges, 1979
   Part 6: Specification for materials and workmanship; steel, 1980
   Part 7: Specification for materials and workmanship; concrete, reinforcement and prestressing tendons, 1978
   Part 8: Recommendations for materials and workmanship; concrete, reinforcement and prestressing tendons, 1978
   Part 9: Bearings, 1983
   Part 10: Fatigue, 1980
   British Standards Institution, London.
2. Departmental Standards:
   BD/13/82, *Design of steel bridges, Use of BS 5400; Part 3; 1982*, October 1982
   BD/16/82, *Design of composite bridges, Use of BS 5400; Part 5; 1979*, November 1982
   Department of Transport, London.
3. **Eurocode No. 2,** *Common unified rules for concrete structures*, (draft), Commission of the European Communities, Brussels, 1984.
4. **Eurocode No. 3**, *Common unified rules for steel structures*, (draft), Commission of the European Communities, Brussels, 1984.
5. **Eurocode No. 4**, *Common unified rules for composite steel and concrete structures*, (draft), Commission of the European Communities, Brussels, 1986.
6. **Basille, G. D.** 'Composite tee beam bridges', *J. Inst. Municipal & County Engrs*, **62**, 1264–72, May 1936.
7. **Viest, I. M.** (Chairman) 'Composite steel-concrete construction', Report of Task Committee, *Proc. A.S.C.E.*, **100**, ST5, 1085–1139, May 1974.
8. **McDevitt, C. F. and Viest, I. M.** (a) 'Interaction of different materials', Introd. Report, 55–79. (b) 'A survey of using steel in combination with other materials', Final Report, 101–17, *Ninth Congress, Int. Assoc. for Bridge and Struct. Eng.*, Amsterdam, 1972.
9. Council on Tall Buildings and Urban Habitat, *Monograph on the Planning and Design of Tall Buildings;* Vol. SB, Structural design of tall steel buildings, chapter A41, Mixed construction, American Society of Civil Engineers, New York, 1978.
10. **Thomas, P. K.** 'A comparative study of highway bridge loadings in different countries', *Supp. Report 135 UC*, Transport and Road Research Laboratory, 1975.
11. **Petignat, J. and Dauner, H.-G.** 'Progress in design and construction of composite bridges of steel and concrete in Switzerland' (in French), *Schweiz. Bauzeitung*, **92**, 5, 89–94, January 1974.
12. **Grelu, H. P.** 'Association du béton précontraint et de la charpente métallique',

Contributions Techniques Françaises, 7th Congress F.I.P., 114–18, Association Française du Béton, Paris, 1974.

13. **Menn, Ch. and Aasheim, P.** 'Le pont sur la Veveyse', *Bull. Tech. de la Suisse Romande*, **97**, 527–31, 1971.

14. *Proceedings, Conference on Steel Bridges*, June 1968. British Constructional Steelwork Association, London, 1969.
    (a) **Sawko, F.** 'Recent developments in the analysis of steel bridges using electronic computers', 39–48.
    (b) **Chapman, J. C. and Teraszkiewicz, J. S.** 'Research on composite construction at Imperial College', 49–58.
    (c) **Gray, S. R.** 'The elevated structure for the Birkenhead Mersey Tunnel approaches', 59–65.
    (d) **Roik, K.** 'Methods of prestressing continuous composite girders', 75–81.
    (e) **Wallace, A. A. C.** 'The White Cart Viaduct', 95–104.
    (f) **Murray, J. and Rigbye, W. W.** 'Design and construction of curved steel bridges in Lancashire', 105–15.
    (g) **Deuce, T. L. G.** 'The Lofthouse Interchange bridges', 117–22.
    (h) **Nicholas, R. J.** 'Development of the preflexion of beams for bridgeworks', 123–32.
    (j) **Elliot, P.** 'Can steel bridges become more competitive?', 199–210.

15. **Jeske, G. and Sontag, H.** 'Emil-Schulz-Brücke, Berlin-Lichterfelde', *Der Bauingenieur*, **39**, 5, 186–93, 1964.

16. **Godfrey, G. B.** *Bridges in Composite Construction*, British Constructional Steelwork Association, London, 1968.

17. **Dubas, P. and Hauri, H.** 'The motorway bridge over the Sarine at Fribourg', *Schweiz. Bauzeitung*, **84**, 3–15, January 1966.

18. **BS 153**, *Steel girder bridges*, British Standards Institution, London, 1958.

19. **Dubas, P.** 'Développements Suisses recents en matière de ponts mixtes acier-béton', *Construzioni Metalliche*, **21**, 1–15, 1969.

20. **Jacquemoud, J., Saluz, R. and Hirt, M. A.** 'Mesures statiques et dynamiques sur le viaduc d'access à la jonction d'Aigle', *Report ICOM 024*, p. 43, Ecole Polytechnique Fédérale de Lausanne, June 1976.

21. **Kavanagh, T. C.** 'Box girder bridge design', *Eng. J. Amer. Inst. Steel Constr.*, **4**, No. 3, 100–106, July 1967.

22. **Branco, F. A. and Green, R.** 'Construction bracing for composite box girder bridges', *Proc. Int. Conf. Short and Medium Span Bridges*, Canadian Society for Civil Engineering, Toronto, 157–69, 1982.

23. **Reed, T. W.** 'The design and construction of the River Stour Bridge, Wimborne bypass', *Struct. Engr*, **61A**, 115–19, April 1983.

24. **Fairhurst, W. A. and Beveridge, A.** 'The superstructure of the Tay Road Bridge', *Struct. Engr*, **43**, 75–82, March 1965.

25. **Daniels, J. H. and Fisher, J. W.** 'Fatigue behaviour of continuous composite beams', *Highway Research Record*, No. 253, 1–20, Highway Research Board, 1968.

26. *Standard Specifications for Highway Bridges*, 10th ed., American Association of State Highway Officials, Washington D.C., 1969.

27. **Sattler, K.** 'Composite construction in theory and practice', *Struct. Engr*, **39**, 124–44, April 1961.

28. **Menzies, J. B.** 'Structural behaviour of the Moat Street flyover, Coventry', *Civ. Eng. and Pub. Wks Review*, **63**, 967–71, September 1968.

29. **Ciolina, F.** 'Ponts route mixtes acier-béton', *Bull. Tech. de la Suisse Romande*, **100**, 200–211, May 1974.

30. **Ward, F. G. and Scott, W. B.** *Composite Construction for Simply-Supported Bridges – Properties of Composite Sections*, Publication BD1, British Constructional Steelwork Association, London, 1967.

31. **Tordoff, D.** *Steel Bridges*, Publication 15/85, British Constructional Steelwork Association, March 1985.

32. **Nash, G. F. J. and Salter, P. R.** *Tables and graphs for simply-supported beam and slab design*, Constrado, London, June 1985.

33. **Frischmann, W.** 'Motorway bridges', *Building with Steel*, No. 14, 14–17, British Steel Corporation, May 1973.

34. **Holland, A. D. and Deuce, T. L. G.** 'A review of small span highway bridge design and standardisation', *J. Instn. Highway Engrs*, **17**, No. 8, 3–27, August 1970.

35. Committee of Inquiry into the basis of design and method of erection of steel box girder bridges. *Interim Design Rules*, Department of Environment, London, 1973.

36. **Ogle, M. H., Watt, A. C. and Baxter, P. T.** 'Saltings Viaduct, Neath (South Wales)'. *Acier*, **39**, 1, 1–8, January 1974.

37. **de Miranda, F.** 'The elevated highway in Genoa', Translation of article in Italian, in *Steel and the Highway Problem*, British Constructional Steelwork Association, London, 1969.

38. **McKay, C.** 'Composite bridge deck construction using steel hollow trapezoidal girders', *J. Instn. Highway Engrs*, **18**, No. 2, 5–16, February 1971.

39. **Smith, D. W.** 'New Scotswood Bridge', *Proc. Instn Civ. Engrs*, **42**, 217–49, February 1969.

40. **Wallace, A. C.** 'Bonar Bridge in Scotland', *Acier*, **9**, 288–96, 1975.

41. **Robinson, J. R.** 'Système nouveau de couverture de ponts-routes métalliques par tôle cintrée et béton armé associés, *Prelim. Report, 4th Congress I.A.B.S.E.*, 649–62, 1952.

42. **Delcamp, A.** 'Le pont suspendu de Tancarville', *Acier*, **25**, No. 4, 149–61, April 1960.

43. **Esquillan, N.** 'Pont-route de Tancarville'. Chap. 7, 'Dalles en béton armé sous chaussée et sous troittoirs', *Annales*, **14**, No. 157, 152–66, January 1961.

44. **Hayward, A. C. G.,** 'Interaction in twin beam and deck systems – a rapid analysis method', *Civ. Eng. and Pub. Wks Review*, **65**, 1164–66, October 1970.

45. **Rockey, K. C., Bannister, J. L. and Evans, H. R.** (eds), *Developments in Bridge Design and Construction*, Crosby Lockwood, UK, 1971.
    (a) **Sawko, F. and Mills, J. H.** 'Design of cantilever slabs for spine beam bridges', 1–26.
    (b) **Loo, Y. C. and Cusens, A. R.** 'Developments of the finite strip method in the analysis of cellular bridge decks', 53–72.
    (c) **Scordelis, A. C.** 'Analytical solutions for box girder bridges', 200–216.
    (d) **Mattock, A. H.** 'Development of design criteria for composite box girder bridges', 371–86.
    (e) **Lee, D. J.** 'The selection of box beam arrangements in bridge design', 400–426.
    (f) **Wex, B. P.** 'Special features in the design of box girders', 427–40.

46. **Tung, D. H. and Fountain, R. S.** 'Approximate torsional analysis of curved box girders by the M/R method', *Eng. J. Amer. Inst. Steel Const.*, **7**, 65–74, July 1970.

47. **Flint, A. R. and Edwards, L. S.** 'Limit state design of highway bridges', *Struct. Engr*, **48**, 93–108, March 1970.

48. **Lowe, P. A. and Flint, A. R.** 'Prediction of collapse loadings for composite highway bridges', *Proc. Instn. Civ. Engrs*, **48**, 645–59, April 1971.

49. **Nash, G. F. J.** *Steel bridge design guide – composite universal beam simply supported span*, p. 63, Constrado, Croydon, January 1984.

50. **Cusens, A. R. and Pama, R. P.** *Bridge Deck Analysis*, Wiley, 1975.

51. **O'Connor, C.** *Design of Bridge Superstructures*, Wiley-Interscience, 1971.

52. **Hambly, E. C.** *Bridge Deck Behaviour*, Chapman and Hall, 1976.

53. **Kollbrunner, C. F. and Basler, K.** *Torsion in Structures*, Springer-Verlag, Berlin, 1969.

54. **Maisel, B. I., Rowe, R. E. and Swann, R. A.** 'Concrete box-girder bridges', *Struct. Engr*, **51**, 363–76, October 1973, and discussion, **52**, 257–72, July 1974.

55. **Moffatt, K. R. and Dowling, P. J.** 'Shear lag in steel box girder bridges', *Struct. Engr*, **53**, 439–48, October 1975.

56. **Pucher, A.** *Influence Surfaces of Elastic Plates*, 3rd ed., Springer, 1964.

57. **Cartledge, P.** (ed.), *Proceedings of Conference on Steel Box Girder Bridges*, Institution of Civil Engineers, London, 1973.
    (a) **Dowling, P. J., Loe, J. A. and Dean, J. A.** 'The behaviour up to collapse of load bearing diaphragms in rectangular and trapezoidal stiffened box girders', 95–117.
    (b) **Henderson, W., Burt, M. E. and Goodearl, K. A.** 'Bridge loading', 193–201.

58. **Maisel, B. I.** 'Review of literature related to the analysis and design of thin-walled beams', *Tech. Rep. TRA 440*, Cement and Concrete Association, July 1970.

59. **Hambly, E. C. and Pennells, E.** 'Grillage analysis applied to cellular bridge decks', *Struct. Engr*, **53**, 267–76, July 1975.

60. **Wright, R. N., Abdel-Samad, S. R. and Robinson, A. R.** 'BEF analogy for analysis of box girders', *Proc. A.S.C.E.*, **94**, ST7, 1719–43, July 1968.

61. **Heilig, R.** 'A contribution to the theory of box-girders of arbitrary cross-sectional shape', *Library Translation No. 145*, Cement and Concrete Association, 1971.

62. **Knittel, G.** 'Analysis of thin-walled box girders of constant symmetrical cross-section' (in German), *Beton u. Stahlbetonbau*, **60**, 205–211, September 1965. English translation (unpublished), Cement and Concrete Association, London.

63. **Kristek, V.** 'Tapered box girders of deformable cross-section', *Proc. A.S.C.E.*, **96**, ST8, 1761–93, August 1970.

64. **Rowe, R. E.** *Concrete Bridge Design*, C.R. Books, 1962.

65. **Johnston, S. B. and Mattock, A. H.** 'Lateral distribution of load in composite box girder bridges', *Highway Research Record*, No. 167, 25–33, Highway Research Board, 1967.

66. **Scordelis, A. C.** 'Folded plates for bridges', *Bull. Int. Ass. for Shell and Spatial Str.*, No. 57, 29–38, March 1975.

67. **Buragohain, D. N.** 'Discrete analysis of curved slab-beam systems', *Publications*, *I.A.B.S.E.*, **34-II**, 19–38, 1974.

68. **West, R.** *Recommendations on the Use of Grillage Analysis for Slab and Pseudo-Slab Bridge Decks*, Cement and Concrete Association and Construction Industry Research and Information Association, 1973.

69. **Wood, R. H.** 'The reinforcement of slabs in accordance with a predetermined field of moments'. *Concrete*, **2**, No. 8, 319–21, August 1968.

70. **Church, J. G. and Clark, L. A.** 'Combination of highway loads and temperature difference on bridges', *Struct. Engr*, **62A**, 177–81, June 1984.

71. **Leonard, D. R.** 'A traffic loading and its use in the fatigue life assessment of highway bridges', *Tech. Note 311*, Transport and Road Research Laboratory, 1968.

72. **BS 4360**, *Specification for weldable structural steels*, British Standards Institution, London, 1979: Amendment 2, 1982.

73. **BS 4449**, *Specification for hot rolled steel bars for the reinforcement of concrete*, British Standards Institution, London, 1978: amended 1983.

74. **BS 4461**, *Specification for cold worked steel bars for the reinforcement of concrete*, British Standards Institution, London, 1978: amended 1983.

75. **Burkhardt, P., Hertig, P. and Aeschlimann, H. U.** 'Experiments on composite steel-concrete girders glued with epoxy resins', (in French). *Materiaux et Constructions*, **8**, 261–77, August-September 1975.

76. **Menzies, J. B.** 'CP 117 and shear connectors in steel-concrete composite beams', *Struct. Engr*, **49**, 137–53, March 1971.

77. **CP 117**, *Composite construction in structural steel and concrete, Part 2: Beams for bridges*, British Standards Institution, London, 1967.

78. **Slutter, R. G. and Driscoll, G. C.** 'Test results and design recommendations for composite beams for buildings', *Fritz Eng. Lab. Report 279.10*, Lehigh University,

USA, January 1962.

79. **Ollgaard, J. G., Slutter, R. G. and Fisher, J. W.** 'Shear strength of stud connectors in lightweight and normal-weight concrete', *Eng. J. Amer. Inst. Steel Construction*, **8**, 55–64, April 1971.

80. **Johnson, R. P. and Oehlers, D. J.** 'Analysis and design for longitudinal shear in composite T-beams', *Proc. Instn Civ. Engrs*, Part 2, **71**, 989–1021, December 1981.

81. **Dallam, L. N.** 'Push-out tests of stud and channel shear connectors in normal-weight and lightweight concrete slabs', *Eng. Expt. Sta. Bull.*, **69**, 66, University of Missouri, April 1968.

82. **Teraszkiewicz, J. S.** 'Tests on stud shear connectors', *Tech. Note 36*, Road Research Laboratory, Crowthorne, December 1965.

83. **Buttry, K. E.** *Behaviour of stud shear connectors in lightweight and normal-weight concrete*, MSc thesis, University of Missouri, Columbia, 1965.

84. **Schlaginhaufen, R.** 'Zur Anwendung der Bolzendübel', *Schweiz. Bauzeitung*, **41**, 83J, 14 October 1965.

85. Department of the Environment. Phase II design studies on composite bridges for BSI Committee B/116/5. *Report on the Appraisal of St Martin's Road Bridge, Kenilworth Bypass*, July 1976.

86. **Hughes, B. D.** '"Cyc-arc" stud welded concrete anchors', *Civ. Eng. and Pub. Wks Review*, **59**, 723–7, June 1964.

87. **McMackin, P. J., Slutter, R. G., and Fisher, J. W.** 'Combined tension and shear tests of headed concrete anchor studs', *Fritz Eng. Lab. Report 200.71.438.2*, Lehigh University, USA, 1971.

88. **CP 117**, *Composite construction in structural steel and concrete, Part 1: Simply-supported beams in building*, British Standards Institution, London, 1965.

89. **Davies, C.** *Steel-Concrete Composite Beams for Buildings*, George Godwin, London, 1975.

90. **Vogel, G. and Skinner, T. A.** *Shear Connector Behaviour from Beam Tests*, two MSc theses, University of Missouri, Columbia, 1971:
    (a) *Composite beams having normal-weight concrete* (G. Vogel)
    (b) *Composite beams having lightweight concrete* (T. A. Skinner).

91. **Teraszkiewicz, J. S.** 'Static tests on stud shear connectors in haunched slabs', *Report LR223*, Road Research Laboratory, Crowthorne, 1968.

92. **Marshall, W. T., Nelson, H. M. and Bannerjee, H. K.** 'An experimental study of the use of HSFG bolts as shear connectors in composite beams', *Struct. Engr*, **49**, 171–8, April 1971.

93. **BS 4395**, *High strength friction grip bolts and associated nuts and washers for structural engineering*. Parts 1 and 2, 1969; Part 3, 1973, British Standards Institution, London.

94. **BS 4604**, *The use of high strength friction grip bolts in structural steelwork:* Parts 1 and 2, 1970; Part 3, 1973, British Standards Institution, London.

95. **Janss, J.** 'Developpement des recherches concernant les constructions mixtes exécutées par le C.R.I.F. a l'Université de Liège', *Prelim. Report, 9th Cong. I.A.B.S.E.*, 125–32, 1972.

96. **Piraprez, E. and Janss, J.** 'Amélioration de la liaison acier-béton par l'emploi de tôles á adherence renforcée', *Report MT 108*, C.R.I.F., Brussels, October 1975.

97. **Fritz, B.** 'Simplified method of calculation for the effects of creep and shrinkage in composite girders' (in German), *Die Bautechnik*, **27**, 37–42, February 1950.

98. **Haensel, J.** *Effects of Creep and Shrinkage in Composite Construction*, Tech. Report 75–12, Institut für Konstruktiven Ingenieurbau, University of Bochum, West Germany, October 1975.

99. *Steel Designers' Manual* (4th ed.), Collins Professional and Technical Books, London, 4th ed. revised, 1983.

100. **Westergaard, H. M.** 'Computation of stresses in bridge slabs due to wheel loads',

Public Roads, **11**, 1–23, March 1930.

101. **Clark, L. A.** *Concrete bridge design to BS 5400*, Construction Press, London, 1983.
102. **Johnson, R. P. and Allison, R. W.** 'Cracking in concrete tension flanges of composite T-beams', *Struct. Engr*, **61B**, 9–16, March 1983.
103. **Johnson, R. P., van Dalen, K. and Kemp, A. R.** 'Ultimate strength of continuous composite beams', *Proc. Conf. on Structural Steelwork, 1966*, 27–35, British Constructional Steelwork Association, London, 1967.
104. **Janss, J. and Lambert, J. C.** 'Tests on composite beams on three supports' (in French), *Report MT85*, C.R.I.F., Brussels, November 1973.
105. **Hamada, S. and Longworth, J.** 'Buckling of composite beams in negative bending', *Proc. A.S.C.E.*, **100**, ST11, 2205–22, November 1974.
106. **Chatterjee, S.** 'New features in the design of beams', *Symp. on BS 5400: Part 3*, 19–32, Institution of Structural Engineers, London, January 1980.
107. **Nethercot, D.** 'Design of beams and plate girders – treatment of overall and local flange buckling', *The Design of Steel Bridges*, eds Rockey, K. C. and Evans, H. R., 243–62, Granada, 1981.
108. **Johnson, R. P. and Bradford, M. A.** 'Distortional lateral buckling of unstiffened composite bridge girders', *Instability and Plastic Collapse of Steel Structures*, ed. Morris, L. J., 569–80, Granada, 1983.
109. **Leonhardt, F.** *Prestressed Concrete Design and Construction* (2nd ed.), Wilhelm Ernst, Berlin and Munich, 1964.
110. **Clarke, J. L.,** 'The fatigue behaviour of stud shear connectors under rotating shear', *Proc. Instn Civ. Engrs*, **53**, 545–56, December 1972.
111. **Johnson, R. P. and Oehlers, D. J.** 'Design for longitudinal shear in composite L-beams', *Proc. Instn Civ. Engrs*, Part 2, **73**, 147–70, March 1982.
112. **Johnson, R. P.** '(1) The Eurocodes, (2) Shrinkage in composite beams', *Proc. US-Japan Joint Seminar, Composite and Mixed Construction*, Seattle, July 1984 (to be published).
113. **Johnson, R. P. and May, I. M.** 'Partial-interaction design of composite beams', *Struct. Engr*, **53**, 305–11, August 1975.
114. **Long, A. E. and Csagoly, P.** 'A note on shrinkage stresses in continuous steel-concrete composite bridges', *Struct. Engr*, **53**, 256–9, June 1975.
115. **Heckel, R.** 'The fourth Danube bridge in Vienna – damage and repair', *Developments in Bridge Design and Construction*, ed. Rockey, K. C. *et al.*, 588–98, Crosby Lockwood, 1971.
116. **Baldwin, J. W. and Guell, D. L.** 'Permanent deflections and loss of camber in steel bridge beams', *Final Report, Projects 12-1 and 12-6*, Engineering Experimental Station, University of Missouri, Columbia, November 1971.
117. **Hallam, M. W.** *The Behaviour of Composite Structural Elements under Repeated Loading*, PhD thesis, Univ. of Sydney, June 1973.
118. **Wright, R. N. and Walker, W. H.** 'Vibration and deflection of steel bridges', *Eng. J. Amer. Inst. Steel Constr.*, **9**, 20–31, January 1972.
119. **Tilly, G. P. et al.,** 'Dynamic behaviour of footbridges', *Survey S-26/84*, 13–24, *Periodica 2/1984*, I.A.B.S.E., Zurich, May 1984.
120. **Gray, S. R., Clark, P. J. and Gent, A. R.** 'Birkenhead-Mersey Tunnel approach viaducts', *Struct. Engr*, **46**, 379–95, December 1968.
121. **Warren, P. A.** *Composite Box Girders Loaded to Collapse*, PhD thesis, Imperial College, University of London, p. 216, July 1980.
122. **Horne, M. R.** 'Basic concepts in the design of webs', *The Design of Steel Bridges*, eds Rockey, K. C. and Evans, H. R., 161–88, Granada, 1981.
123. **Moffatt, K. R.** 'An analytical study of the longitudinal bending behaviour of composite box girder bridges having incomplete interaction', *CESLIC Report CBI*, Imperial College, University of London, February 1976.
124. **Casillas, J., Siess, C. P. and Khachaturian, N.** 'Studies of reinforced concrete beams

and slabs reinforced with steel plates', *Civ. Eng. Studies, Str. Research Series*, No. 134, University of Illinois, April 1957.

125. **Maeda, Y. and Matsui, S.** 'Prefabricated steel deck plates sandwiching concrete', *Prelim. Report, I.A.B.S.E. Symp. on Mass-Produced Steel Struct.*, 335–42, Prague, 1971.

126. **Solomon, S. K., Smith, D. W. and Cusens, A. R.** 'Flexural tests of steel-concrete-steel sandwiches', *Mag. Conc. Research*, **28**, 94, 13–20, March 1976.

127. **Clarke, J. L. and Morley, C. T.** 'Steel-concrete composite plates with flexible shear connectors', *Proc. Instn. Civ. Engrs Part 2*, **53**, 557–68, December 1972.

128. **Moffatt, K. R.** *Finite Element Analysis of Box Girder Bridges*, PhD thesis, Imperial College, University of London, January 1974.

129. **Lou, C. T.** *Some Aspects of the Behaviour of Composite Plates*, M.Sc thesis, Imperial College, University of London, January 1974.

130. *Specification for the Design, Fabrication, and Erection of Structural Steel for Buildings*, American Institute for Steel Construction, 1969.

131. **Moffatt, K. R. and Dowling, P. J.** 'Interface shear stresses in composite plates under wheel loads', *Struct. Engr*, **62B**, 1–5, March 1984.

132. **McKay, C.** 'Composite bridge deck construction using steel hollow trapezoidal girders', *J. Instn Highway Engrs*, **18**, 5–16, February 1971.

133. **Yam, L. C. P. and Chapman, J. C.** 'The inelastic behaviour of simply-supported composite beams of steel and concrete', *Proc. Instn Civ. Engrs*, **41**, 651–83, December 1968.

134. **Johnson, R. P.** 'Loss of interaction in short-span composite beams and plates', *J. Const. Steel Research*, **1**, No. 2, 11–16, January 1981.

135. **Johnson, R. P. and Arnaouti, C.** 'Punching shear strength of concrete slabs subjected to in-plane biaxial tension', *Mag. Conc. Research*, **32**, 45–50, March 1980.

136. **Rockey, K. C. and Evans, H. R.** (eds), *The Design of Steel Bridges*, Granada, 1981.
    (a) **Ogle, M.** 'The new fatigue loading for highway bridges', 433–53.
    (b) **Halse, W. I.** 'The fatigue assessment of bridges to BS 5400', 455–73.

137. **Oehlers, D. J. and Foley, L.** 'The fatigue strength of stud shear connections in composite beams', *Proc. Instn Civ. Engrs*, Part 2, **79**, 365–81, June 1985.

138. **Climenhaga, J. J. and Johnson, R. P.** 'Fatigue strength of form-reinforced composite slabs for bridge decks', *Publications, I.A.B.S.E.*, **35-1**, 89–101, 1975.

139. **BS 5975**, *Code of practice for falsework*, British Standards Institution, London, 1982.

140. **Kempster, E.** 'GRC as permanent formwork', *Civ. Engrg*, 57–63, August 1981.

141. **True, G. F.** 'Glassfibre reinforced cement permanent formwork', Concrete Society current practice sheet No. 97, *Concrete*, **19**, 31–3, February 1985.

142. *Handbook No. 1, Permanent formwork*, Glassfibre Reinforced Cement Association, 1985.

143. **Dowling, P. J. and Labib, F.** 'Precast concrete bridge shutters', *CESLIC Report PS 2*, Imperial College, University of London, April 1972.

144. **Burton, K. T. and Hognestad, E.** 'Fatigue tests of reinforcing bars – tack welding of stirrups', *J. Amer. Conc. Inst.*, **64**, 244–52, May 1967.

145. **Kerensky, O. A. and Dallard, N. J.** 'The four-level interchange at Almondsbury', *Proc. Instn Civ. Engrs*, **40**, 295–322, July 1968.

146. **Morice, P. B.** 'Discussion of Paper by A. D. Holland', *Proc. Inst. Civ. Engrs*, Part 2, **4**, 269–70, June 1955.

147. **Johnson, R. P. and Smith, D. G. E.** 'Design of cased beams', *Tech. Paper 213*, B.S.I. Sub-Committee B/20/5 – Composite structures for buildings, January 1976 (unpublished).

148. **Johnson, R. P.** 'Composite beam design', *The Consulting Engr*, **32**, No. 11, 40–45, November 1968.

149. **Hàwkins, N. M.** 'Strength of concrete-encased steel beams', *Civ. Eng. Trans.*, Institution of Engineers, Australia, **CE15**, Nos. 1 and 2, 39–44, 1973.

150. **Chatterjee, S.** 'BS 5400: Part 3, Design of steel bridges', Paper 19, *Proc. National Struct. Steel Conf.*, British Constructional Steelwork Association, Dec. 1984.

151. **Boon, A. and Decoppet, J. P.** 'Recent examples of steel bridges with reinforced concrete slabs' (in French) *Schweiz. Bauzeitung*, **92**, 5, 73–7, January 1974.

152. **Presig, P.** 'The motorway bridge over the Veveyse', *Schweiz. Bauzeitung*, **87**, 18, 340–41, May 1969.

153. **Sedlacek, H.** 'Pedestrian bridges', *Krupp Tech. Review*, **24**, 1, 5–7, 1966.

154. **Climenhaga, J. J. and Johnson, R. P.** 'Local buckling in continuous composite beams', *Struct. Engr*, **50**, 367–74, September 1972.

155. **Base, G. D., Read, J. A., Beeby, A. W. and Taylor, H. P. J.** 'An investigation of the crack control characteristics of various types of bar in reinforced concrete beams', *Res. Report 18, Parts I and II*, Cement and Concrete Association, London, 1966.

156. **Beeby, A. W.** 'An investigation of cracking in slabs spanning one way', *Tech. Report TRA 433*, Cement and Concrete Association, London, 1970.

157. Department of the Environment, 'Rules for the design and construction of Preflex beams in highway bridges', *Tech. Memo BE 7/73*, October 1973.

158. **Baes, L. and Lipski, A.** *Preflex beams – principles, note on calculations and descriptive notes*, Sections I, II and III, Preflex S.A., Brussels, Belgium, June 1953, May 1954, April 1958.

159. **Evans, R. H. and White, A. D.** 'Characteristics of preflexed prestressed concrete beams', *Proc. Instn Civ. Engrs*, **30**, 709–29, April 1965.

160. **Stanger, R. H.** *Report of Loading Tests on Horseshoe Connectors for Messrs Boulton and Paul Ltd*, Elstree Laboratories, England (unpublished), July 1964.

161. **Menzies, J. B.** 'Fatigue tests on horseshoe connectors', *Tech. Paper No. 3A*, BSI Sub-committee B/116/5 – Composite Bridges, London, July 1968 (unpublished).

162. **Tyler, R. G.** *Creep and Shrinkage of Concrete Bridge Structures*, PhD thesis, King's College, University of London, January 1973.

163. **Bondale, D. S. and Clark, P. J.** 'Composite construction in the Almondsbury interchange', *Proc. Conf. Structural Steelwork*, 91–100, British Constructional Steelwork Association, 1967.

164. **Bondale, D. S.** *The Effect of Concrete Encasement on Eccentrically Loaded Steel Columns*, PhD thesis, Imperial College, University of London, 1962.

165. **Basu, A. K.** 'Computation of failure loads of composite columns', *Proc. Instn Civ. Engrs*, **36**, 305–12, October 1967.

166. **Basu, A. K. and Sommerville, W.** 'Derivation of formulae for the design of rectangular composite columns', *Proc. Instn Civ. Engrs*, Supplementary Volume, 233–80, 1969.

167. **Virdi, K. S. and Dowling, P. J.** 'The ultimate strength of composite columns in biaxial bending', *Proc. Instn. Civ. Engrs*, **55**, Part 2, 251–72, March 1973.

168. **May, I. M. and Johnson, R. P.** 'Inelastic analysis of biaxially restrained columns', *Proc. Instn Civ. Engrs*, Part 2, **65**, 323–37, June 1978.

169. **Johnson, R. P. and May, I. M.** 'Tests on restrained composite columns', *Struct. Engr*, **56B**, 21–8, June 1978.

170. **Johnson, R. P. and Anderson, D.** 'Design studies for composite columns', *Tech. Papers 83 and 85*, BSI Sub-Committee B/20/5, London, January 1974.

171. **Anderson, D.** 'Design methods for composite columns', *Tech. Paper 130*, BSI Sub-Committee B/116/5, London, August 1976.

172. **Virdi, K. S. and Dowling, P. J.** 'A unified design method for composite columns', *Publications, I.A.B.S.E.*, **36-II**, 165–84, 1976.

173. **Buckby, R. J.** 'Effects of eccentricities due to construction tolerances on the failure load of composite columns under nominally axial loading', *Tech. Papers 118 and 128*, BSI Sub-Committee B/116/5, London, November 1974 and July 1976.

174. **Dowling, P. J., Chu, H. F. and Virdi, K. S.** 'The design of composite columns for biaxial bending', *Prelim. Report, Second Int. Colloq. on Stability*, Liege, April 1977.

175. **Sen, H. K.** *Biaxial Effects in Concrete-filled Tubular Steel Columns*, PhD Thesis,

Imperial College, University of London, 1969.
176. **Bridge, R. Q.** *Factors Affecting the Behaviour of Composite Steel and Concrete Columns*, PhD Thesis, University of Sydney, Australia, 1975.
177. **CP 114,** *Structural use of reinforced concrete in buildings*, British Standards Institution, London, 1965.
178. **BS 449: Part 2.** *The use of structural steel in building, (Metric units)*, British Standards Institution, London 1969.
179. **Wood, R. H.** 'Effective lengths of columns in multi-storey buildings', *Struct. Engr.* **52**, 235–44, 295–302, 341–6, July–September 1974.
180. **Hughes, B. P.** 'Early thermal movement and cracking of concrete', Current Practice Sheet 3PC/06/2 No. 6, *Concrete*, **7**, 43–4, May 1973.
181. **Hughes, B. P.** 'Controlling shrinkage and thermal cracking', *Concrete*, **6**, 39–42, May 1972.
182. **Johnson, R. P.** 'Shrinkage and shear lag in concrete flanges of composite bridge beams', *Research Report CE/16*, University of Warwick, September 1984: published in *Festschrift Roik* (ed. Albrecht, G.), Mitteilung 83–4, Institut für Konstruktiven Ingenieurbau, 426–38, Ruhr-Universität Bochum, September 1984.
183. **Hayward, A. C. G.** 'Alternative designs of shorter span bridges', Paper 22, *Proc. National Structural Steel Conf.*, British Constructional Steelwork Association, December 1984.
184. **Allison, R. W., Johnson, R. P. and May, I. M.** 'Tension-field action in composite plate girders', *Proc. Instn Civ. Engrs*, Part 2, **73**, 255–76, June 1982.
185. **BS 8110**, *Structural use of concrete*, Parts 1 and 2, British Standards Institution, London 1985.
186. Departmental Standards: *Interim revised loading specification*, August 1982 (revised Nov. 1982); and *Tech. Memo. BD 24/84*, Department of Transport, London, 1984.
187. **Horne, M. R.** 'Structural action in steel box girder bridges, *Guide No. 3*, Construction Industry Research and Information Association, London, April 1977.
188. **Johnson, R. P. and Bradford, M. A.** 'Inelastic buckling of composite bridge girders near internal supports', to be published, *Proc. Instn Civ. Engrs*, Part 2, 1986 or 1987.
189. **CP 110,** *The structural use of concrete*, Parts 1, 2, and 3, British Standards Institution, London 1972.
190. 'The design of highway bridge parapets', *Tech. Memo. (Bridges) BE5*, 4th revision, Department of Transport, London, July 1982.
191. **Kato, B.** *et al.,* 'Composite beams using newly developed H-shaped steel with protrusions,' *IABSE-ECCS Symp., Steel in Buildings*, Luxembourg, *Reports*, **48**, 309–16, International Association for Bridge and Structural Engineering, 1985.
192. **Heins, C. P.** *et al.,* 'Composite beams in torsion', *Proc. A.S.C.E.*, **98**, ST5, 1105–117, May 1972.
193. *Ontario Highway Bridge Design Code*, Ministry of Transportation and Communications, Ontario, Canada, 1979.
194. **Dorton, R. A.,** 'The Conestogo River Bridge – design and testing', *Proc. Canadian Struct Eng. Conf.*, Montreal, 18–39, February 1976.
195. **Johnson, R. P.,** 'Desk study of the ultimate strength and failure modes of composite bridges of short-to-medium span', to be published, Transport and Road Research Laboratory, 1986.
196. **Moffatt, K. R. and Dowling, P. J.** 'Distribution of longitudinal stresses in cracked reinforced concrete tension flanges of composite girders', *Proc. Instn Civ. Engrs*, Part 2, **67**, 185–90, March 1979.

Index